PROJECTIVE DIFFERENTIAL GEOMETRY OF SUBMANIFOLDS

North-Holland Mathematical Library

Board of Advisory Editors:

M. Artin, H. Bass, J. Eells, W. Feit, P.J. Freyd, F.W. Gehring, H. Halberstam, L.V. Hörmander, J.H.B. Kemperman, H.A. Lauwerier, W.A.J. Luxemburg, L. Nachbin, F.P. Peterson, I.M. Singer and A.C. Zaanen

VOLUME 49

NORTH-HOLLAND
AMSTERDAM • LONDON • NEW YORK • TOKYO

Projective Differential Geometry of Submanifolds

M.A. AKIVIS
Moscow Institute of Steel and Alloys
Department of Mathematics
Moscow, Russia

V.V. GOLDBERG
New Jersey Institute of Technology
Department of Mathematics
Newark, NJ, USA

1993
NORTH-HOLLAND
AMSTERDAM • LONDON • NEW YORK • TOKYO

ELSEVIER SCIENCE PUBLISHERS B.V.
Sara Burgerhartstraat 25
P.O. Box 211, 1000 AE Amsterdam, The Netherlands

Library of Congress Cataloging-in-Publication Data

Akivis, M. A. (Maks Aĭzikovich)
Projective differential geometry of submanifolds M.A. Akivis, V.V. Gol'dberg.
p. cm. -- (North-Holland mathematical library ; 49)
Includes bibliographical references.
ISBN 0-444-89771-2
1. Submanifolds. 2. Projective differential geometry.
I. Gol'dberg, V. V., (Vladislav Viktorovich) II. Title.
III. Series.
QA649.A38 1993
516.3'62--dc20 93-10725
CIP

ISBN: 0 444 89771 2

This book is printed on acid-free paper

Transferred to digital printing 2005

Preface

1. The projective differential geometry of submanifolds of *three-dimensional projective* space was intensively developed during the first half of this century. Its main results were presented in the monographs [Wi 06] by Wilczynski, [Tz 24] by Tzitzéica, [Če 26] by Čech, [FČ 26], [FČ 31] by Fubini and Čech, [La 32], [La 42] by Lane, [Fi 37], [Fi 50], [Fi 56] by Finikov, [Bo 50a] by Bol, [H 45] by Hlavatý, [Mi 58] by Mihailescu, [Shc 60] by Shcherbakov, [Ko 63] by Kovantsov, [Go 64] by Godeaux, [Shu 64] by Shulikovskii, [Šv 65] by Švec, [Su 73] by Su Buchin, [Kr 80] by Kruglyakov, and [Sto 92a] and [Sto 92b] by Stolyarov. Many papers of É. Cartan were also devoted to projective differential geometry (see [Ca 16], [Ca 18], [Ca 19], [Ca 20a], [Ca 20b], [Ca 20c], [Ca 20d], [Ca 24], [Ca 27], [Ca 31], [Ca 44] and [Ca 45b]). Moreover, É. Cartan was among the first geometers to study multidimensional projective differential geometry systematically.

Following É. Cartan, in the 1930's—1940's many geometers (Bol, Bompiani, Bortolotti, Brauner, Chern, Finikov, Fubini, Hlavatý, Kanitani, Muracchini, Norden, Vangeldère, Villa, Vaona and others) also considered some problems in multidimensional projective differential geometry. During these years, many papers on multidimensional projective differential geometry were published in the Soviet Union. However, these papers remained unnoticed by western geometers. Many of these papers were authored by the participants in the seminar on classical differential geometry at the Moscow State University. The seminar was under the supervision of S.P. Finikov (1883–1964), G.F. Laptev (1911–1972) and A.M. Vasilyev (1923–1987).

In 1978 S.S. Chern in his scientific autobiography [C 78] emphasized the importance of projective differential geometry. He wrote: "I wish to say that I believe that projective differential geometry will be of increasing importance. In several complex variables and in the transcendental theory of algebraic varieties the importance of the Kähler metric cannot be over-emphasized. On the other hand, projective properties are in the holomorphic category. They will appear when the problems involve, directly or indirectly, the linear subspaces or their generalizations."

Note that projective differential geometry is a basis for Euclidean and non-Euclidean differential geometries since metric properties of submanifolds of Euclidean and non-Euclidean spaces should only be added to their projective properties. This fact was noted by É. Cartan in his paper [Ca 19].

In recent years the interest in multidimensional projective differential geometry increased again. Many interesting works devoted to different problems of multidimensional projective differential geometry were published (see [A 82c], [A 83b], [A 84a], [A 84b], [A 88], [A 92a], [A 92b], [Cha 90], [Gr 74], [GH 79], [JM 92], [Kr 80], [P 90], [Sto 92a], [Sto 92b], [Sas 88], [Sas 91], [Wo 84], [Ya 85], [Ya 92], etc.).

However, there is as yet no book in which the *multidimensional* projective differential geometry has been systematically presented. The present book will fill the indicated gap in the literature on differential geometry. In particular, this book reflects the content of many papers by Soviet geometers in multidimensional projective differential geometry.

In this book we give the foundations of the local projective differential geometry of submanifolds and present many results obtained after World War II. In particular, we investigate here a series of special types of submanifolds with a "special projective structure". The problem of studying such submanifolds was posed by É. Cartan in [Ca 19].

The authors of this book were very much influenced by the paper [GH 79] of Griffiths and Harris where the relationship between local differential geometry and algebraic geometry was stressed.

In our exposition we emphasize a projective base of some problems which were usually considered in affine, Euclidean and non-Euclidean geometries. Among these problems are: the theory of conic conjugate nets (Section **3.6**), the theory of parabolic submanifolds without singularities (Section **4.7**), the construction of the generalized Koenigs nets on a hypersurface (Section **5.5**), some aspects of the theory of normal connections on submanifolds (Sections **6.3** and **6.4**), the projective interpretation of the notion of affine normal (Section **7.3**) and hypersurfaces with the parallel second fundamental form (Section **7.2**) and the projective theory of space hyperbands (Section **7.6**). A series of other problems of this nature, for example, the projective interpretation of the Egorov transformations (see [A 84b]) is out of the scope of this book.

As a rule, we use the index notations in our presentation. In our opinion, this allowed us to obtain a deeper understanding of the essence of problems of the local differential geometry.

As a rule, we also do not distinguish between the presentation of material in a real domain and in a complex domain. This is the reason that in the book we use the notation $\mathbf{GL}(n)$ for the general linear group instead of $\mathbf{GL}(n, \mathbf{R})$ and $\mathbf{GL}(n, \mathbf{C})$ as is often done.

However, in some places, namely in those where we are forced to solve algebraic equations or find intersections of algebraic images, the assumption that the objects under consideration be complex becomes essential.

Note also that if we impose a restriction on a submanifold, then, as a rule, we assume that this condition holds at all points of this submanifold. More precisely, we consider only a domain on a submanifold where this restriction holds.

2. We will make a few general remarks for readers of this book. First of all, note that the book is intended for graduate students whose field is differential geometry, and for mathematicians and teachers conducting research in this subject. This book can also be used for a few special courses for graduate students in Mathematics.

In our presentation of material we use the tensorial methods in combination with the methods of exterior differential forms and moving frames of Élie Cartan. The reader is assumed to be familiar with these methods as well as with the basics of modern differential geometry. Many notions of differential geometry are explained briefly in the text and some are given without any explanation. As references, the books [KN 63], [St 64], [Ca 37], [CaH 67] and [BCGGG 91] are recommended. For Russian readers the books [A 77] and [Vas 87] can be also recommended.

All functions, vector and tensor fields and differential forms are considered to be differentiable sufficiently many times.

The book consists of eight chapters whose subjects are clear from the Table of Contents. Sections, formulas and figures are numbered within each chapter. Each chapter is accompanied by a set of notes with remarks of historical and bibliographical nature. A sufficiently complete bibliography, a list of notations and index are given at the end of the book.

A large portion of the book was written during the summers of 1991 and 1992 in the Mathematisches Forschungsinstitut Oberwolfach (MFO), Germany. We express our deep gratitude to Professor Dr. M. Barner, the Director of MFO, for giving us the opportunity to work in MFO using its excellent conditions and facilities.

We also are grateful to the Mathematics Departments of the Moscow Institute of Steel and Alloys, Russia, and New Jersey Institute of Technology, USA, where we are working, and to the Department of Mathematics and Computer Science of Ben-Gurion University of the Negev for the assistance provided in the process of our writing the book.

We express our sincere gratitude to G.R. Jensen, V.V. Konnov, J.B. Little and J. Vilms for reading some of the book chapters and making many useful suggestions, and to L.V. Goldstein for her invaluable assistance in preparing the manuscript for publication.

Moscow, Russia *Maks A. Akivis*

Livingston, New Jersey, USA *Vladislav V. Goldberg*

Table of Contents

CHAPTER 4

Tangentially Degenerate Submanifolds

CHAPTER 5

Submanifolds with Asymptotic and Conjugate Distributions

CHAPTER 6

Normalized Submanifolds in a Projective Space

CHAPTER 7

Projective Differential Geometry of Hypersurfaces

CHAPTER 8

Algebraization Problems in Projective Differential Geometry

Chapter 1

Preliminaries

1.1 Vector Space

1. In what follows, the notion of a finite-dimensional vector space L^n over the field of real or complex numbers will play an important role. We will not state here the basic axioms and properties of a vector space— they can be found in any textbook on linear algebra. Note only that a *frame* (or a *basis*) of an n-dimensional vector space L^n is a system consisting of n linearly independent vectors $e_1, e_2, \ldots, e_n$. A transition from one frame $R = \{e_1, e_2, \ldots, e_n\}$ to another frame $R' = \{e'_1, e'_2, \ldots, e'_n\}$ is determined by the relation

$$e'_i = a_i^j e_j, \quad i, j = 1, \ldots, n, \tag{1.1}$$

where (a_i^j) is a non-singular square matrix. (In these formulas as well as everywhere in the sequel it is understood that summation is carried out over the indices which appear twice: once above and once below.) Thus, the family $\mathcal{R}(L^n)$ of frames in the space L^n depends on n^2 parameters.

The transformations of type (1.1) of frames form a group which is isomorphic to the general linear group $\mathbf{GL}(n)$ of non-singular square matrices (a_i^j) of order n. As was the case for the family $\mathcal{R}(L^n)$ of frames, this group depends on n^2 parameters.

Let us fix a frame R_0, and let R_a be an arbitrary frame in the space L^n, where a is a set of parameters determining the location of the frame R_a relative to the frame R_0. Denote by S_a a transformation of type (1.1) which sends the frame R_0 to the frame R_a:

$$R_a = S_a R_0. \tag{1.2}$$

Equation (1.1) implies that the transformation S_a is a differentiable function (in fact, linear) of parameters a. Let R_{a+da} be a frame near the frame R_a. Then, the transition from the frame R_a to the frame R_{a+da} is determined by the transformation $S_{a+da}S_a^{-1}$. Since $S_a S_a^{-1} = I$, where I is the identity transformation of the space L^n, and S_a is a differentiable function of a, the transformation $S_{a+da}S_a^{-1}$ can be represented in the form:

$$S_{a+da}S_a^{-1} = I + S_\omega + o(da), \tag{1.3}$$

where $o(da)$ are infinitesimals of higher orders than da. The transformation S_ω defining the principal linear part of transformation (1.3) is called an *infinitesimal transformation* of a frame in the space L^n.

The *derivational formulas* or the *equations of infinitesimal displacement* of a frame are equations of the form:

$$dR_a = S_\omega R_a. \tag{1.4}$$

Since the frame R_a in the space L^n consists of n linearly independent vectors e_i, in the coordinate form, formula (1.4) can be written as follows:

$$de_i = \omega_i^j e_j, \tag{1.5}$$

where ω_i^j are linear differential forms which depend on the parameters a (defining the location of the frame R_a) and their differentials da.

Let us find the explicit expressions for the forms ω_i^j in terms of the parameters a_i^j of the group $\mathbf{GL}(n)$. For the vectors e_i and e_i' of the frames R_a and R_{a+da}, we have:

$$e_i = a_i^j e_j^0, \quad e_i' = (a + da + o(da))_i^j e_j^0,$$

where e_j^0 are the vectors of the fixed frame R_0. It follows that

$$e_i^0 = \widetilde{a}_i^j e_j,$$

where $(\widetilde{a}_i^j)$ is the inverse matrix of the matrix (a_i^j). The last two formulas imply

$$e_i' = e_i + da_i^k \widetilde{a}_k^j e_j + o_i(da). \tag{1.6}$$

Comparing formulas (1.5) and (1.6), we obtain:

$$\omega_i^j = \widetilde{a}_k^j da_i^k = -a_i^k d\widetilde{a}_k^j, \tag{1.7}$$

or in matrix notation

$$\omega = a^{-1} da. \tag{1.8}$$

The matrix form $\omega = (\omega_i^j)$ is called an *invariant linear form of the general linear group* $\mathbf{GL}(n)$.

2. Let us find the law of transformation of the coordinates of a vector under transformations of a frame in the space L^n. Suppose we have two frames R and R' whose vectors are connected by relations (1.1). An arbitrary vector x can be represented in the form of linear combinations of the vectors of these two frames:

$$x = x^i e_i = {}'x^i e'_i. \tag{1.9}$$

Using formulas (1.1), we find from (1.9) that

$$x^i = a^i_j \, {}'x^j, \quad {}'x^i = \widetilde{a}^i_j x^j, \tag{1.10}$$

where $(\widetilde{a}^j_i)$ is the inverse matrix of the matrix (a^j_i).

In what follows, it will be more convenient for us to replace equations (1.10) by equivalent differential equations. We assume that the vector x is unchanged under transformations of a frame, i.e. we assume that $dx = 0$. If we differentiate the first equation of (1.9) and apply formulas (1.5), we obtain:

$$0 = dx^i e_i + x^i de_i = (dx^i + x^j \omega^i_j) e_i.$$

The linear independence of the vectors e_i implies that

$$dx^i + x^j \omega^i_j = 0. \tag{1.11}$$

These equations are the desired differential equations which are equivalent to equations (1.10). The latter equations can be recovered by integrating equations (1.11).

Next, let us find the differential equations for coordinates of a covector ξ_i. It follows from the definition of a covector that its contraction $\xi_i x^i$ with coordinates x^i of an arbitrary vector x is constant, i.e. this contraction does not depend on a choice of a frame:

$$\xi_i x^i = \text{const}.$$

Differentiating this relation and using formulas (1.11), we find that

$$d\xi_i x^i + \xi_i dx^i = (d\xi_i - \xi_j \omega^j_i) x^i = 0.$$

Since this holds for any vector x^i, it follows that

$$d\xi_i - \xi_j \omega^j_i = 0. \tag{1.12}$$

Similar equations can be derived for a tensor of any type. For example, let us consider a tensor t of the type (1, 2) with components t^i_{jk}. It follows from the definition of such a tensor that its contraction with coordinates x^j, y^k and ξ_i of arbitrary vectors x, y and an arbitrary covector ξ does not depend on a choice of a frame:

$$t^i_{jk} x^j y^k \xi_i = \text{const}.$$

Differentiating this relation and using formulas (1.11) and (1.12), we find the differential equations which the components t^i_{jk} of the tensor t satisfy:

$$dt^i_{jk} - t^i_{lk} \omega^l_j - t^i_{jl} \omega^l_k + t^l_{jk} \omega^i_l = 0. \tag{1.13}$$

By integrating equations (1.12) and (1.13), we can get the laws of transformation of the coordinates of a covector ξ_i and the tensor t^i_{jk} under transformation (1.11) of a frame:

$$'\xi_i = a^j_i \xi_j, \quad 't^i_{jk} = a^l_j a^m_k \tilde{a}^i_p t^p_{lm}, \tag{1.14}$$

To simplify the form of equations (1.11), (1.12) and (1.13) and similar equations, it is convenient to introduce a differential operator ∇ defined by the following formulas:

$$\nabla x^i = dx^i + x^j \omega^i_j, \quad \nabla \xi_i = d\xi_i - \xi^j \omega^i_j. \tag{1.15}$$

Using this operator, we can write equations (1.11), (1.12) and (1.13) in the form:

$$\nabla x^i = 0, \quad \nabla \xi_i = 0, \quad \nabla t^i_{jk} = 0. \tag{1.16}$$

In addition to the vectors and tensors which were considered above and which were invariant under transformations of a frame, we will encounter objects that get multiplied by some number under transformations of a frame. This number depends on a choice of a basis and on some other factors. Such objects are called *relative vectors* and *relative tensors*. Their coordinates satisfy equations which are slightly different from equations (1.16). For example, for a relative tensor of type (1, 2), these equations have the form:

$$\nabla t^i_{jk} = \theta t^i_{jk}, \tag{1.17}$$

where θ is a linear differential form. The following law of transformation:

$$'t^i_{jk} = \lambda a^l_j a^m_k \tilde{a}^i_p t^p_{lm}. \tag{1.18}$$

corresponds to equations (1.17).

The simplest tensor is the tensor of type (0, 0) or an *absolute invariant*, i.e. a quantity K which does not depend on a choice of a frame. For this quantity, equation (1.17) becomes

$$dK = 0. \tag{1.19}$$

A *relative invariant* is a quantity K which is multiplied by a factor under transformations of a frame. For this quantity, equation (1.17) becomes

$$dK = \theta K. \tag{1.20}$$

1.2 Differentiable Manifolds

1. The second basic notion which will be needed is the notion of a *differentiable manifold*. We give here only the main points of the definition. For more detail, we refer the reader to other books (see, for example, [KN 63] or [D 71]).

A neighborhood of any point of a differentiable manifold M is homeomorphic to an open simply connected domain of the coordinate space $\mathbf{R}^n$ (or $\mathbf{C}^n$ if the manifold M is complex). This allows us to introduce coordinates in the neighborhood of any point of the manifold. The number n is the *dimension* of the manifold M.

If neighborhoods of two points of the manifold M have a non-empty intersection, then the two coordinate systems defined in this intersection are connected by means of invertible differentiable functions. The differentiability class of these functions is called the *class* of the differentiable manifold. Coordinates defined in a neighborhood of a point of a differentiable manifold admit invertible transformations of the same class of differentiability. In what follows, we will assume the differentiable manifolds under considerations to be of class C^∞, and in the complex case we will assume them to be analytic.

Consider a point x of an n-dimensional differentiable manifold M^n. In a neighborhood of the point x, we introduce coordinates in such a way that the point x itself has zero coordinates. Let $x^i = x^i(t)$ be a smooth curve passing through the point x. We parameterize this curve in so that $x^i(0) = 0$. The quantities $\left.\frac{dx^i}{dt}\right|_{t=0} = \xi^i$ are called the *coordinates of the tangent vector* ξ to the curve under consideration, at the point x. The parametric equations of the curve can be written as $x^i(t) = \xi^i t + o^i(t)$, where $o^i(t)$ are infinitesimals of orders higher than t.

The set of tangent vectors to all curves passing through a point $x \in M^n$ forms an n-dimensional vector space. This space is called the *tangent space* to the manifold M^n at the point x and is denoted by $T_x(M^n)$. The set of all tangent spaces of the manifold M^n is called its *tangent bundle* and is denoted by $T(M^n)$. An element of the tangent bundle is a pair (x, ξ), where $x \in M^n$ and $\xi \in T_x(M^n)$. This explains why the tangent bundle is also a differentiable manifold of dimension $2n$, dim $T(M^n) = 2n$.

Next, we consider the set of all possible frames $R_x = \{e_i\}$ in each tangent space. This set can be viewed as a fiber of a fibration $\mathcal{R}(M^n)$ which is called the *frame bundle* over the manifold M^n. Since the family of frames at a fixed point x depends on n^2 parameters, the dimension of the fiber bundle $\mathcal{R}(M^n)$ is equal to $n^2 + n$: dim $\mathcal{R}(M^n) = n^2 + n$.

Let ξ be a vector of the space $T_x(M^n) : \xi \in T_x(M^n)$. The decomposition of this vector relative to the basis $\{e_i\}$ has the form:

$$\xi = \omega^i(\xi) e_i, \tag{1.21}$$

where $\omega^i(\xi)$ are the coordinates of the vector ξ with respect to the basis $\{e_i\}$. These coordinates are linear forms constituting a *cobasis* (a dual basis) in the space $T_x(M^n)$. This cobasis is a basis in the dual space $T_x^*(M^n)$. An element of the dual space is a linear form over $T_x(M^n)$. It follows from formula (1.21) that

$$\omega^i(e_j) = \delta^i_j. \tag{1.22}$$

The set of spaces $T_x^*(M^n)$ forms the *cotangent bundle* $T^*(M^n)$ over the manifold M^n.

Since every tangent space $T_x(M^n)$ is an n-dimensional vector space, we can consider tensors of different types in this space. A *tensor field* $t(x)$ is a function that assigns to each point $x \in M^n$ the value of the tensor t at this point. We will assume that the function $t(x)$ is differentiable as many times as we need.

In each space $T_x(M^n)$, the frames $\{e_i\}$ admit transformations whose differentials can be written in the form (1.5). Since further on we will also consider displacements of the point x along the manifold M^n, we will rewrite formulas (1.5) in the form:

$$\delta e_i = \pi^j_i e_j, \tag{1.23}$$

where δ denotes differentiation under condition that the point x is fixed, i.e. δ is the restriction of the operator of differentiation d to the fiber $R_x(M^n)$ of the frame bundle $\mathcal{R}(M^n)$, and the forms π^j_i are invariant forms of the general linear group $\mathbf{GL}(n)$ of frame transformations in the space $T_x(M^n)$. Parameters defining the location of a frame in the space $T_x(M^n)$ are called *secondary parameters*, in contrast to *principal parameters* which define the location of the point x in the manifold M^n. This is the reason why the symbol δ is called the *operator of differentiation with respect to the secondary parameters* and the 1-forms π^i_j are called the *secondary forms.*

If a tensor field is given on the manifold M^n, then the coordinates of this field must satisfy equations of type (1.13) at any point of this field. For example, for the field $t^i_{jk}(x)$, these equations have the form:

$$\delta t^i_{jk} - t^i_{lk}\pi^l_j - t^i_{jl}\pi^l_k + t^l_{jk}\pi^i_l = 0. \tag{1.24}$$

If, in accordance with formulas (1.15), we denote the left-hand side of this equation by $\nabla_\delta t^i_{jk}$, then this equation takes the form:

$$\nabla_\delta t^i_{jk} = 0. \tag{1.25}$$

2. Let M and N be two manifolds of dimension m and n respectively, and let $f : M \to N$ be a differentiable mapping M into N. Consider a point $a \in M$, its image $b = f(a) \in N$ under the mapping f and coordinate neighborhoods U_a and U_b of the points a and b. The mapping f defines a correspondence

$$y^u = f^u(x^i), \quad i = 1, \dots, m, \ u = 1, \dots, n,$$

between coordinates of points $x \in U_a$ and $y \in U_b$. A mapping f is differentiable of class r, $f \in C^r$, if and only if the functions f^u are differentiable scalar functions of the same class. If the functions f^u are infinitely differentiable functions, then the mapping f belongs to the class C^∞, and if the the functions f^u are analytic functions, then $f \in C^\omega$.

Consider the matrix

$$\mathcal{M} = \left(\frac{\partial y^u}{\partial x^i}\right)$$

having n rows and m columns. This matrix is called the *Jacobi matrix* of the mapping f. It is obvious that the rank r of this matrix satisfies the condition

$$r \leq \min(m, n).$$

It is also obvious that the rank r depends on a point $x \in U_a$. If the rank reaches its maximal value at a point x, i.e. $r = \min(m, n)$, then a mapping f is said to be *nondegenerate* at the point x, and the point x itself is called a *regular* point of a mapping f. If $r < \min(m, n)$ at a point x, then the point x is called a *singular* point of a mapping f.

The following relations can exist between the dimensions m and n:

a) $m < n$. In this case a mapping f is called *injective.* In a neighborhood of a regular point a, the image $f(M) = V$ of a manifold M is an m-dimensional submanifold of the manifold N, and the point $b = f(a)$ is a regular point of the submanifold V. Moreover, the tangent subspace $T_b(V)$ at a regular point b is an m-dimensional submanifold of the tangent subspace $T_b(N)$ whose dimension is equal to n. In particular, as we indicated earlier, if $m = 1$, the submanifold V is a *curve* in N, and if $m = n - 1$, the submanifold V is a *hypersurface* in N.

b) $m > n$. In this case a mapping f is called *surjective.* In a neighborhood U_a of a regular point a, this mapping defines a foliation whose leaves F_y are the complete preimages $f^{-1}(y)$ of the points $y \in U_b$, where $b = f(a)$. The dimension of a leaf is equal to $m - n$, and the dimension of the subspace tangent to the leaf F_y is also $m - n$. If $\dim N = 1$, then we may assume that $N \subset \mathbf{R}$, and the leaves F_y are the level hypersurfaces of the function

$$y = f(x^1, \ldots, x^m)$$

defining the mapping $M \to \mathbf{R}$.

c) $m = n$. In this case, in a neighborhood of a regular point a, a mapping f is *bijective.* The tangent subspaces $T_a(M)$ and $T_b(N)$ to the manifolds M and N at the points a and b are of the same dimension, and the mapping f defines a nondegenerate linear map $f_* : T_a(M) \to T_b(N)$ with the matrix $\mathcal{M}$.

Note also that if $m < n$, in a neighborhood of a regular point a the correspondence between the manifolds M and $f(M)$ is bijective.

3. Let x^i be coordinates in a neighborhood of a point x of the manifold M^n and let $f(x)$ be a function defined in this neighborhood. Then the differential of this function can be written in the form:

$$df = \frac{\partial f}{\partial x^i} dx^i. \tag{1.26}$$

The latter expression is a linear differential form in a coordinate neighborhood of the manifold M^n. However, this form is a form of special type since its coefficients are partial derivatives of the function $f(x)$. A linear differential form of general type can be written in the form:

$$\theta = a_i dx^i. \tag{1.27}$$

Its coefficients $a_i = a_i(x)$ are coordinates of a differentiable covector field which is defined on the manifold M^n. The set of all linear forms on the manifold M^n is denoted by $\Lambda^1(M^n)$.

For the linear forms, the operations of addition and multiplication by a function can be defined in a natural way. In addition, for two linear forms θ_1 and θ_2, the operation of exterior multiplication $\theta_1 \wedge \theta_2$ can be defined. This operation is linear with respect to each factor and is anti-commutative: $\theta_2 \wedge \theta_1 = -\theta_1 \wedge \theta_2$. The product $\theta_1 \wedge \theta_2$ is an *exterior quadratic form*. The exterior quadratic forms of general type are obtained by means of linear combinations of the exterior products of linear forms. The linear operations can be defined in a natural way in the set of exterior quadratic forms, and this set is a module over the ring of smooth functions on the manifold M^n. This module is denoted by $\Lambda^2(M^n)$ (see, for example, [KN 63], pp. 5–7). The localization of this module over each coordinate neighborhood $U \subset M^n$ is a free module with $\binom{n}{2} = \frac{n(n-1)}{2}$ generators. At each point the exterior quadratic forms form a vector space Λ^2 of dimension $\frac{n(n-1)}{2}$ over the field of real or complex numbers.

In a similar manner, one can define the exterior differential forms of degree p, $p \leq n$ on the manifold M^n, and these forms generate a module $\Lambda^p(M^n)$ over the same ring. The localization of this module over each neighborhood $U \subset M^n$ is a free module of dimension $\binom{n}{p}$.

The multiplication of exterior forms of different degrees can be also defined. If θ_1 and θ_2 are exterior forms of degrees p and q, respectively, then their exterior product $\theta_1 \wedge \theta_2$ is an exterior form of degree $p+q$. This product satisfies the following property:

$$\theta_1 \wedge \theta_2 = (-1)^{pq} \theta_2 \wedge \theta_1. \tag{1.28}$$

By the skew-symmetry, the exterior forms of degree greater than n vanish.

The exterior forms of different degrees form the *Grassmann algebra* on the manifold M^n:

$$\Lambda = \Lambda^0 + \Lambda^1 + \Lambda^2 + \ldots + \Lambda^n; \tag{1.29}$$

here Λ^p is the module of exterior forms of degree p. In particular, Λ^0 is the ring of differentiable functions on the manifold M^n. Exterior forms of degree p are also called *p-forms*, and 1-forms are also called the *Pfaffian forms*.

We now consider an exterior differential form of degree two on a manifold M^n. In terms of the coordinates x^i, this form can be written as

$$\theta = a_{ij} dx^i \wedge dx^j, \quad i, j = 1, \ldots, n,$$

where $a_{ij} = -a_{ji}$, and $dx^i \wedge dx^j$ are the basis 2-forms. A skew-symmetric bilinear form is associated with the form θ. This bilinear form is defined by the formula:

$$\theta(\xi, \eta) = a_{ij} \xi^i \eta^j,$$

where ξ and η are vector fields defined in $T(M^n)$. If these two vector fields satisfy the equation

$$\theta(\xi, \eta) = 0,$$

then we say that they are *in involution* with respect to the exterior quadratic form θ. The notion of the value of an exterior p-form on a system consisting of p vector fields given on the manifold M^n can be defined in a similar manner.

Note further the following proposition of algebraic nature, which is called the *Cartan lemma*:

Lemma 1.1 (*Cartan*) *Suppose the linearly independent* 1*-forms* $\omega^1, \omega^2, \ldots, \omega^p$ *and the* 1*-forms* $\theta_1, \theta_2, \ldots, \theta_p$ *are connected by the relation:*

$$\theta_1 \wedge \omega^1 + \ldots + \theta_p \wedge \omega^p = 0. \tag{1.30}$$

Then the forms θ_a *are linearly expressed in terms of the forms* ω^a *as follows:*

$$\theta_a = l_{ab} \omega^b, \tag{1.31}$$

where

$$l_{ab} = l_{ba}. \tag{1.32}$$

Proof. Since the forms ω^a, $a = 1, \ldots, p$, are linearly independent forms in a vector space T^*, by adding the forms ω^ξ, $\xi = p + 1, \ldots, n$ we complete $\omega^1, \ldots, \omega^p$ to a basis for T^*. Then

$$\theta_a = l_{ab} \omega^b + l_{a\xi} \omega^\xi.$$

Substituting this into relation (1.30), we obtain

$$l_{ab} \omega^a \wedge \omega^b + \omega^a \wedge l_{a\xi} \omega^\xi = 0$$

which implies $l_{a\xi} = 0$ and $l_{ab} = l_{ba}$. ■

In the algebra of differential forms, another operation—the *exterior differentiation* can be defined. For functions, i.e. exterior forms of degree zero, this operation coincides with ordinary differentiation, and for exterior forms of type

$$\theta = a dx^{i_1} \wedge \ldots \wedge dx^{i_p}, \tag{1.33}$$

this operation is defined by means of the formula:

$$d\theta = da \wedge dx^{i_1} \wedge \ldots \wedge dx^{i_p} \tag{1.34}$$

The operation of exterior differentiation is a linear operation; it is a linear mapping of the space $\Lambda^p(M^n)$ into the space $\Lambda^{p+1}(M^n)$:

$$d : \Lambda^p \to \Lambda^{p+1}. \tag{1.35}$$

Using formula (1.34), the formula for differentiation of a product of two exterior forms can be proved. Namely, if the forms θ_1 and θ_2 have degrees p and q, respectively, then

$$d(\theta_1 \wedge \theta_2) = d\theta_1 \wedge \theta_2 + (-1)^p \theta_1 \wedge d\theta_2. \tag{1.36}$$

In addition, the following formula holds:

$$d(d\theta) = 0 \tag{1.37}$$

This formula is called the *Poincaré lemma.* In particular, for a function f on M^n we have $d(df) = 0$.

Conversely, if ω is an 1-form given in a simply connected domain of a manifold M^n and such that $d\omega = 0$, then $\omega = df$. A 1-form ω satisfying the condition $d\omega = 0$ is called *closed*, and a form ω satisfying the condition $\omega = df$ is called *exact.*

Note also that in fact, the operation of exterior differentiation defined by formula (1.34) by means of coordinates, does not depend on the choice of coordinates on the manifold M^n, i.e. this operation is invariant: it commutes with the operation of coordinate change on the manifold M^n.

4. As an example, we will apply the operation of exterior differentiation to derive the structure equations of the general linear group $\mathbf{GL}(n)$. In subsection **1**, invariant forms for this group were determined for the frame bundle $\mathcal{R}(L^n)$ of a vector space L^n and were written in the form (1.7) or in the matrix form (1.8). Applying exterior differentiation to equations (1.8) and using equations (1.34), we obtain

$$d\omega = da^{-1} \wedge da. \tag{1.38}$$

From relation (1.8) we find that

$$da = a\omega, \tag{1.39}$$

and since $aa^{-1} = I$, we have

$$da^{-1} = -a^{-1}daa^{-1} = -\omega a^{-1}. \tag{1.40}$$

Substituting expressions (1.39) and (1.40) into equation (1.38), we arrive at the equation

$$d\omega = -\omega \wedge \omega. \tag{1.41}$$

In coordinate form, this equation is written as

$$d\omega_j^i = -\omega_k^i \wedge \omega_j^k,$$

or, more often, as

$$d\omega_j^i = \omega_j^k \wedge \omega_k^i. \tag{1.42}$$

Equations (1.41) and (1.42) are called the *structure equations* or the *Maurer–Cartan equations* of the general linear group $\mathbf{GL}(n)$.

5. Suppose that a system of linearly independent 1-forms θ^a, $a = p + 1, \ldots, n$, is given on a manifold M^n. At each point x of the manifold M^n, this system determines a linear subspace Δ_x of the space $T_x(M^n)$ via the equations

$$\theta^a(\xi) = 0. \tag{1.43}$$

The dimension of this subspace is equal to p. A set of such p-dimensional subspaces Δ_x given at every point x of the manifold M^n is called a *p-dimensional distribution* and is denoted by $\Delta^p(M^n)$.

An *integral manifold* of a system of Pfaffian equations

$$\theta^a = 0 \tag{1.44}$$

is a submanifold V^q of dimension q, $q \leq p$, which, at any of its points x, is tangent to the element Δ_x^p of the distribution $\Delta^p(M^n)$ defined by the system of forms θ^a.

It is easy to prove that the system (1.44) always possesses one-dimensional integral manifolds. If the system has integral manifolds of maximal possible dimension p which form a foliation on the manifold M^n, we will say that the system is *completely integrable*. This means that through any point $x \in M^n$, there passes a unique p-dimensional integral manifold V^p of the system (1.44). A necessary and sufficient condition for a system (1.44) to be completely integrable is given by the Frobenius theorem (see [KN 63], vol. 2, p. 323).

Theorem 1.2 (*Frobenius*) *The system* (1.44) *is completely integrable if and only if the exterior differentials of the forms* θ^a *vanish by means of the equations of this system.* ■

Analytically this can be written as follows:

$$d\theta^a = \theta^b \wedge \theta_b^a, \tag{1.45}$$

where θ_b^a are some new 1-forms.

Note that the structure equations (1.42) of the general linear group $\mathbf{GL}(n)$, which we found earlier, are conditions of complete integrability for the equations (1.5) defining the infinitesimal displacement of a frame of the space L^n.

Note also that if a system of forms ω_j^i is given, and it depends on $\rho \leq n^2$ parameters and satisfies structure equations (1.42), then by Frobenius' theorem, this system uniquely (up to a transformation of the general linear group $\mathbf{GL}(n)$) determines a ρ-parameter family of frames $\mathcal{R}^\rho$ in the space L^n.

6. If the system (1.44) is not completely integrable, then it could still possess integral manifolds of dimension $q < p$. We will say that the system of Pfaffian equations (1.44) is *in involution* if at least one two-dimensional integral manifold V^2 passes through each one-dimensional integral manifold V^1 of this system, at least one three-dimensional integral manifold V^3 passes through each of its two-dimensional integral manifold V^2, etc., and finally, at least one integral manifold V^q of dimension q passes through each integral manifold V^{q-1} of dimension $q - 1$.

Later on we will often apply the *Cartan test* for the system of Pfaffian equations (1.44) to be in involution.

To formulate the Cartan test, first of all note that if V^q is an integral manifold of system (1.44), then on this manifold not only the system (1.44) vanishes but also the system

$$d\theta^a = 0. \tag{1.46}$$

A q-dimensional subspace Δ_x^q tangent to the integral manifold V^q is characterized by the fact that each of its vectors satisfies each equation of system (1.44), and each pair of its vectors is in involution relative to the exterior quadratic forms $d\theta^a$, i.e. the pair satisfies the system (1.46). These vectors are called the *one-dimensional integral elements* of system (1.44), and the subspaces of dimension $k \leq q$ spanned by some or all of these vectors are called the *k-dimensional integral elements* of system (1.44).

Let ξ_1 be a one-dimensional integral element of system (1.44). A two-dimensional integral element passing through the element ξ_1 is determined by a vector ξ_2 which, in addition to the system of equations (1.43), together with ξ_1 satisfy the system:

$$d\theta^a(\xi_1, \xi_2) = 0. \tag{1.47}$$

If the vector ξ_1 is fixed, the system (1.47) is a linear homogeneous system for finding ξ_2. Denote by r_1 the rank of this system. Suppose that ξ_2 is a solution of system (1.47). The vectors ξ_1 and ξ_2 determine a two-dimensional integral element E_2 of system (1.44). To find a three-dimensional integral element of this system, we should consider the system:

$$d\theta^a(\xi_1, \xi_3) = 0, \quad d\theta^a(\xi_2, \xi_3) = 0. \tag{1.48}$$

Each vector ξ_3 satisfying equations (1.48), together with the vectors ξ_1 and ξ_2 determines a three-dimensional integral element E_3. Denote by r_3 the rank of system (1.48). Similarly we can construct integral elements $E_4, \ldots, E_q$. They are by the relation:

$$\xi_1 = E_1 \subset E_2 \subset E_3 \subset \ldots \subset E_q.$$

Denote by r_k the rank of the system of type (1.48) defining a vector ξ_{k+1}, which is in involution with the previously defined vectors $\xi_1, \ldots, \xi_k$, and let

$$s_1 = r_1, \quad s_2 = r_2 - r_1, \quad \ldots, s_{q-1} = r_{q-1} - r_{q-2}.$$

Let s_q be the dimension of the subspace defined by a system of type (1.48) for finding a vector ξ_q. The integers $s_1, s_2, \ldots, s_q$ are called the *characters* of system (1.44), and the integer

$$Q = s_1 + 2s_2 + \ldots + qs_q$$

is called its *Cartan number.* The characters of the Pfaffian system (1.44) are connected by the inequalities:

$$s_1 \geq s_2 \geq \ldots \geq s_q. \tag{1.49}$$

The left-hand sides of equations (1.46) are exterior products of some linear forms from which q forms are linearly independent and are the basis forms of the integral manifold V^q. Let us call these 1-forms ω^a, $a = 1, \ldots, q$. In addition, the equations (1.46) contains forms ω^u whose number is equal to $s_1 + s_2 + \ldots + s_q$. Applying the procedure outlined in the proof of the Cartan lemma, one can express the forms ω^u as linear combinations of the forms ω^a. The number of independent coefficients in these linear combinations is called the *arbitrariness of the general integral element* and is denoted by the letter N.

We can now formulate the Cartan test.

Theorem 1.3 (*Cartan's test*) *For a system of Pfaffian equations* (1.44) *to be in involution, it is necessary and sufficient that the condition $Q = N$ holds. Moreover, its q-dimensional integral manifold V^q depends on s_k functions of k variables where s_k is the last nonvanishing character in sequence* (1.49). ■

Note also that if the system (1.44) of Pfaffian equations is not in involution, this does not mean that this system has no solution. The further investigation of this system is connected with its successive differential prolongations. Moreover, it can be proved that after a finite number of prolongations one obtains either a system which is in involution—and in this case there exists a solution of system (1.44)—or he arrives at a contradiction which proves that the system has no solution.

The reader can find a more detailed exposition of the theory of systems of Pfaffian equations in involution in the books [BCGGG 91], [Ca 45], [Fi 48], [Gr 83] and [GJ 87]. Examples of application of Cartan's test can be found further on.

7. Let us find the structure equations of a differentiable manifold M^n. As we have already noted, if a function $f(x)$ is given on the manifold M^n, then in local coordinates x^i, the differential of this function can be written in form (1.26). The operators $\frac{\partial}{\partial x^i}$ of differentiation with respect to the coordinates x^i form a basis of the tangent space $T_x(M^n)$, called the *natural basis*. We view the differentials dx^i as the coordinates of a tangent vector $d = \frac{\partial}{\partial x^i} dx^i$ with respect to this basis. If we replace the natural basis $\{\frac{\partial}{\partial x^i}\}$ by an arbitrary basis $\{e_i\}$ of the space $T_x(M^n)$:

$$e_i = x_i^j \frac{\partial}{\partial x^j}, \quad \frac{\partial}{\partial x^i} = \widetilde{x}_i^j e_j, \tag{1.50}$$

where (x_i^j) and $(\widetilde{x}_i^j)$ are mutually inverse matrices, then we can expand the vector d as

$$d = e_j \widetilde{x}_i^j dx^i = \omega^j e_j, \tag{1.51}$$

where we used the notation:

$$\omega^j = \widetilde{x}_i^j dx^i, \quad i, j = 1, \dots, n. \tag{1.52}$$

The forms ω^j are called the *base forms* of the manifold M^n.

Taking exterior derivatives of equations (1.52), we obtain

$$d\omega^i = d\widetilde{x}_j^i \wedge dx^j. \tag{1.53}$$

Eliminating the differentials dx^j by means of relations (1.52) from equations (1.53), we arrive at the equations:

$$d\omega^i = d\widetilde{x}_k^i \wedge x_j^k \omega^j. \tag{1.54}$$

Equation (1.54) implies that

$$d\omega^i = \omega^j \wedge \omega_j^i, \tag{1.55}$$

where the forms ω_j^i are not uniquely defined. In fact, subtracting (1.54) from (1.55), we find that

$$\omega^j \wedge (\omega_j^i + x_j^k d\widetilde{x}_k^i) = 0.$$

Applying the Cartan lemma to these equations, we obtain the equations

$$\omega_j^i + x_j^k d\widetilde{x}_k^i = x_{jk}^i \omega^k,$$

or

$$\omega_j^i = -x_j^k d\widetilde{x}_k^i + x_{jk}^i \omega^k, \tag{1.56}$$

where $x_{jk}^i = x_{kj}^i$.

Equations (1.55) are the first set of structure equations of the manifold M^n. By the Frobenius theorem, it follows from equations (1.55) that the system of equations $\omega^i = 0$ is completely integrable. The first integrals of this system are the coordinates x^i of a point x of the manifold M^n.

Let us find the second set of the structure equations of the manifold M^n, which are satisfied by the forms ω_j^i. Exterior differentiation of equations (1.55) leads to the equations:

$$d\omega_j^i = -dx_j^k \wedge d\widetilde{x}_k^i + dx_{jk}^i \wedge \omega^k + x_{jk}^i \omega^l \wedge \omega_l^k. \tag{1.57}$$

The entries of the matrices (x_i^j) and $(\widetilde{x}_i^j)$ are connected by the relation:

$$x_j^k \widetilde{x}_k^i = \delta_j^i.$$

If we differentiate this relation, we find that

$$dx_j^k = -x_q^k x_j^l d\widetilde{x}_l^q.$$

Substituting these expressions for dx_j^k into equations (1.57) and using relations (1.55), we find that

$$d\omega_j^i = \omega_j^k \wedge \omega_k^i + (\nabla x_{jk}^i + x_{jl}^p x_{pk}^i \omega^l) \wedge \omega^k, \tag{1.58}$$

where ∇x_{jk}^i are defined according to the rule (1.15). Define also the 1-forms:

$$\omega_{jk}^i = \nabla x_{jk}^i + x_{jl}^p x_{pk}^i \omega^l + x_{jkl}^i \omega^l, \tag{1.59}$$

where $x_{jkl}^i = x_{jlk}^i$. Using these equations, we can write equations (1.58) as

$$d\omega_j^i = \omega_j^k \wedge \omega_k^i + \omega_{jk}^i \wedge \omega^k. \tag{1.60}$$

These equations form the second set of structure equations of the manifold M^n.

Using the same procedure which we just used to define the forms ω^i, ω_j^i and ω_{jk}^i on the differentiable manifold M^n, and to find structure equations for these forms, we can define higher-order forms $\omega_{jkl}^i, \ldots$ and find structure equations for them (see [Lap 66]). However, in this book we will not need these higher-order forms and equations.

As we already noted above, the forms ω^i are basis forms of the manifold M^n. The forms ω_j^i are the fiber forms of the bundle $\mathcal{R}^1(M^n)$ of frames of first order over M^n, and the forms ω_{jk}^i together with the forms ω_j^i are the fiber forms of the bundle $\mathcal{R}^2(M^n)$ of frames of second order over M^n. The fibers $\mathcal{R}_x^1$ and $\mathcal{R}_x^2$ of these two fibrations are defined on the manifolds $\mathcal{R}^1(M^n)$ and $\mathcal{R}^2(M^n)$ by the equations $\omega^i = 0$.

We denote by δ the restriction of the differential d to the fibers $\mathcal{R}_x^1$ and $\mathcal{R}_x^2$ of the frame bundles under consideration. Let us also denote the restrictions of the forms ω_j^i and ω_{jk}^i to these bundles by $\pi_j^i = \omega_j^i(\delta)$ and $\pi_{jk}^i = \omega_{jk}^i(\delta)$ respectively. Then it follows from equations (1.60) that

$$\delta\pi_j^i = \pi_j^k \wedge \pi_k^i. \tag{1.61}$$

Equations (1.61) coincide with the structure equations (1.42) of the general linear group $\mathbf{GL}(n)$. Thus, the forms π_j^i are invariant forms of the group $\mathbf{GL}(n)$ of admissible transformations of the first order frames $\{e_i\}$ associated with the point x of the manifold M^n, and the fiber $\mathcal{R}_x^1$ is diffeomorphic to this group. This fiber is an orbit of a point of a representation space of the group $\mathbf{GL}(n)$.

This and relations (1.5) show that if $\omega^i = 0$, the vectors e_i composing a frame in the space $T_x(M^n)$ satisfy the equations

$$\delta e_i = \pi_i^j e_j,$$

and the forms ω^i composing a coframe satisfy the equations

$$\delta\omega^i = -\pi_j^i\omega^j. \tag{1.62}$$

Next, consider the forms $\pi_{jk}^i = \omega_{jk}^i(\delta)$. Relations (1.59) imply that

$$\pi_{jk}^i = \nabla_\delta x_{jk}^i,$$

and thus $\pi_{jk}^i = \pi_{kj}^i$. It is not so difficult to show that the forms π_{jk}^i satisfy the following structure equations

$$\delta\pi_{jk}^i = \pi_{jk}^l \wedge \pi_l^i + \pi_j^l \wedge \pi_{lk}^i + \pi_k^l \wedge \pi_{jl}^i$$

(see [Lap 66]) and that these forms together with the forms π_j^i are invariant forms of the group $\mathbf{GL}^2(n)$ of admissible transformations of the second order frames associated with the point $x \in M^n$. The group $\mathbf{GL}^2(n)$ is diffeomorphic to the fiber $\mathcal{R}_x^2$.

8. In what follows we will use the notion of an affine connection in a frame bundle. The forms ω_j^i in equations (1.55) and (1.60) are invariantly defined in the frame bundle $\mathcal{R}^2(M^n)$ of second order. An affine connection γ on a manifold M^n is defined in the frame bundle $\mathcal{R}^2(M^n)$ by means of an invariant horizontal distribution Δ which is defined by a system of Pfaffian forms

$$\theta_j^i = \omega_j^i - \Gamma_{jk}^i\omega^k \tag{1.63}$$

vanishing on Δ. The distribution Δ is invariant with respect to the group of affine transformations acting in $\mathcal{R}^1(M^n)$.

Using equations (1.63), we eliminate the forms ω_j^i from equations (1.55). As a result, we obtain

$$d\omega^i = \omega^j \wedge \theta^i_j + R^i_{jk}\omega^j \wedge \omega^k, \tag{1.64}$$

where $R^i_{jk} = \Gamma^i_{[jk]}$. The condition for the distribution Δ to be invariant leads to the following equations:

$$d\theta^i_j = \theta^i_k \wedge \theta^k_j + R^i_{jkl}\omega^k \wedge \omega^l. \tag{1.65}$$

The Pfaffian forms $\theta = (\theta^i_j)$ with their values in the Lie algebra gl(n) of the group $\mathbf{GL(n)}$ are called the *connection forms* of the connection γ.

The quantities R^i_{jk} and R^i_{jkl} form tensors which are called the *torsion tensor* and the *curvature tensor* of the connection γ respectively.

Conversely, one can prove that if in the frame bundle $\mathcal{R}^2(M^n)$, the forms θ^i_j are given, and these forms together with the forms ω^i satisfy equations (1.64) and (1.65), then the forms θ^i_j define an affine connection γ on M^n, and the tensors R^i_{jk} and R^i_{jkl} are the torsion and curvature tensors of this connection γ.

As a rule, in our considerations the torsion-free affine connections will arise for which $R^i_{jk} = 0$. For these connections the form $\omega = (\omega^i_j)$ can be chosen as a connection form. Under this assumption, the structure equations (1.64) and (1.65) can be written in the form:

$$d\omega^i = \omega^j \wedge \omega^i_j, \quad d\omega^i_j = \omega^i_k \wedge \omega^k_j + R^i_{jkl}\omega^k \wedge \omega^l. \tag{1.66}$$

A more detailed presentation of the foundations of the theory of affine connections can be found in the books [KN 63] and [Lich 55] (see also the papers [Lap 66] and [Lap 69]).

1.3 Projective Space

1. We will assume that the reader is familiar with the notions of projective plane and three-dimensional projective space. These notions can be generalized for the multidimensional case in a natural way (see [D 64]).

Consider an $(n+1)$-dimensional vector space L^{n+1}. Denote by $\widetilde{L}^{n+1} = L^{n+1}\backslash\{0\}$ the set of all nonzero vectors of the space L^{n+1}. We will consider collinear vectors of L^{n+1} to be equivalent and define the *n-dimensional projective space* P^n as the quotient of the set $\widetilde{L}^{n+1}$ by this equivalence relation. This means that a point of P^n is a collection of nonzero collinear vectors λx of L^{n+1}, i.e. a point of P^n is a one-dimensional subspace of L^{n+1} (minus the origin). A straight line of P^n is a two-dimensional subspace of L^{n+1}, etc. If in L^{n+1} a basis defined by the vectors $e_0, e_1, \ldots, e_n$, is given, then any vector $x \neq 0$ of L^{n+1} can be decomposed relative to this basis:

$$x = x^0 e_0 + x^1 e_1 + \ldots + x^n e_n,$$

where the numbers $x^0, x^1, \dots, x^n$ are the coordinates of the vector x relative to the basis $\{e_i\}$. In the space L^{n+1}, a set of collinear vectors corresponds to the point x of P^n, and the coordinates of this set are the numbers $(\lambda x^0, \lambda x^1, \dots, \lambda x^n)$. These numbers are called the *homogeneous coordinates* of the point $x \in P^n$. Note that they are unique up to multiplicative factor.

Linear transformations of the space L^{n+1} give rise to corresponding *projective transformations* of the space P^n. Under these transformations, straight lines are transformed into straight lines, planes into planes etc. Since a point in P^n is defined by homogeneous coordinates, transformations of the form

$$y^u = \rho x^u, \quad u = 0, 1, \dots, n,$$

define the identity transformation of the space P^n. Thus, the projective transformations can be written as

$$\rho y^u = a_v^u x^v, \quad u, v = 0, 1, \dots, n,$$

where $\det(a_v^u) \neq 0$. Therefore, the *group of projective transformations* of the space P^n depends on $(n+1)^2 - 1 = n^2 + 2n$ parameters. This group is denoted by $\mathbf{PGL}(n)$.

A *projective frame* in the space P^n is a system consisting of $n+1$ points A_u, $u = 0, 1, \dots, n$, and the *unity point* E which are in general position. In the space L^{n+1}, to the points A_u there correspond linearly independent vectors e_u, and the vector $e = \sum_{u=0}^n e_u$ corresponds to the point E. These vectors are defined in L^{n+1} up to a common factor. It follows that the set of projective frames $\{A_u\}$ depends on $n^2 + 2n$ parameters. We will always assume below that the unity point E is given along with the basis points A_u, although we might not mention it on every occasion.

We will perform the linear operations on points of a projective space P^n via the corresponding vectors in the space L^{n+1}. These operations will be invariant in P^n if we multiply all corresponding vectors in L^{n+1} by a common factor.

In some instances, we will assume that a vectorial frame in L^{n+1} is normalized by the condition:

$$e_0 \wedge e_1 \wedge \dots \wedge e_n = 1, \tag{1.67}$$

where the wedge denotes the exterior product. Condition (1.67) can be always achieved by multiplying all vectors of the frame by an appropriate factor. When such a normalization has been done, the vectors of a frame in L^{n+1} corresponding to the point of a projective frame $\{A_u\}$ are uniquely determined. Thus, the group of projective transformations of the space P^n is isomorphic to the special linear group $\mathbf{SL}(n+1)$ of transformations of L^{n+1}. Sometimes we will write the normalization condition (1.67) in the form:

$$A_0 \wedge A_1 \wedge \dots \wedge A_n = 1. \tag{1.68}$$

The equations of infinitesimal displacement of a frame in P^n have the same form (1.5) as they had in L^n:

$$dA_u = \omega_u^v A_v, \tag{1.69}$$

but now the indices u and v take the values from 0 to n, and by condition (1.68), the forms ω_u^v in equations (1.69) are connected by the relation:

$$\omega_0^0 + \omega_1^1 + \ldots + \omega_n^n = 0. \tag{1.70}$$

This condition shows that the number of linearly independent forms ω_u^v becomes equal to the number of parameters on which the group $\mathbf{PGL}(n)$ of projective transformations of the space P^n depends.

The structure equations of the space P^n have the same form as they had in the space L^n:

$$d\omega_v^u = \omega_v^w \wedge \omega_w^u, \tag{1.71}$$

but now we have a new range for the indices u, v and w: $u, v, w = 0, 1 \ldots, n$.

It is well-known that a projective space P^n is a differentiable manifold. Let us show that equations (1.71) are a particular case of the structure equations (1.56) and (1.60) of a differentiable manifold. To show this, first of all, we write equations (1.71) for $v = 0$ and $u = i$, where $i = 1, \ldots, n$, in the form:

$$d\omega_0^i = \omega_0^j \wedge \theta_j^i,$$

where $\theta_j^i = \omega_j^i - \delta_j^i \omega_0^0$. These equations differ from equations (1.55) only in notation. Next, taking exterior derivatives of the forms θ_j^i and applying equations (1.71), we find that

$$d\theta_j^i = \theta_j^k \wedge \theta_k^i + (\delta_k^i \omega_j^0 + \delta_j^i \omega_k^0) \wedge \omega_0^k.$$

Comparing these equations with equations (1.60), we observe that they coincide if

$$\omega_{jk}^i = \delta_k^i \omega_j^0 + \delta_j^i \omega_k^0.$$

The latter relations prove that if a first order frame is fixed, the second order frames of a projective space P^n depend on n parameters while on a general differentiable manifold they depend on n^3 parameters.

Note that the forms ω^i constitute a basis in the cotangent space $T_x^*(P^n)$ of a projective space P^n. The corresponding basis in the tangent space $T_x(P^n)$ is formed by the vectors v_i which are directed along the lines A_0A_i (see [GH 79]).

Consider a hyperplane ξ in P^n. The equations of this hyperplane can be written in the form:

$$\xi_u x^u = 0, \quad u = 0, 1, \ldots, n, \tag{1.72}$$

where the coefficients ξ_u are defined up to a constant factor. These coefficients can be viewed as homogeneous coordinates of the hyperplane ξ. They are called the *tangential coordinates* of the hyperplane ξ. This consideration shows that the collection of hyperplanes of a projective space P^n is a new projective space of the same dimension n. This space is denoted by P^{n*} and is called the *dual space* of P^n.

Let us denote the left-hand side of equation (1.72) by (ξ, x). Then the condition of the incidence of a point x with coordinates x^u and a hyperplane $\xi = (\xi_u)$ has the form:

$$(\xi, x) = 0. \tag{1.73}$$

In the space P^{n*}, let us choose a coframe consisting of $n+1$ hyperplanes α^u which are connected with the points of the frame $\{A_u\}$ by the following condition:

$$(\alpha^u, A_v) = \delta_v^u. \tag{1.74}$$

This coframe is called *dual* to the frame $\{A_u\}$. Condition (1.74) means that the hyperplane α^u contains all points A_v, $v \neq u$, and that the condition of normalization $(\alpha^u, A_u) = 1$ holds.

We write the equations of infinitesimal displacement of the tangential frame $\{\alpha_u\}$ in the form:

$$d\alpha_u = \widetilde{\omega}_v^u \alpha^v, \quad u, v = 0, 1, \ldots, n. \tag{1.75}$$

Differentiating relations (1.74) and using equations (1.69) and (1.74), we arrive at the equation

$$\omega_v^u + \widetilde{\omega}_v^u = 0,$$

from which it follows that equations (1.75) take the form:

$$d\alpha_u = -\omega_v^u \alpha^v. \tag{1.76}$$

The structure equations (1.71) are the conditions for complete integrability of both equations (1.69) of infinitesimal displacement of a point frame and equations (1.76) of infinitesimal displacement of a tangential frame. Thus, if the forms ω_v^u depending on some number ρ, $\rho \leq n^2 + n$ parameters, are given, then, up to a projective transformation of the space P^n, they define in P^n a family of projective frames, and this family depends on ρ parameters. The location of this family of frames is completely determined by the location of a frame corresponding to initial values of parameters. Conversely, if in P^n a family of projective frames which depends on ρ parameters is given, then the components ω_v^u of infinitesimal displacement of this family are unchanged under its projective transformation. Hence, the forms ω_v^u are invariant forms with respect to transformations of the projective group.

2. As was already noted in Preface, a projective space can be used to represent all classical homogeneous spaces: affine, Euclidean, non-Euclidean, conformal and other spaces. To do this, one fixes certain invariant objects in a projective space P^n and reduces the group of transformations of the space by requiring that they be invariant. We will now show how this is carried out for the basic homogeneous spaces.

An *affine space* A^n is a projective space P^n in which a hyperplane α is fixed. This hyperplane is called the *ideal hyperplane* or the *hyperplane at infinity* (or the *improper hyperplane*). The *affine transformations* are those projective transformations that transform this hyperplane into itself. Straight lines of the space P^n which intersect the ideal hyperplane at the same point are called *parallel straight lines* of the space A^n. Two-dimensional planes of P^n intersecting the ideal hyperplane along the same straight line are called *parallel 2-planes* of the space A^n, etc.

As a frame in the space A^n, it is natural to take a projective frame whose points $A_1, \ldots, A_n$ lie in the ideal hyperplane α. The equations of infinitesimal displacement of such a frame have the form:

$$\begin{aligned} dA_0 &= \omega_0^0 A_0 + \omega_0^i A_i, \\ dA_i &= \omega_i^j A_j, \quad i, j = 1, \ldots, n. \end{aligned} \tag{1.77}$$

Equations (1.77) show that in this case the forms ω_i^0 in equations (1.69) are equal to zero: $\omega_i^0 = 0$. This and structure equations (1.71) imply that $d\omega_0^0 = 0$. Thus, the form ω_0^0 is a total differential: $\omega_0^0 = d \ln |\lambda|$. Substituting this value of the form ω_0^0 into the first equation of (1.77), we find that

$$dA_0 = \frac{d\lambda}{\lambda} A_0 + \omega_0^i A_i.$$

It follows that

$$d\frac{A_0}{\lambda} = \omega_0^i \frac{A_i}{\lambda}. \tag{1.78}$$

If we set

$$\frac{A_0}{\lambda} = x, \quad \frac{A_i}{\lambda} = e_i, \tag{1.79}$$

then equation (1.78) can be written as

$$dx = \omega_0^i e_i. \tag{1.80}$$

Differentiating the second equation of (1.79), we obtain

$$de_i = \theta_i^j e_j, \tag{1.81}$$

where

$$\theta_i^j = \omega_i^j - \delta_i^j \, d \ln |\lambda|.$$

We may consider the point x as the vertex of an affine frame and the vectors e_i as its basis vectors. Equations (1.80) and (1.81) are the equations of infinitesimal displacement of this affine frame $\{x, e_i\}$. These equations contain $n + n^2$ linearly independent forms ω_0^i and θ_i^j. This corresponds to the fact that the group of affine transformations of the space A^n depends on $n + n^2$ parameters. The forms ω_0^i determine a parallel displacement of the frame, and the forms θ_i^j determine the isotropy transformations of this frame which keep the point x invariant.

The structure equations of the space A^n can be obtained from equations (1.71). In fact, we derive from those equations that

$$d\omega_0^i = \omega_0^0 \wedge \omega_0^i + \omega_0^j \wedge \omega_j^i = \omega_0^j \wedge \theta_j^i,$$
$$d\theta_j^i = d\omega_j^i = \omega_j^k \wedge \omega_k^i = \theta_j^k \wedge \theta_k^i.$$

As a result, the structure equations of the affine space A^n have the form:

$$d\omega_0^i = \omega_0^j \wedge \theta_j^i, \quad d\theta_j^i = \theta_j^k \wedge \theta_k^i. \tag{1.82}$$

These equations imply that the isotropy transformations form an invariant subgroup in the group of affine transformations of the space A^n, and this subgroup is isomorphic to the general linear group $\mathbf{GL}(n)$. In addition, equations (1.82) imply that the parallel displacements form a subgroup which is not an invariant subgroup.

A *Euclidean space* E^n is obtained from an affine space A^n if in the ideal hyperplane of the latter space a nondegenerate imaginary quadric Q of dimension $n - 2$ is fixed. The equations of this quadric Q can be written in the form:

$$x^0 = 0, \quad \sum_{i=1}^{n} (x^i)^2 = 0. \tag{1.83}$$

The *Euclidean transformations* are those affine transformations that transform this quadric into itself.

The quadric Q allows us to define the scalar product (a, b) of vectors a and b in the Euclidean space E^n. If we take vectors of a frame in such a way that

$$(e_i, e_j) = \delta_{ij} \tag{1.84}$$

(here δ_{ij} is the Kronecker symbol: $\delta_{ii} = 1$ and $\delta_{ij} = 0$ if $i \neq j$), then the forms θ_j^i from equations (1.81) are connected by the relations:

$$\theta_i^j + \theta_j^i = 0, \tag{1.85}$$

which are obtained by differentiating equations (1.84). The number of independent forms in equations (1.80) and (1.81) is now equal to $n + \frac{1}{2}n(n-1) = \frac{1}{2}n(n+1)$. This number coincides with the number of parameters on which the group of motions of space E^n depends. The structure equations of the space E^n still have the form (1.82).

A *non-Euclidean space* is a projective space P^n in which a nondegenerate invariant hyperquadric

$$Q(X, X) = g_{uv} x^u x^v = 0, \quad u, v = 0, 1, \ldots, n, \tag{1.86}$$

is given. Suppose for definiteness that a non-Euclidean space is elliptic, i.e. the hyperquadric $Q(X, X)$ is positive definite. Then we may choose the points of a projective frame $\{A_u\}$ in such a way that they form an autopolar simplex with respect to this hyperquadric, and in addition, we normalize the vertices of this simplex. This means that we have

$$Q(A_u, A_v) = \delta_{uv}, \tag{1.87}$$

and the forms ω_u^v from equations (1.69) will satisfy

$$\omega_u^v + \omega_v^u = 0. \tag{1.88}$$

The *elliptic transformations* are those projective transformations of the space P^n that preserve the hyperquadric Q. These transformations depend on $\frac{1}{2}n(n+1)$ parameters, and the latter number coincides with the number of independent forms among the forms ω_v^u.

The geometry of a *conformal space* C^n can be realized as the geometry on a nondegenerate real hyperquadric Q of a projective space P^{n+1} of dimension $n+1$, and the group of conformal transformations of the space C^n is isomorphic to the subgroup of the group of projective transformations of the space P^n that keep the hyperquadric Q invariant.

3. In what follows, we will often use a special construction called the *projectivization.*

Let P^n be a projective space of dimension n, and let P^m be a subspace of dimension m, where $0 \leq m < n$. We say that two points x and y of P^n are *in the relation* P^m and will write this as $x P^m y$ if there exists the straight line xy intersecting the subspace P^m. It is easy to check that the introduced relation is an equivalence relation. Thus, the points which are in the relation P^m are called *P^m-equivalent.* This equivalence relation divides all points of the space P^n into the equivalence classes in such a way that all points of an $(m+1)$-plane P^{m+1} containing the subspace P^m belong to one class.

The equivalence relation introduced above allows us to factorize the space P^n by this relation. The resulting factor space P^n/P^m is called the *projectivization* of P^n with the *center* P^m and is denoted by $\mathcal{P}_m(P^n)$:

$$\mathcal{P}_m(P^n) = P^n/P^m.$$

The projectivization $\mathcal{P}_m(P^n)$ is a projective space of dimension $n-m-1$: $\mathcal{P}_m(P^n) = \widetilde{P}^{n-m-1}$. Let us take a basis in P^n in such a way that its points A_i, $i = 0, 1, \ldots, m$, belong to the center P^m of projectivization. Then the basis of the space $\widetilde{P}^{n-m-1}$ is formed by the points $\mathcal{P}_m(A_\alpha) = \widetilde{A}_\alpha$, $\alpha = m+1, \ldots, n$. Since the center P^m is unchanged under projectivization, the equations of infinitesimal displacement of the frame $\{A_i, A_\alpha\}$ of the space P^n can be written in the form:

$$dA_i = \omega_i^j A_j, \quad dA_\alpha = \omega_\alpha^\beta A_\beta + \omega_\alpha^i A_i.$$

Thus, in this family of frames we have $\omega_i^\alpha = 0$. Hence, the structure equations (1.71) of a projective space P^n imply that

$$d\omega_\alpha^\beta = \omega_\alpha^\gamma \wedge \omega_\gamma^\beta. \tag{1.89}$$

This allows us to consider the forms ω_α^β as the components of infinitesimal displacement of the frame $\{\widetilde{A}_\alpha\}$ of the projectivization $\widetilde{P}^{n-m-1} = \mathcal{P}_m(P^n)$, so that

$$d\widetilde{A}_\alpha = \omega_\alpha^\beta \widetilde{A}_\beta.$$

On some occasions we will identify the points $\widetilde{A}_\alpha$ of the projectivization $\mathcal{P}_m(P^n)$ with the points A_α of the projective space P^n.

Note that one can also consider the projectivization of a vector space L^m by its 0-dimensional subspace $\{0\}$. The result of this projectivization is $P^{m-1} = L^m \backslash \{0\}$. Actually, in the definition of projective space P^n itself (see subsection **1.3.1**), we already used the projectivization, so that $P^n = L^{n+1} \backslash \{0\}$.

4. We consider a point x in the space P^n and associate with this point a family of frames such that $A_0 = x$. If the point x is fixed, then this family depends on $n(n+2) - n = n(n+1)$ parameters. Analytically, this family of frames is defined by the equations

$$\omega_0^i = 0, \quad i = 1, \ldots, n, \tag{1.90}$$

and the equations of infinitesimal displacement of this family of frames have the form:

$$\delta A_0 = \pi_0^0 A_0, \quad \delta A_i = \pi_i^0 A_0 + \pi_i^j A_j, \tag{1.91}$$

where δ is the operator of differentiation with the point x fixed, i.e. when $\omega_0^i = 0$, and

$$\pi_u^v = \omega_u^v(\delta) = \omega_u^v \Big|_{\omega_0^i = 0}. \tag{1.92}$$

The equations (1.91) are the equations of infinitesimal displacement of a frame of the stationary subgroup of the point A_0, and the forms in these equations are invariant forms of this subgroup.

Since by equations (1.71) we have

$$d\omega_0^i = \omega_0^0 \wedge \omega_0^i + \omega_0^j \wedge \omega_j^i,$$

then by the Frobenius theorem, the system (1.90) is completely integrable, and its independent first integrals are the nonhomogeneous coordinates of the point $x = A_0$.

Equations (1.90) define a subbundle of the bundle of frames $\{A_u\}$ of the space P^n. The base of this subbundle is the space P^n itself, and its fiber over a point x is the family of frames with $A_0 = x$. This subbundle can be denoted by

$$\pi : \mathcal{R}(P^n) \to P^n \tag{1.93}$$

where π is given by

$$\pi\{A_0, A_1, \ldots, A_n\} = A_0.$$

The forms ω_0^i, $i = 1, \ldots, n$, which for brevity we will denote simply by ω^i, are the *base forms* of this subbundle. These forms are also called the *horizontal forms* for the subbundle (1.93). The forms π_0^0, π_i^0 and π_i^j in equations (1.91) are its *fiber forms*.

In a similar manner, we can find the stationary subgroup of a hyperplane ξ of P^n. We associate a family of frames with this hyperplane in such a way that the points $A_1, \ldots, A_n$ belong to the hyperplane ξ and the normalization condition $(\xi, A_0) = 1$ holds. Then the hyperplane ξ coincides with the basis hyperplane α^0 from the coframe $\{\alpha^u\}$. By means of equations (1.76), the condition for this hyperplane to be fixed is

$$\omega_i^0 = 0, \quad i = 1, \ldots, n, \tag{1.94}$$

and the equations of infinitesimal displacement in the family of frames associated with the hyperplane α^0 are

$$\delta\alpha^0 = -\pi_0^0\alpha^0, \quad \delta\alpha^i = -\pi_0^i\alpha^0 - \pi_j^i\alpha^j. \tag{1.95}$$

The 1-forms on the right-hand sides of these equations are invariant forms of the stationary subgroup of hyperplane $\alpha^0 = \xi$.

The system of equations (1.94) is completely integrable and determines a subbundle

$$\pi^* : \mathcal{R}(P^{n*}) \to P^{n*} \tag{1.96}$$

in the bundle of coframes, where

$$\pi^*\{\alpha^0, \alpha^1, \ldots, \alpha^n\} = \alpha^0.$$

The forms ω_i^0 are the base forms of this subbundle, and the forms π_0^0, π_0^i and π_j^i are its fiber forms.

We further consider an m-dimensional subspace P^m of a projective space P^n and find its stationary subgroup. We place the points A_i, $i = 0, 1, \ldots, m$, in this subspace P^m. If the subspace P^m is fixed, then the equations of infinitesimal displacement of frames associated with this subspace have the form:

$$\delta A_i = \pi_i^j A_j, \quad \delta A_\alpha = \pi_\alpha^i A_i + \pi_\alpha^\beta A_\beta, \tag{1.97}$$

where $i, j = 0, 1, \ldots, m$ and $\alpha, \beta = m+1, \ldots, n$. It follows that this family of frames is defined by

$$\omega_i^\alpha = 0. \tag{1.98}$$

The symbol δ in equations (1.97) is the operator of differentiation when the subspace P^m is fixed (i.e. when $\omega_i^\alpha = 0$), and $\pi_v^u = \omega_v^u(\delta) = \omega_v^u\Big|_{\omega_i^\alpha=0}$. Since the number of forms ω_i^α is equal to $(m+1)(n-m)$, the stationary subgroup of the subspace P^m depends on $n(n+2)-(m+1)(n-m)$ parameters. The complete integrability of equations (1.98) can be verified directly. The independent first integrals of this system are nonhomogeneous coordinates of the subspace P^m in the space P^n. Their number is $(m+1)(n-m)$, and this number coincides with the dimension of the manifold of m-dimensional subspaces of P^n. This manifold is called the *Grassmannian* of m-dimensional subspaces in P^n and is denoted by $G(m,n)$.

The system of equations (1.98) is completely integrable and defines a subbundle of the set $\mathcal{R}(P^n)$ of frames in P^n. The forms ω_i^α are base forms of this subbundle, and the forms π_i^j, π_α^i and π_α^β are its fiber forms.

The subspace P^m is the intersection of the hyperplanes α^β, $\beta = m+1, \ldots, n$, of the tangential frame. It follows from equations (1.76) that the invariance condition for the subspace P^m leads to the same equations (1.98), and the stationary subgroup of the subspace P^m can be written as

$$\delta\alpha^\beta = -\pi_\gamma^\beta \alpha^\gamma, \quad \delta\alpha^i = -\pi_j^i \alpha^j - \pi_\beta^i \alpha^\beta. \tag{1.99}$$

This implies that the Grassmannian $G(m,n)$ of m-dimensional subspaces in a projective space P^n can be also defined as the Grassmannian $G(n-m, n)$ of $(n-m)$-dimensional subspaces in the dual space P^{n*}.

1.4 Some Algebraic Manifolds

1. We now consider some algebraic varieties in a projective space, which we will need in our considerations.

First of all, we will study in more detail the Grassmannian $G(m,n)$ of m-dimensional subspaces in a projective space P^n. Consider a fixed frame $\{E_u\}$ in P^n and denote the coordinates of a point X relative to this frame by x^u. Thus, we have $X = x^u E_u$. Let P^m be an m-dimensional subspace in P^n. Let us take $m+1$ linearly independent points X_i, $i = 0, 1, \ldots, m$, in the subspace P^m. We will call them *basis points* of the P^m. Let us write the coordinates of the points X_i relative to the frame $\{E_u\}$ in the form of a matrix:

$$(x_i^u) = \begin{pmatrix} x_0^0 & x_0^1 & \dots & x_0^n \\ x_1^0 & x_1^1 & \dots & x_1^n \\ \dots & \dots & \dots & \dots \\ x_m^0 & x_m^1 & \dots & x_m^n \end{pmatrix}. \tag{1.100}$$

Consider the minors $p^{i_0 i_1 \dots i_m}$ of order $m+1$ of this matrix:

$$p^{i_0 i_1 \dots i_m} = \begin{vmatrix} x_0^{i_0} & x_0^{i_1} & \dots & x_0^{i_m} \\ x_1^{i_0} & x_1^{i_1} & \dots & x_1^{i_m} \\ \dots & \dots & \dots & \dots \\ x_m^{i_0} & x_m^{i_1} & \dots & x_m^{i_m} \end{vmatrix}. \tag{1.101}$$

Since the matrix has $m+1$ rows and $n+1$ columns, the total number of such minors is equal to $\binom{n+1}{m+1}$. If we change the basis in the subspace P^m, the matrix (1.100) will also change but all its minors will be multiplied by the same factor, namely the determinant of the basis change matrix. Thus, these minors can be taken as homogeneous projective coordinates of a point in the projective space P^N of dimension $N = \binom{n+1}{m+1} - 1$. These coordinates are called the *Grassmann coordinates* of the $P^m \subset P^n$. It is easy to see that these coordinates are skew-symmetric, and that not proportional sets of Grassmann coordinates correspond to different m-dimensional subspaces.

The Grassmann coordinates $p^{i_0 i_1 \dots i_m}$ are not independent—they satisfy the sequence of quadratic relations:

$$p^{i_0 i_1 \dots i_{m-1}[i_m} p^{j_0 j_1 \dots j_m]} = 0, \tag{1.102}$$

which follows from equations (1.100) and (1.101) (see, for example, [HP 47]). In formulas (1.102) (and many other formulas of this book), the square brackets enclosing some (or all) upper (or lower) indices denote the alternation with respect to the enclosed indices while the parentheses in the indices denote the symmetrization. For example,

$$\begin{aligned} &t^{[ij]} = \tfrac{1}{2}(t^{ij} - t^{j,i}), \quad t^{(ij)} = \tfrac{1}{2}(t^{ij} + t^{j,i}), \\ &t^{[ijk]} = \tfrac{1}{3!}(t^{ijk} + t^{jki} + t^{kij} - t^{jik} - t^{kji} - t^{ikj}), \\ &t^{(ijk)} = \tfrac{1}{3!}(t^{ijk} + t^{jki} + t^{kij} + t^{jik} + t^{kji} + t^{ikj}). \end{aligned}$$

Relations (1.102) define in the space P^N an algebraic variety of dimension $(m+1)(n-m)$, which is the number of linearly independent basis forms ω_i^α on the Grassmannian. We will denote this algebraic variety by $\Omega(m,n)$. There is a one-to-one correspondence between m-dimensional subspaces P^m of P^n and the points of the variety $\Omega(m,n)$. This correspondence defines the mapping $\varphi : G(m,n) \to \Omega(m,n)$, called the *Grassmann mapping*.

As an example, we consider the Grassmannian $G(1,3)$ – the manifold of straight lines of the three-dimensional projective space P^3. In this case the matrix (1.100) takes the form:

$$\begin{pmatrix} x_0^0 & x_0^1 & x_0^2 & x_0^3 \\ x_1^0 & x_1^1 & x_1^2 & x_1^3 \end{pmatrix}.$$

Its minors

$$p^{i_0 i_1} = \begin{vmatrix} x_0^{i_0} & x_0^{i_1} \\ x_1^{i_0} & x_1^{i_1} \end{vmatrix}$$

are usually called the *Plücker coordinates* of the straight line l defined by the points X_0 and X_1. Since $\binom{4}{2} = 6$, the minors are homogeneous projective coordinates of a point in the space P^5. It is easy to prove that these coordinates satisfy the single quadratic equation:

$$p^{01}p^{23} + p^{02}p^{31} + p^{03}p^{12} = 0.$$

Therefore, the variety $\Omega(1,3)$ is a hyperquadric in P^5, called the *Plücker hyperquadric.*

Let us study the structure of the Grassmannian $G(m,n)$ and its image $\Omega(m,n)$ in the space P^N, where $N = \binom{n+1}{m+1} - 1$. Let p and q be two m-dimensional subspaces of P^n having in common an $(m-1)$dimensional subspace P^{m-1}. These two subspaces generate a linear pencil $\lambda p + \mu q$ of m-dimensional subspaces. A straight line of the variety $\Omega(m,n)$ corresponds to this pencil. All subspaces of the pencil belong to the same subspace P^{m+1} of dimension $m+1$, and a pair of subspaces $P^{m-1} \subset P^{m+1}$ completely defines the pencil and therefore a straight line on $\Omega(m,n)$.

Consider further an $(n-m)$-bundle of m-dimensional subspaces passing through a fixed subspace P^{m-1}. An $(n-m)$-dimensional plane generator ξ^{n-m} of the variety $\Omega(m,n)$ corresponds to this bundle. Since the space P^n contains the $m(n-m+1)$-dimensional family of subspaces P^{m-1}, the variety $\Omega(m,n)$ carries a family of $(n-m)$-dimensional plane generators ξ^{n-m}, and the latter family depends on $m(n-m+1)$ parameters.

Let P^{m+1} be a fixed $(m+1)$-dimensional subspace in P^n. Consider all its m-dimensional subspaces P^m. They form a plane field of dimension $m+1$. An $(m+1)$-dimensional plane generator η^{m+1} of the variety $\Omega(m,n)$ corresponds to this field. Since P^n contains the $(m+2)(n-m-1)$-parameter family of subspaces P^{m+1}, the variety $\Omega(m,n)$ carries an $(m+2)(n-m-1)$-parameter family of plane generators η^{m+1}.

If $P^{m-1} \subset P^{m+1}$, then the plane generators ξ^{n-m} and η^{m+1} of the variety $\Omega(m,n)$ corresponding to these subspaces intersect each other along a straight line. Otherwise, they do not have common points.

Next, consider in P^n a fixed subspace P^m. It contains an m-parameter family of subspaces P^{m-1}. Thus, an m-parameter family of generators ξ^{n-m} passes through the point $p \in \Omega(m,n)$ corresponding to the P^m. There is also an $(n-m-1)$-parameter family of subspaces P^{m+1} passing through the same subspace P^m. Thus, an $(n-m-1)$-parameter family of generators η^{m+1} passes through the point $p \in \Omega(m,n)$. Moreover, any two generators ξ^{n-m} and η^{m+1} passing through the point p have a straight line as their intersection. It follows that all plane generators ξ^{n-m} and η^{m+1} passing through the point $p \in \Omega(m,n)$ are generators of a cone with its vertex at the point p, and this cone is located on the variety $\Omega(m, n)$. We will denote this cone by $C_p(n-m, m+1)$ and call it the *Segre cone.* The projectivization of the Segre cone with the center at a point p is the *Segre variety* $S(n-m-1, m)$ which we will study later.

In the space P^n, to the Segre cone $C_p(n-m, m+1)$ there corresponds the set of all m-dimensional subspaces intersecting a fixed subspace P^m along the subspace of dimension $m-1$. It follows that the dimension of the Segre cone $C_p(n-m, m+1)$ is equal to n.

2. The so-called determinant submanifolds are interesting examples of submanifolds in a projective space.

Consider a projective space P^N of dimension $N = ml + m + l$ in which projective coordinates are matrices (x_i^α) with $i = 0, 1, \ldots, m$; $\alpha = 0, 1, \ldots, l$, and we suppose $m \leq l$. A *determinant manifold* is defined by the condition

$$1 \leq \operatorname{rank}(x_i^\alpha) \leq r, \quad r \leq m. \tag{1.103}$$

Consider first the extreme case $r = 1$. In this case, the matrix (x_i^α) has the form of a simple dyad:

$$x_i^\alpha = t^\alpha s_i, \tag{1.104}$$

where t^α and s_i are homogeneous parameters which can be taken as coordinates of points in the spaces P^l and P^{m*}.

The determinant manifold defined by equation (1.104) is called the *Segre variety* and is denoted by $S(m, l)$. This variety carries two families of plane generators $s_i = \lambda c_i$ and $t^\alpha = \mu c^\alpha$ where c_i and c^α are constants. The generators of these two families are of dimensions l and m, respectively. The Segre variety is an embedding

$$P^l \times P^{m*} \rightarrow P^N \tag{1.105}$$

of the direct product of the spaces P^l and P^{m*} into the space P^N, and the dimension of the Segre variety is $l + m$.

Suppose now that in relation (1.103) the rank $r = 2$. In this case the entries of the matrix (x_i^α) can be written in the form:

$$x_i^\alpha = \lambda \ {}'t^\alpha s_i' + \mu \ {}''t^\alpha s_i'', \tag{1.106}$$

i.e. the matrix (x_i^α) is a linear combination of two simple dyads. Each of these dyads determines a point on the Segre variety $S(m, l)$. If the parameters λ and μ vary, the point of the space P^N with coordinates x_i^α describes a straight line —the *bisecant* of the Segre variety $S(m, l)$. Thus, if $r = 2$, equation (1.103) defines the *bisecant variety* for the Segre variety $S(m, l)$.

Similarly, for any r, the determinant manifold (1.103) is a family of $(r-1)$-secant subspaces for the Segre variety $S(m, l)$.

For example, if $m = l = 1$, then $N = 3$, and the equations of the Segre variety $S(1, 1)$ can be written as

$$x_0^0 = t^0 s_0, \quad x_1^0 = t^0 s_1,$$
$$x_0^1 = t^1 s_0, \quad x_1^1 = t^1 s_1. \tag{1.107}$$

Eliminating the parameters t^α and s_i from these equations, we arrive at the quadratic equation:

$$x_0^0 x_1^1 - x_0^1 x_1^0 = 0, \tag{1.108}$$

defining in the space P^3 a ruled surface of second order which carries two families of rectilinear generators: $s_i = \text{const}$ and $t^\alpha = \text{const}$. This surface is an embedding of the direct product $P^1 \times P^{1*}$ into the space P^3.

Next, we consider another type of determinant manifold defined in a projective space P^N of dimension $N = \frac{1}{2}(m+1)(m+2) - 1$, where projective coordinates are symmetric matrices (x^{ij}), $i, j = 0, 1, \ldots, m$, by the equation:

$$\operatorname{rank}(x^{ij}) = r, \quad r \leq m. \tag{1.109}$$

If $r = 1$, then each entry of a matrix (x^{ij}) is the tensorial square of a vector t^i:

$$x^{ij} = t^i t^j. \tag{1.110}$$

The parameters t^i can be considered as homogeneous coordinates of a point in a projective space P^m. Thus, the manifold defined by equations (1.110) is a symmetric embedding of the P^m into the P^N:

$$\nu : P^m \longrightarrow P^N.$$

The manifold (1.110) is called the *Veronese variety* and is denoted by $V(m)$. Its dimension is m.

If $r > 1$, the determinant manifold (1.109) is the variety of $(r-1)$-secant subspaces for the Veronese variety $V(m)$.

As an example of the Veronese variety, we consider the case $m = 2$. Then $N = 5$, and the variety $V(2)$ defined by equation (1.110) for $i, j = 0, 1, 2$, is a symmetric embedding of the two-dimensional projective plane into the space P^5. The variety $V(2)$ is a two-dimensional surface of fourth order in P^5 (see, for example, [SR 85]).

Note some properties of the Veronese surface $V(2)$. To each straight line of the plane P^2 there corresponds a curve of second order on the Veronese surface $V(2)$, and this surface carries a two-parameter family of second order curves. Through each point of the surface $V(2)$ there passes a one-parameter family of such curves, and through any pair of points of the surface $V(2)$ there passes a unique curve of this family. Two-dimensional planes in P^5 containing these curves are called *conisecant planes* of the surface $V(2)$.

To the second order curve defined by the equation

$$a_{ij} t^i t^j = 0 \tag{1.111}$$

in the plane P^2, there corresponds a fourth order curve on the Veronese surface $V(2)$. This curve is the intersection of the Veronese surface $V(2)$ with the hyperplane

$$a_{ij}x^{ij} = 0. \tag{1.112}$$

If the curve (1.111) is composed of two straight lines, then the corresponding fourth order curve is decomposed into two conics. For curves of this type, we have $\det(a_{ij}) = 0$, and the hyperplane (1.112) defining this curve is tangent to the Veronese surface $V(2)$ at a point of intersection of these two conics. If the curve (1.111) is a double straight line, then $a_{ij} = a_i a_j$, and the hyperplane (1.112) is tangent to the Veronese surface $V(2)$ along a double conic.

If $r = 2$, the manifold defined in the space P^5 by equation (1.109) is a hypercubic defined by the equation

$$\begin{vmatrix} x^{00} & x^{01} & x^{02} \\ x^{10} & x^{11} & x^{12} \\ x^{20} & x^{21} & x^{22} \end{vmatrix} = 0, \quad x^{ij} = x^{ji}, \tag{1.113}$$

and called the *cubic symmetroid.* This hypercubic is a bisecant variety for the Veronese surface $V(2)$. It carries two families of two-dimensional plane generators. One of these families consists of conisecant planes of the surface $V(2)$, and the second consists of two-dimensional planes tangent to this surface. The Veronese surface $V(2)$ is the manifold of singular points of the cubic symmetroid (1.113).

NOTES

1.2. For more detail on differentiable manifolds see, for example, the books [KN 63] or [D 71] and on the theory of systems of Pfaffian equations in involution the books [BCGGG 91], [Ca 45], [Fi 48], [Gr 83] and [GJ 87].

A more detailed presentation of the foundations of the theory of affine connections can be found in the books [KN 63] and [Lich 55] (see also the papers [Lap 66] and [Lap 69]).

1.3. For more detail on the notion of a multidimensional projective space see the book [D 64] by Dieudonné and the paper [GH 79] by Griffiths and Harris.

1.4 On Grassmann coordinates see, for example, the book [HP 47].

On Veronese variety see the book [SR 85]. The embedding (1.110) generating the Veronese variety was considered in many papers and books from different points of view (see [CDK 70], [EH 87], [GH 78], [GH 79], [J 89], [LP 71], [Nom 76], [NY 74], [Sas 91], [SegC 21a], [SegC 21b], [SegC 22] and [Sev 01]).

Chapter 2

The Foundations of Projective Differential Geometry of Submanifolds

2.1 Submanifolds in a Projective Space and Their Tangent Subspaces

1. Let M be a connected m-dimensional differentiable manifold, and let f be a nondegenerate differentiable mapping (an immersion) of M into a projective space P^n:

$$f : M \to P^n$$

where $m < n$. The image $V^m = f(M)$ of this mapping is called an *m-dimensional submanifold* in the projective space P^n. If t^i, $i = 1, \ldots, m$, are differentiable coordinates on the manifold M, then the submanifold V^m can be given locally by the equations

$$x^u = x^u(t^i), \quad u = 0, 1, \ldots, n, \tag{2.1}$$

where $x^u(t^i)$ are differentiable functions of the variables t^i and the rank of the matrix $(\frac{\partial x^u}{\partial t^i})$ is equal to m. Since x^u are homogeneous coordinates of a point x of the space P^n, the functions x^u admit multiplication by a common factor.

The submanifold V^m can be also given locally by a system consisting of $n - m$ independent equations of the form:

$$F^\alpha(x^0, x^1, \ldots, x^n) = 0, \quad \alpha = m + 1, \ldots, n, \tag{2.2}$$

where F^α are homogeneous differentiable functions. In a neighborhood of a nonsingular point x, the Jacobi matrix $\left(\frac{\partial F^\alpha}{\partial x^u}\right)$ is of rank $n - m$. Hence we may assume that equations (2.2) can be solved for the variables x^α:

$$x^\alpha = x^\alpha(x^0, x^1, \ldots, x^m) = 0, \quad \alpha = m+1, \ldots, n. \tag{2.3}$$

Here the right-hand sides are homogeneous functions of first degree. Therefore, these right-hand sides as well as the right-hand sides of equations (2.1) contain m essential variables which determine the location of a point on the submanifold V^m. If we set $x^i/x^0 = t^i$, we reduce equations (2.3) to the form (2.1).

In Section **1.4** we considered some algebraic submanifolds in a projective space. Certainly, those are differentiable manifolds. Moreover, equations (1.98) defining the image $\Omega(m,n)$ of the Grassmannian $G(m,n)$ in the space P^N, where $N = \binom{n+1}{m+1} - 1$, are of form (2.2), and equations (1.100) and (1.106), defining the Segre and Veronese varieties respectively, are of form (2.1). However, the parameters in equations (1.100) and (1.106) are homogeneous while the parameters in equations (2.1) are nonhomogeneous. But as we indicated above for equation (2.3), in a neighborhood of a nonsingular point, it is easy to change homogeneous parameters for nonhomogeneous ones.

2. Let V^m be a differentiable submanifold in the projective space P^n and let x be its nonsingular point. In what follows we will always assume that a point $x \in V^m$ under consideration is nonsingular without specifying this additionally. Consider all smooth curves passing through a point $x \in V^m$. The tangent lines to these curves at the point x, lie in an m-dimensional subspace $T_x(V^m)$ of the space P^n, called the *tangent subspace to the submanifold* V^m *at the point* x. For brevity, we will also use the symbol $T_x^{(1)}$ for the subspace $T_x(V^m)$.

We associate a family of moving frames $\{A_u\}$, $u = 0, 1, \ldots, n$, with each point $x \in V^m$, and we assume that for all these frames the point A_0 coincides with the point x, and the points A_i, $i = 1, \ldots, m$, lie in the tangent subspace $T_x^{(1)}$. The frames of this family are called *first order frames*. Since the differential $dx = dA_0$ of the point x belongs to the tangent subspace $T_x^{(1)}$, its decomposition with respect to the vertices of the frame $\{A_u\}$ can be written as:

$$dA_0 = \omega_0^0 A_0 + \omega_0^i A_i. \tag{2.4}$$

Thus, in the space P^n, the submanifold V^m along with the family of first order frames is defined by the following system of Pfaffian equations:

$$\omega_0^\alpha = 0, \ \alpha = m+1, \ldots, n, \tag{2.5}$$

and the forms ω_0^i in equation (2.4) are linearly independent and form a cobasis in the tangent subspace $T_x^{(1)}$. For brevity, we denote these forms by ω^i:

$$\omega_0^i = \omega^i.$$

By the structure equations (1.67) of a projective space P^n and by equations (2.5), the exterior differentials of the forms ω^i can be written as

$$d\omega^i = \omega^j \wedge (\omega^i_j - \delta^i_j \omega^0_0).$$

This implies that the 1-forms

$$\theta^i_j = \omega^i_j - \delta^i_j \omega^0_0 \tag{2.6}$$

are the base forms of the frame bundle $\mathcal{R}^1(M)$ of first order frames on the manifold M of parameters of the submanifold V^m. The forms ω^i are the basis forms of the manifold M as well as of the submanifold V^m. By relation (1.62), if the point x is fixed, the forms ω^i satisfy the differential equations:

$$\delta\omega^i + \omega^j(\pi^i_j - \delta^i_j \pi^0_0) = 0, \tag{2.7}$$

where, as in Chapter 1, the symbol δ denotes the restriction of the differential d to the fiber $\mathcal{R}^1_x$ of the frame bundle $\mathcal{R}^1(M)$, and $\pi^u_v = \omega^u_v(\delta)$.

If the point x is fixed on the submanifold V^m, then the forms ω^i vanish: $\omega^i = 0$. In this case, the tangent subspace $T^{(1)}_x$ is also fixed. Hence the forms ω^α_i also vanish. Thus, if the point x is fixed, then the admissible transformations of the moving frames are determined by the following derivational equations:

$$\delta A_0 = \pi^0_0 A_0, \quad \delta A_i = \pi^0_i A_0 + \pi^j_i A_j, \quad \delta A_\alpha = \pi^0_\alpha A_0 + \pi^i_\alpha A_i + \pi^\beta_\alpha A_\beta. \tag{2.8}$$

The 1-forms $\pi^0_0, \pi^0_i, \pi^j_i, \pi^0_\alpha, \pi^i_\alpha$ and π^β_α in (2.8) define the group of transformations of first order frames associated with the point $x = A_0$. This group is called the ***stationary subgroup*** of the plane element $(x, T^{(1)}_x)$ of V^m.

Since the family of first order frames is associated with each point x of the submanifold V^m, the *bundle* $\mathcal{R}^1(V^m)$ *of frames of first order* is defined on the whole submanifold V^m. The base of this bundle is the submanifold V^m itself, its base forms are the forms ω^i, its typical fiber is a set of first order frames associated with a point $x = A_0$, and its fiber forms are the forms $\omega^0_0, \omega^0_i, \omega^j_i, \omega^0_\alpha, \omega^i_\alpha$ and ω^β_α.

Consider the projectivization $\mathcal{P}_0 T^{(1)}_x$ of the tangent subspace $T^{(1)}_x$ with the center $A_0 = x$ (see subsection **1.3.3**). This projectivization is a projective space $\widetilde{P}^{m-1}$ whose elements are the straight lines of the space $T^{(1)}_x$ passing through the point x.

As we indicated in Section **1.3**, this projectivization defines an equivalence relation in the set of points of the space $T^{(1)}_x$. This explains why it is natural to denote this projectivization by $T^{(1)}_x / A_0$:

$$\mathcal{P}_0 T^{(1)}_x = T^{(1)}_x / A_0.$$

A frame in the space $\mathcal{P}_0 T^{(1)}_x = \widetilde{P}^{m-1}$ is formed by the points $\widetilde{A}_i = \mathcal{P}_0(A_i)$, and the forms ω^i become homogeneous coordinates of the point $\widetilde{Y} \in \widetilde{P}^{m-1}$, i.e.

$$\widetilde{Y} = \omega^i \widetilde{A}_i. \tag{2.9}$$

Consider also the projectivization of the space P^n with the tangent subspace $T_x^{(1)}$ as the center of projectivization. The elements of this projectivization are $(m+1)$-dimensional subspaces of the space P^n containing the m-dimensional subspace $T_x^{(1)}$. We denote this projectivization by $\mathcal{P}_m P^n = P^n/T_x^{(1)}$; it is a projective space $\widetilde{P}^{n-m-1}$. The basis points of the space $\widetilde{P}^{n-m-1}$ are the points $\widetilde{A}_\alpha = \mathcal{P}_m(A_\alpha)$, determined by $(m+1)$-dimensional subspaces passing through the points A_α and the center $T_x^{(1)}$ of projectivization. The space $\widetilde{P}^{n-m-1} = P^n/T_x^{(1)}$ is called the *first normal subspace* of the submanifold V^m at its point x and is denoted by $N_x(V^m) = N_x^{(1)}$.

3. We will illustrate the notions introduced above by considering some examples. Consider first the Grassmannian $G(m,n)$. As we did in Section **1.4**, with each element $p = P^m$ of $G(m,n)$ we associate a family of moving frames whose points A_i, $i = 0, 1, \dots, m$, span the subspace P^m. Then we have

$$dA_i = \omega_i^j A_j + \omega_i^\alpha A_\alpha, \tag{2.10}$$

where ω_i^α are the basis forms of $G(m,n)$.

The subspace P^m can be represented as

$$p = A_0 \wedge A_1 \wedge \dots \wedge A_m,$$

where the symbol $\wedge$ denotes the exterior product. Differentiating this equation and using formulas (2.10), we obtain

$$dp = \omega p + \omega_i^\alpha p_\alpha^i, \tag{2.11}$$

where $\omega = \omega_0^0 + \omega_1^1 + \dots + \omega_m^m$, and

$$p_\alpha^i = A_0 \wedge \dots \wedge A_{i-1} \wedge A_\alpha \wedge A_{i+1} \wedge \dots \wedge A_m \tag{2.12}$$

are m-dimensional subspaces in P^n which are those faces of the coordinate simplex that are adjacent to the face p.

Under the Grassmann mapping, to the subspaces p and p_α^i there correspond points on the manifold $\Omega(m,n)$, where $N = \binom{n+1}{m+1} - 1$. We will denote these points by the same letters as the subspaces corresponding to these points. These points define the tangent subspace to the manifold $\Omega(m,n)$ at the point p. It is easy to see from (2.11) that the dimension of this subspace is equal to $\rho = (m+1)(n-m)$, i.e. this dimension coincides with the dimension of the Grassmannian. Since this dimension is the same at all points of the manifold $\Omega(m,n)$, this manifold does not have singular points.

Formula (2.11) can be written as

$$dp/p = \omega_i^\alpha \widetilde{p}_\alpha^i.$$

Thus, the projectivization of the tangent subspace $T_p^{(1)}$ to the manifold $\Omega(m,n)$ by the point p is isomorphic to the space of homeomorphisms $\mathrm{Hom}(\widetilde{P}^{m-1}, \widetilde{P}^{n-m-1})$ (see [GH 79]).

As our next example, we will take the Segre variety $S(m,l)$ belonging to a projective space P^N of dimension $N = lm + l + m$. Since this variety is an embedding $P^l \times P^{m*} \to P^N$, its element is the tensor product $Y \otimes \eta$, where $Y \in P^l$ and $\eta \in P^{m*}$ (see formula (1.100)). To construct a moving frame associated with the Segre variety $S(m,l)$, we consider the frames $\{A_\alpha\}$, $\alpha = 0, 1, \ldots, l$, in P^l and $\{\xi^i\}$, $i = 0, 1, \ldots m$, in P^{m*} whose equations of infinitesimal displacement have the form:

$$dA_\alpha = \omega_\alpha^\beta A_\beta, \quad d\xi^i = \theta_j^i \xi^j. \tag{2.13}$$

A moving frame associated with $S(m,l)$ is formed by the points

$$p_\alpha^i = A_\alpha \otimes \xi^i. \tag{2.14}$$

Differentiating these points, we find the equations of infinitesimal displacement of frames in the space P^N:

$$dp_\alpha^i = \omega_\alpha^\beta p_\beta^i + \theta_j^i p_\alpha^j. \tag{2.15}$$

If we take an arbitrary point y of the space P^l as the point A_0 and an arbitrary hyperplane η of the space P^{m*} as the hyperplane ξ^0, $A_0 = Y$, $\xi^0 = \eta$, then for an arbitrary point $p = Y \otimes \eta$ of the Segre variety $S(m,l)$ we have

$$p = Y \otimes \eta = A_0 \otimes \xi^0 = p_0^0.$$

Therefore,

$$dp_0^0 = (\omega_0^0 + \theta_0^0) p_0^0 + \omega_0^{\hat\beta} p_{\hat\beta}^0 + \theta_{\hat j}^0 p_0^{\hat j}, \tag{2.16}$$

where $\hat\beta = 1, \ldots, l$ and $\hat j = 1, \ldots, m$. These relations show that the tangent space $T_p^{(1)}$ to the Segre variety $S(m,l)$ at each of its points p is determined by the linearly independent points $p_0^0, p_{\hat\beta}^0$ and $p_0^{\hat j}$ and has the dimension $m + l$, which is equal to the dimension of the variety $S(m,l)$. Thus, this manifold also does not have singularities.

Finally, we consider the Veronese variety $V(m)$ embedded in the space P^N of dimension $N = \frac{1}{2}(m+1)(m+2) - 1$. The element of this variety is $p = X \otimes X$, where the point $X \in P^m$ (see formula (1.106)). Suppose that the points A_i, $i = 0, 1, \ldots, m$, form a frame in the space P^m. We write the infinitesimal displacement of this frame in the form:

$$dA_i = \omega_i^j A_j. \tag{2.17}$$

Let $p_{ij} = \frac{1}{2}(A_i \otimes A_j + A_j \otimes A_i)$ be a symmetric tensor product of the points A_i and A_j. The points p_{ij} form a frame associated with the Veronese variety $V(m)$ in the space P^N. If $i = j$, the points belong to $V(m)$, and if $i \neq j$, this is not the case. By (2.17), differentiation of the points p_{ij} gives

$$dp_{ij} = \omega_i^k p_{kj} + \omega_j^k p_{ik}. \tag{2.18}$$

Then, for the point p_{00}, we find from (2.18) that

$$dp_{00} = 2\omega_0^0 p_{00} + 2\omega_0^{\hat{i}} p_{\hat{i}0}. \tag{2.19}$$

where $\hat{i} = 1, \ldots, m$. This formula shows that the tangent space $T_p^{(1)}$ to the variety $V(m)$ is determined by the linearly independent points p_{00} and $p_{\hat{i}0}$ and has the dimension m, which is equal to the dimension of this variety. This proves that the Veronese variety $V(m)$ does not have singular points.

We will encounter manifolds with singular points when we will study the determinant submanifolds defined by equations (1.99) and (1.105) in the case $r > 1$. We will study such manifolds in detail in Chapter 4.

2.2 The Second Fundamental Form of a Submanifold

The further investigation of a submanifold V^m in a projective space P^n is concerned with differential prolongations of the equations (2.5) defining this submanifold along with the family of first order moving frames associated with it. Exterior differentiation of these equations gives:

$$\omega^i \wedge \omega_i^\alpha = 0. \tag{2.20}$$

Applying the Cartan lemma to these exterior equations, we obtain the expressions of the forms ω_i^α in terms of the basis forms ω^i of the submanifold V^m:

$$\omega_i^\alpha = b_{ij}^\alpha \omega^j, \quad b_{ij}^\alpha = b_{ji}^\alpha. \tag{2.21}$$

As we noted earlier, the forms $\{\omega_0^\alpha, \omega_i^\alpha\}$ are the basis forms of the Grassmannian $G(m,n)$. Equations (2.5) and (2.21) define a mapping of V^m into this Grassmannian:

$$\gamma : V^m \to G(m,n), \tag{2.22}$$

called the *Gauss mapping*. We will denote the image of the submanifold V^m under the mapping γ by $\gamma(V^m)$. If the Gauss mapping γ is nondegenerate, i.e. if the tangent subspace $T_x^{(1)}$ to the submanifold V^m depends on m parameters, then the submanifold V^m is called *tangentially nondegenerate*. In this case, the forms ω_i^α in equations (2.21) cannot be expressed in terms of fewer than m linearly independent forms ω^i. Otherwise, the submanifold V^m is called *tangentially degenerate*.

To establish the nature of the geometric object with the components b_{ij}^α, we evaluate the exterior differentials of equations (2.21) by means of structure equations (1.67) of the space P^n. This results in the following exterior equations:

$$\nabla b^\alpha_{ij} \wedge \omega^j = 0, \tag{2.23}$$

where

$$\nabla b^\alpha_{ij} = db^\alpha_{ij} - b^\alpha_{kj}\theta^k_i - b^\alpha_{ik}\theta^k_j + b^\beta_{ij}\theta^\alpha_\beta, \tag{2.24}$$

and the forms θ^j_i are determined by formulas (2.6). As we noted above, these forms are connected with transformations of the first order frames in the subspace $T_x(M)$ tangent to the manifold M of parameters of the submanifold V^m. Similarly, the forms

$$\theta^\alpha_\beta = \omega^\alpha_\beta - \delta^\alpha_\beta\omega^0_0 \tag{2.25}$$

determine admissible transformations of moving frames in the space $N_x(V^m)$.

Applying the Cartan lemma to exterior quadratic equation (2.23), we obtain

$$\nabla b^\alpha_{ij} = b^\alpha_{ijk}\omega^k, \tag{2.26}$$

where the coefficients b^α_{ijk} are symmetric in all lower indices. It follows from these equations that if $\omega^i = 0$, we have

$$\nabla_\delta b^\alpha_{ij} = \delta b^\alpha_{ij} - b^\alpha_{kj}\sigma^k_i - b^\alpha_{ik}\sigma^k_j + b^\beta_{ij}\sigma^\alpha_\beta = 0, \tag{2.27}$$

where

$$\sigma^j_i = \pi^j_i - \delta^j_i\pi^0_0, \quad \sigma^\alpha_\beta = \pi^\alpha_\beta - \delta^\alpha_\beta\pi^0_0.$$

Comparing equations (2.27) with equations (1.13), we see that the quantities b^α_{ij} form a tensor relative to the indices i and j. They also form a tensor relative to the index α under transformations of moving frames in the space $N_x(V^m)$. Tensors of this kind are called *mixed tensors.*

The tensor b^α_{ij} is connected with the second order differential neighborhood of a point x of the submanifold V^m. This is the reason that this tensor is called the *second fundamental tensor of the submanifold* V^m. Let us clarify the geometric meaning of this tensor. To do this, we compute the second differential of the point $x = A_0$ by differentiating the relation (2.4):

$$d^2A_0 = (d\omega^0_0 + (\omega^0_0)^2 + \omega^i_0\omega^0_i)A_0 + (\omega^0_0\omega^i_0 + \omega^j_0\omega^i_j)A_i + \omega^i_0\omega^\alpha_i A_\alpha. \tag{2.28}$$

Factorizing the latter relation by the tangent subspace $T^{(1)}_x$, we obtain

$$d^2A_0/T^{(1)}_x = \omega^i_0\omega^\alpha_i\widetilde{A}_\alpha. \tag{2.29}$$

Substituting the values of ω^α_i from equations (2.21) into equation (2.29) and denoting the left-hand side by $\Phi_{(2)}$, we find that

$$\Phi_{(2)} = b^\alpha_{ij}\omega^i\omega^j\widetilde{A}_\alpha. \tag{2.30}$$

This expression is a quadratic form with respect to the coordinates ω^i, having values in the normal subspace $N_x^{(1)}$. The form $\Phi_{(2)}$ is called the *second fundamental form* of the submanifold V^m. Thus, the second fundamental form defines a mapping of the tangent subspace $T_x^{(1)}(V^m)$ into the first normal subspace $N_x^{(1)}(V^m)$:

$$\Phi_{(2)} : \mathrm{Sym}^2 T_x^{(1)}(V^m) \to N_x^{(1)}(V^m).$$

This mapping is called the *Meusnier–Euler mapping* (see [GH 79]).

Note that *a submanifold V^m is an m-plane or a part of an m-plane if and only if the second fundamental form $\Phi_{(2)}$ vanishes on V^m*. In fact, if $\Phi_{(2)} \equiv 0$, then it follows from formula (2.29) that $\omega_i^\alpha = 0$ on V^m. This implies that the equations of infinitesimal displacement of a moving frame become:

$$dA_0 = \omega_0^0 A_0 + \omega^i A_i, \quad dA_i = \omega_i^0 A_0 + \omega_i^j A_j,$$

and as a result, the m-plane $A_0 \wedge A_1 \wedge \ldots \wedge A_m$ is fixed, and the point A_0 moves in this m-plane.

The scalar forms

$$\Phi_{(2)}^\alpha = b_{ij}^\alpha \omega^i \omega^j \tag{2.31}$$

are the coordinates of the form $\Phi_{(2)}$ with respect to the moving frame $\{\widetilde{A}_\alpha\}$ in the space $N_x^{(1)}$. Let us denote the maximal number of linearly independent forms $\Phi_{(2)}^\alpha$ by m_1. In some instances, it is convenient to consider the bundle of second fundamental forms of the submanifold V^m defined by the relation:

$$\Phi_{(2)}(\lambda) = \lambda_\alpha b_{ij}^\alpha \omega^i \omega^j, \tag{2.32}$$

where $\lambda = (\lambda_\alpha)$. The number m_1 is the dimension of this bundle.

In the space $N_x^{(1)}$, consider the points

$$\widetilde{B}_{ij} = b_{ij}^\alpha \widetilde{A}_\alpha. \tag{2.33}$$

Since $\widetilde{B}_{ij} = \widetilde{B}_{ji}$, the number of these points is equal to $\frac{1}{2}m(m+1)$. However, it is not necessarily the case that all these points are linearly independent. The maximal number of linearly independent points $\widetilde{B}_{ij}$ coincides with the maximal number of linearly independent forms $\Phi_{(2)}^\alpha$, which we denoted by m_1. Note that according to our general point of view (see Preface), we suppose that the integer m_1 is the same on the entire submanifold V^m in question, and we will make similar assumptions relative to all other integer-valued invariants arising in our further considerations.

It is obvious that the number m_1 satisfies the following inequalities:

$$0 \leq m_1 \leq \frac{m(m+1)}{2}, \quad \text{and} \quad m_1 \leq n - m. \tag{2.34}$$

In the space $N_x^{(1)}$, the points $\widetilde{B}_{ij}$ span the subspace $\widetilde{P}^{m_1-1}$.

Next, in the space P^n, we consider the subspace $\mathcal{P}_m^{-1}(\widetilde{P}^{m_1-1})$ which is the linear span of the subspace $T_x^{(1)}$ and the points $B_{ij} = b_{ij}^\alpha A_\alpha$. By relation (2.28), this subspace is also the linear span of all two-dimensional osculating planes of all curves of the submanifold V^m passing through the point x. By this reason, this subspace is called the *second osculating subspace* of the submanifold V^m at its point x, and is denoted by $T_x^{(2)}$. We consider the tangent subspace $T_x^{(1)}$ as the *first osculating subspace* of the submanifold V^m at a point x.

As examples, we find the second fundamental forms of the Segre variety $S(m,l)$ and the Veronese variety $V(m)$ introduced in Section **1.4**.

To find the second differential of the point $p = p_0^0$ of the Segre variety $S(m,l)$, we will first write the projectivization with the center $T_p^{(1)}$ of the differentials of the points $p_{\hat{\beta}}^0$ and $p_0^{\hat{j}}$ which along with the point p_0^0, determine the tangent subspace of the Segre variety $S(m,l)$ at the point p_0^0:

$$dp_{\hat{\beta}}^0/T_p^{(1)} = \theta_{\hat{i}}^0 p_{\hat{\beta}}^{\hat{i}}, \quad dp_0^{\hat{j}}/T_p^{(1)} = \omega_0^{\hat{\alpha}} p_{\hat{\alpha}}^{\hat{j}}. \tag{2.35}$$

Using these relations, we obtain the projectivization with the center $T_p^{(1)}$ of the second differential of the point $p = p_0^0$:

$$d^2p/T_p^{(1)} = 2\omega_0^{\hat{\beta}} \theta_{\hat{j}}^0 p_{\hat{\beta}}^{\hat{j}}. \tag{2.36}$$

The right-hand side of this expression is the second fundamental form $\Phi_{(2)}$ of the Segre variety $S(m,l)$. Its coordinates have the form:

$$\Phi_{(2)\hat{j}}^{\hat{\beta}} = 2\omega_0^{\hat{\beta}} \theta_{\hat{j}}^0. \tag{2.37}$$

The number m_1 of these coordinates is equal to ml, and this number is significantly less than the maximal possible value for a manifold of dimension $m+l$, which is equal to $\frac{1}{2}(m+l)(m+l+1)$.

The second osculating subspace $T_p^{(2)}$ of the Segre variety $S(m,l)$ is spanned by the points $p_0^0, p_{\hat{\beta}}^0, p_0^{\hat{j}}$ and $p_{\hat{\beta}}^{\hat{j}}$. These points form a moving frame of the space P^{ml+m+l} in which the variety $S(m,l)$ lies. Thus, the second osculating subspace $T_p^{(2)}$ of the Segre variety $S(m,l)$ coincides with the space P^{ml+m+l}.

Similarly, for the Veronese variety $V(m)$ we obtain:

$$d^2p_{00}/T_p^{(1)} = 2\omega_0^{\hat{i}}\omega_0^{\hat{j}} p_{\hat{i}\hat{j}}, \tag{2.38}$$

where $\hat{i}, \hat{j} = 1, \ldots, m$, and $p_{\hat{i}\hat{j}} = p_{\hat{j}\hat{i}}$. The right-hand side of expression (2.38) is the second fundamental form $\Phi_{(2)}$ of the Veronese variety $V(m)$, and the coordinates of this form are written as

$$\Phi_{(2)}^{\hat{i}\hat{j}} = 2\omega_0^{\hat{i}}\omega_0^{\hat{j}}. \tag{2.39}$$

The number m_1 of these coordinates is equal to $\frac{1}{2}m(m+1)$, i.e. this number assumes the maximal possible value. The points $p_{00}, p_{0\hat{i}}$ and $p_{\hat{i}\hat{j}}$ form a moving frame of the space P^N, $N = \frac{1}{2}(m+1)(m+2) - 1$, in which the Veronese variety $V(m)$ lies. Thus, as for the Segre variety $S(m, l)$, the second osculating subspace $T_p^{(2)}$ of the Veronese variety $V(m)$ coincides with the space P^N.

2.3 Osculating Subspaces and Fundamental Forms of Higher Orders of a Submanifold

1. We will make a further specialization of moving frames $\{A_u\}$ associated with a point $x \in V^m$. To do this, we place the vertices $A_{m+1}, \ldots, A_{m+m_1}$ of the frames into the second osculating subspace $T_x^{(2)}$, whose dimension is equal to $m + m_1$. The frames thus obtained are called the *frames of second order.*

With this specialization, the points B_{ij}, which together with the points A_0 and A_i define the second osculating subspace $T_x^{(2)}$, are expressed in terms of the points A_{i_1} alone: $B_{ij} = b_{ij}^{i_1} A_{i_1}$, $i_1 = m+1, \ldots, m+m_1$. So, we have

$$b_{ij}^{\alpha_1} = 0, \quad \alpha_1 = m + m_1 + 1, \ldots, n, \tag{2.40}$$

and therefore formulas (2.21) break up into two groups:

$$\omega_i^{i_1} = b_{ij}^{i_1} \omega^j, \tag{2.41}$$

$$\omega_i^{\alpha_1} = 0. \tag{2.42}$$

Therefore the second fundamental forms $\Phi_{(2)}^{\alpha}$ of the submanifold V^m can be written as follows:

$$\Phi_{(2)}^{i_1} = b_{ij}^{i_1} \omega^i \omega^j, \quad \Phi_{(2)}^{\alpha_1} = 0, \tag{2.43}$$

and formula (2.29) becomes

$$d^2 A_0 / T_x^{(1)} = \omega^i \omega_i^{i_1} \widetilde{A}_{i_1}. \tag{2.44}$$

The forms $\Phi_{(2)}^{i_1}$ are linearly independent, and the matrix $(b_{ij}^{i_1})$ of coefficients of these forms, having m_1 rows and $\frac{1}{2}m(m+1)$ columns, is of rank m_1.

Consider now the projectivization with the center $T_x^{(1)}$ of the projective space $T_x^{(2)}$. This projectivization is a projective space of dimension $m_1 - 1$. We will call this space the *reduced first normal subspace* of the submanifold V^m and denote it by $\widetilde{N}_x^{(1)}$:

$$\widetilde{N}_x^{(1)} = T_x^{(2)} / T_x^{(1)}. \tag{2.45}$$

If $n > m + m_1$, then at the point $x \in V^m$ it is also possible to define the *second normal subspace*

$$N_x^{(2)} = P^n / T_x^{(2)}, \tag{2.46}$$

whose dimension is equal to $n - m - m_1 - 1$ and whose basis is formed by the points $\widetilde{A}_{\alpha_1} = \mathcal{P}_{m+m_1} A_{\alpha_1}$.

Another result of the specialization of moving frames indicated above is that the stationary subgroup associated with a point x of the submanifold V^m is reduced. The last equation of system (2.8) breaks up into two groups:

$$\begin{aligned} \delta A_{i_1} &= \pi_{i_1}^0 A_0 + \pi_{i_1}^j A_j + \pi_{i_1}^{j_1} A_{j_1}, \\ \delta A_{\alpha_1} &= \pi_{\alpha_1}^0 A_0 + \pi_{\alpha_1}^j A_j + \pi_{\alpha_1}^{j_1} A_{j_1} + \pi_{\alpha_1}^{\beta_1} A_{\beta_1}, \end{aligned} \tag{2.47}$$

and the points $\widetilde{A}_{i_1} = \mathcal{P}_m A_{i_1}$ and $\widetilde{A}_{\alpha_1} = \mathcal{P}_{m+m_1} A_{\alpha_1}$, which form a basis of the normal spaces $\widetilde{N}_x^{(1)}$ and $N_x^{(2)}$, are transformed according to the formulas:

$$\delta \widetilde{A}_{i_1} = \pi_{i_1}^{j_1} \widetilde{A}_{j_1}, \quad \delta \widetilde{A}_{\alpha_1} = \pi_{\alpha_1}^{\beta_1} \widetilde{A}_{\beta_1}.$$

Let us now establish the form of equations (2.26) after the specialization of moving frames indicated above. These equations also break up into two groups:

$$\nabla b_{ij}^{i_1} = d b_{ij}^{i_1} - b_{kj}^{i_1} \theta_i^k - b_{ik}^{i_1} \theta_j^k + b_{ij}^{j_1} \theta_{j_1}^{i_1} = b_{ijk}^{i_1} \omega^k, \tag{2.48}$$

$$\nabla b_{ij}^{\alpha_1} = b_{ij}^{i_1} \omega_{i_1}^{\alpha_1} = b_{ijk}^{\alpha_1} \omega^k. \tag{2.49}$$

Equations (2.48) show that, under transformations (2.47), the quantities $b_{ij}^{i_1}$ form a tensor relative to the indices i, j and i_1. Since the matrix $(b_{ij}^{i_1})$ is of rank m_1, equations (2.49) can be solved with respect to the forms $\omega_{i_1}^{\alpha_1}$:

$$\omega_{i_1}^{\alpha_1} = c_{i_1 k}^{\alpha_1} \omega^k. \tag{2.50}$$

Substituting these expressions of the forms $\omega_{i_1}^{\alpha_1}$ into equations (2.49), we obtain

$$b_{ij}^{i_1} c_{i_1 k}^{\alpha_1} = b_{ijk}^{\alpha_1}. \tag{2.51}$$

Since the quantities $b_{ijk}^{\alpha_1}$ are symmetric in the indices j and k, we find from (2.51) that

$$b_{ij}^{i_1} c_{i_1 k}^{\alpha_1} = b_{ik}^{i_1} c_{i_1 j}^{\alpha_1}. \tag{2.52}$$

This equation can also be obtained as a result of exterior differentiation of equations (2.42). We will use equation (2.52) many times in our further considerations.

In the same manner as we did in Section **2.2** for the tensor b_{ij}^{α}, we can prove that the quantities $b_{ijk}^{\alpha_1}$ form a tensor relative to the indices i, j, k and α_1. This and the relations (2.51) imply that the quantities $c_{i_1 k}^{\alpha_1}$ also form a tensor relative to the indices k, i_1 and α_1. As to the quantities $b_{ijk}^{i_1}$ in relations (2.48), it is easy to verify that they do not form a tensor, but depend on the choice of the subspace $A_0 \wedge A_{m+1} \wedge \ldots \wedge A_{m+m_1}$ which is complementary to the subspace $T_x^{(1)}$ in the osculating subspace $T_x^{(2)}$.

2. Consider the third differential of the point $x = A_0$ of the submanifold V^m. Differentiating equations (2.28) and making the projectivization with the center $T_x^{(2)}$, we obtain

$$d^3 A_0/T_x^{(2)} = \omega^i \omega_i^{i_1} \omega_{i_1}^{\alpha_1} \widetilde{A}_{\alpha_1}. \tag{2.53}$$

where $i = 1, \ldots, m$; $i_1 = m+1, \ldots, m+m_1$, and $\alpha_1 = m+m_1+1, \ldots, n$. The expression on the right-hand side of equation (2.53) is a cubic form with respect to coordinates ω^i, with values in the space $N_x^{(2)} = P^n/T_x^{(2)}$. This form is called the *third fundamental form* of the submanifold V^m at the point x, and is denoted by $\Phi_{(3)}$:

$$\Phi_{(3)} = \omega^i \omega_i^{i_1} \omega_{i_1}^{\alpha_1} \widetilde{A}_{\alpha_1}.$$

This form defines a mapping of the tangent subspace $T_x^{(1)}(V^m)$ into the second normal subspace $N_x^{(2)}(V^m)$:

$$\Phi_{(3)} : \mathrm{Sym}^3 T_x^{(1)}(V^m) \to N_x^{(2)}(V^m).$$

The coordinates of the form $\Phi_{(3)}$ are the scalar forms

$$\Phi_{(3)}^{\alpha_1} = \omega^i \omega_i^{i_1} \omega_{i_1}^{\alpha_1}. \tag{2.54}$$

It follows from (2.41) and (2.43) that the latter forms can be written as:

$$\Phi_{(3)}^{\alpha_1} = \Phi_{(2)}^{i_1} \omega_{i_1}^{\alpha_1}, \tag{2.55}$$

and (2.43), (2.50) and (2.51) imply that these forms can also be written as

$$\Phi_{(3)}^{\alpha_1} = b_{ijk}^{\alpha_1} \omega^i \omega^j \omega^k. \tag{2.56}$$

Differentiating equations (2.56) with respect to the variables ω^k, we obtain:

$$\frac{\partial \Phi_{(3)}^{\alpha_1}}{\partial \omega^k} = 3 b_{ijk}^{\alpha_1} \omega^i \omega^j.$$

Substituting expressions (2.51) of $b_{ijk}^{\alpha_1}$ into this equation, we find that

$$\frac{\partial \Phi_{(3)}^{\alpha_1}}{\partial \omega^k} = 3 b_{ij}^{i_1} c_{i_1 k}^{\alpha_1} \omega^i \omega^j = 3 c_{i_1 k}^{\alpha_1} \Phi_{(2)}^{i_1}. \tag{2.57}$$

The tensor $b_{ijk}^{\alpha_1}$, which defines the third fundamental forms of the submanifold V^m, and is connected with the third differential neighborhood of this submanifold, is called the *third fundamental tensor* of the submanifold V^m.

In some instances, it is more convenient to consider a bundle of third fundamental forms of the submanifold V^m, defined by the formula:

$$\Phi_{(3)}(\lambda_1) = \lambda_{\alpha_1} \Phi^{\alpha_1}_{(3)}, \tag{2.58}$$

where $\lambda_1 = (\lambda_{\alpha_1})$.

In the normal subspace $N_x^{(2)}$, consider the points

$$\widetilde{B}_{ijk} = b^{\alpha_1}_{ijk} \widetilde{A}_{\alpha_1}. \tag{2.59}$$

Since the tensor $b^{\alpha_1}_{ijk}$ is symmetric in the lower indices, the maximal number of linearly independent points $\widetilde{B}_{ijk}$ does not exceed the number $\frac{1}{6}m(m+1)(m+2)$. Denote by m_2 the maximal number of linearly independent points $\widetilde{B}_{ijk}$. Then it is obvious that this number satisfies the inequalities:

$$0 \leq m_2 \leq \frac{1}{6}m(m+1)(m+2), \quad m_2 \leq n - m - m_1. \tag{2.60}$$

The number m_2 is equal to the maximal number of linearly independent forms $\Phi^{\alpha_1}_{(3)}$ defined by (2.56) and coincides with the dimension of the bundle (2.58).

The points $\widetilde{B}_{ijk}$ define a subspace of dimension $m_2 - 1$ in the normal space $N_x^{(2)}$. We denote this subspace by $\widetilde{N}_x^{(2)}$ and call it the *second reduced normal subspace*. The subspace $\mathcal{P}^{-1}_{m+m_1}(\widetilde{N}_x^{(2)})$ of the space P^n is of dimension $m + m_1 + m_2$ and is the linear span of the subspace $T_x^{(2)}$ and the points $B_{ijk} = b^{\alpha_1}_{ijk} A_{\alpha_1}$. By equation (2.53), the subspace $\mathcal{P}^{-1}_{m+m_1}(\widetilde{N}_x^{(2)})$ is the linear span of three-dimensional osculating planes of all curves of the submanifold V^m passing through the point x. For this reason, this subspace is called the *third osculating subspace* of the submanifold V^m at the point x and is denoted by $T_x^{(3)}$.

We place the points $A_{m+m_1+1}, \ldots, A_{m+m_1+m_2}$ into the third osculating subspace $T_x^{(3)}$. Then the cubic forms $\Phi^{i_2}_{(3)}$, $i_2 = m + m_1 + 1, \ldots, m + m_1 + m_2$, defined by formulas (2.56), become linearly independent, and the forms $\Phi^{\alpha_2}_{(3)}$, $\alpha_2 = m+m_1+m_2+1, \ldots, n$, vanish. This implies the following equations:

$$b^{\alpha_2}_{ijk} = 0, \tag{2.61}$$

$$b^{i_1}_{ij} \omega^{i_2}_{i_1} = b^{i_2}_{ijk} \omega^k, \quad b^{i_1}_{ij} \omega^{\alpha_2}_{i_1} = 0, \tag{2.62}$$

$$\Phi^{i_2}_{(3)} = b^{i_2}_{ijk} \omega^i \omega^j \omega^k, \quad \Phi^{\alpha_2}_{(3)} = 0. \tag{2.63}$$

Since the rank of the rectangular matrix $(b^{i_1}_{ij})$ is equal to m_1, it follows from the second of equations (2.62) that

$$\omega^{\alpha_2}_{i_1} = 0. \tag{2.64}$$

These equations show that the differentials of the points A_{i_1}, belonging to the second osculating subspace $T_x^{(2)}$ of the submanifold V^m, are contained in its third osculating subspace $T_x^{(3)}$.

The construction of osculating subspaces of the submanifold V^m can be continued. In this construction, the osculating subspace $T_x^{(q)}$ of order q of the submanifold V^m at a point x is the linear span of osculating planes of order q of all curves of the submanifold V^m passing through the point x.

While doing this construction, we can encounter two possibilities:

1. The osculating subspace of a certain order p coincides with the ambient space P^n. Then the sequence of osculating subspaces of the submanifold V^m has the form:
$$x = A_0 \in T_x^{(1)} \subset T_x^{(2)} \subset \ldots \subset T_x^{(p)} = P^n.$$
In this sequence, the dimension of each preceding subspace is less than the dimension of the following one.

2. Starting from some order p, the dimension of osculating subspaces is stabilized, i.e. $T_x^{(p+1)} = T_x^{(p)}$.

In the latter case we can prove the following theorem.

Theorem 2.1 *If starting from some order p, the dimension of osculating subspaces at every point $x \in V^m$ is stabilized, then the subspace $T_x^{(p)}$ is the same for all points x of the submanifold V^m, and this submanifold lies in this fixed osculating subspace $T_x^{(p)}$.*

Proof. If we make the specialization of moving frames similar to the specializations which we made earlier for the subspaces $T_x^{(2)}$ and $T_x^{(3)}$, it follows that the osculating subspace $T_x^{(p)}$ is defined by the points $A_0, A_i, A_{i_1}, \ldots, A_{i_{p-1}}$. The differentials of these points have the form:

$$\begin{aligned} dA_0 &= \omega_0^0 A_0 + \omega_0^i A_i, \\ dA_i &= \omega_i^0 A_0 + \omega_i^j A_j + \omega_i^{j_1} A_{j_1}, \\ dA_{i_1} &= \omega_{i_1}^0 A_0 + \omega_{i_1}^j A_j + \omega_{i_1}^{j_1} A_{j_1} + \omega_{i_1}^{j_2} A_{j_2}, \\ &\ldots\ldots\ldots\ldots\ldots\ldots\ldots\ldots\ldots\ldots\ldots\ldots \\ dA_{i_{p-1}} &= \omega_{i_{p-1}}^0 A_0 + \omega_{i_{p-1}}^j A_j + \ldots + \omega_{i_{p-1}}^{j_{p-1}} A_{j_{p-1}} + \omega_{i_{p-1}}^{j_p} A_{j_p}. \end{aligned} \tag{2.65}$$

Since, by hypothesis, we have $T_x^{(p+1)} = T_x^{(p)}$, the last term in the latter equation of (2.65) vanishes, i.e. we have
$$\omega_{i_{p-1}}^{j_p} = 0.$$
Hence, the subspace $T_x^{(p)}$ is fixed, and the submanifold V^m entirely lies in this subspace. ■

In what follows, we will always assume that the ambient space P^n coincides with the fixed osculating subspace $T_x^{(p)}$, i.e. that we are in the situation of the first case indicated above.

2.4 Asymptotic and Conjugate Directions of Different Orders on a Submanifold

1. A curve on a two-dimensional surface V^2 of a Euclidean space E^3 is called *asymptotic* if its osculating planes coincide with the tangent planes to the surface V^2 or are undetermined (see for example, [Bl 21], p. 52 or [Bl 50], p. 65). This definition is projectively invariant and can be generalized to the case where we have a submanifold of any dimension m in a projective space P^n. Namely, a curve l on a submanifold V^m is said to be *asymptotic* if its two-dimensional osculating plane at any of its points x belongs to the tangent subspace $T_x^{(1)}$ to the submanifold V^m at this point or is undetermined.

If a curve l is given on the submanifold V^m by a parametric equation $x = x(t)$, then its osculating plane is determined by the points $x(t), x'(t)$ and $x''(t)$. But since $x = A_0$, this plane can also be defined by the points A_0, dA_0 and d^2A_0. Since for an asymptotic line the second differential of its point belongs to the tangent subspace $T_x^{(1)}$, it follows from equation (2.28) that on this curve we have

$$\Phi_{(2)} = \omega^i\omega_i^\alpha A_\alpha = 0, \tag{2.66}$$

i.e. the second fundamental form of the submanifold V^m vanishes on l. In coordinate form, this condition can be written as follows:

$$b_{ij}^\alpha \omega^i\omega^j = 0. \tag{2.67}$$

On the curve l the basis forms ω^i have the form: $\omega^i = \xi^i dt$, where ξ^i are coordinates of a tangent vector to the curve. Substituting these expressions into equations (2.67), we obtain

$$b_{ij}^\alpha \xi^i\xi^j = 0. \tag{2.68}$$

These equations define a cone of directions at any point x. This cone is called the *asymptotic cone.*

If we place the points A_{i_1}, $i_1 = m+1, \ldots, m+m_1$, of our moving frames into the second osculating subspace $T_x^{(2)}$, as we did in the beginning of Section **2.3**, then by (2.43) , the equations of the asymptotic cone at the point x can be written in the form:

$$b_{ij}^{i_1}\xi^i\xi^j = 0, \;\; i_1 = m+1, \ldots, m+m_1. \tag{2.69}$$

The problem of existence of asymptotic directions at the point x of the submanifold V^m is reduced to the finding of nontrivial solutions of the system of equations (2.69). This is an algebraic problem. In general, nontrivial solutions exist if $m_1 \leq m-1$. However, in some special cases, nontrivial solutions of equations (2.69) can exist even if $m_1 > m-1$. Some of such special cases will be considered later.

Since asymptotic directions defined by equations (2.68) are connected with the second order neighborhood of a point $x \in V^m$, they are also called the *asymptotic directions of second order.* Similarly, the asymptotic curves defined by differential equations (2.67) are called *asymptotic lines of second order.*

On the submanifold V^m, asymptotic curves and asymptotic directions of any order q, $q \leq p$, where p is the order of the osculating subspace of the submanifold V^m coinciding with the ambient space P^n (see Section **2.3**), can also be defined. A curve l on the submanifold V^m is said to be an *asymptotic line of order* q if at any of its points x, its osculating plane of order q belongs to the osculating subspace $T_x^{(q-1)}$ of order $q-1$ of the submanifold V^m.

In the same manner as it was done in Sections **1.2** and **1.3** for the differentials of the second and third order, we can prove that

$$d^q A_0 / T_x^{(q-1)} = \Phi_{(q)}^{i_{q-1}} \widetilde{A}_{i_{q-1}}, \tag{2.70}$$

where the points $\widetilde{A}_{i_{q-1}}$, $i_{q-1} = m+m_1+\ldots+m_{q-2}+1, \ldots, m+m_1+\ldots+m_{q-1}$, form a basis of the reduced normal subspace $\widetilde{N}_x^{(q-1)} = T_x^{(q)}/T_x^{(q-1)}$, and $\Phi_{(q)}^{i_{q-1}}$ are the fundamental differential forms of order q connected with a point x of the submanifold V^m. These forms have the following expressions:

$$\Phi_{(q)}^{i_{q-1}} = \omega^i \omega_i^{i_1} \omega_{i_1}^{i_2} \ldots \omega_{i_{q-2}}^{i_{q-1}} = b_{ij\ldots k}^{i_{q-1}} \omega^i \omega^j \ldots \omega^k. \tag{2.71}$$

The forms $\Phi_{(q)}^{i_{q-1}}$ have degree q, and this explains why the number of the lower indices in the coefficients $b_{ij\ldots k}^{i_{q-1}}$ is equal to q. These coefficients are symmetric in all lower indices and form the qth fundamental tensor of the submanifold V^m. Formulas (2.71) imply that for forms of any order q, $q \leq p$, the following relations similar to relations (2.55) hold:

$$\Phi_{(q)}^{i_{q-1}} = \Phi_{(q-1)}^{i_{q-2}} \omega_{(q-2)}^{i_{q-2}}. \tag{2.72}$$

From the definition of asymptotic lines of order q and relations (2.70) it follows that these lines satisfy the differential equations:

$$\Phi_{(q)}^{i_{q-1}} = 0,$$

or more specifically, the equations

$$b_{ij\ldots k}^{i_{q-1}} \omega^i \omega^j \ldots \omega^k = 0. \tag{2.73}$$

If ξ^i are coordinates of the tangent vector to an asymptotic line of order q passing through a fixed point x of the submanifold V^m, then we can see from equations (2.73) that these coordinates satisfy the equations:

$$b_{ij\ldots k}^{i_{q-1}} \xi^i \xi^j \ldots \xi^k = 0. \tag{2.74}$$

These equations define a cone in the tangent subspace T_x. This cone is called the *asymptotic cone of order* q and is denoted by $C_{(q)}$.

Equation (2.72) implies that the asymptotic cone $C_{(q-1)}$ of order $q-1$ belongs to the asymptotic cone $C_{(q)}$ of order q. Thus, at any point $x \in V^m$, there exists a sequence of asymptotic cones which belong to the tangent space T_x and have the property:

$$C_{(2)} \subset C_{(3)} \subset \ldots \subset C_{(p)} \subset T_x, \tag{2.75}$$

i.e. the asymptotic cones form a filtration.

We can now explain the geometric meaning of relations (2.32) and (2.58). In the first of them the quantities $\lambda = (\lambda_\alpha)$, which form a covector, are tangential coordinates of the hyperplane $\lambda_\alpha x^\alpha = 0$ which is tangent to the submanifold V^m at the point x. If the covector λ_α is fixed, equation (2.32) gives the expression of the second fundamental form of the submanifold V^m relative to the hyperplane $\lambda_\alpha x^\alpha = 0$. The equation $\Phi_{(2)}(\lambda) = 0$ defines on V^m a cone of directions which are asymptotic relative to this hyperplane. The two-dimensional osculating planes of the curves that are tangent to these directions at the point x belong to the hyperplane $\lambda_\alpha x^\alpha = 0$.

In the same way, the quantities $\lambda_1 = (\lambda_{\alpha_1})$ in equation (2.58) are tangential coordinates of the osculating hyperplane $\lambda_{\alpha_1} x^{\alpha_1} = 0$ of the submanifold V^m at a point x, and the cubic form (2.58) is the third fundamental form of the submanifold V^m relative to this hyperplane. The equation $\Phi_{(3)}(\lambda_1) = 0$ defines at the point $x \in V^m$ a cubic cone of directions which are asymptotic relative to the hyperplane $\lambda_{\alpha_1} x^{\alpha_1} = 0$.

2. Another important notion of classical differential geometry—the notion of conjugate directions and conjugate lines—can also be generalized to higher dimensions.

Consider first the fundamental forms $\Phi^{i_1}_{(2)}$ of second order of a submanifold V^m. Two directions $\xi = (\xi^i)$ and $\eta = (\eta^i)$ tangent to the submanifold V^m at the point x are said to be *conjugate directions of second order* if they annihilate all bilinear forms associated with the quadratic forms $\Phi^{i_1}_{(2)}$, i.e. if they satisfy the conditions:

$$\Phi^{i_1}_{(2)}(\xi, \eta) = b^{i_1}_{ij} \xi^i \eta^j = 0. \tag{2.76}$$

For an arbitrary vector ξ, the system (2.76) is a system of homogeneous equations relative to the coordinates of the vector η. If this system has a nontrivial solution, then it determines the directions conjugate to the direction ξ. According to this definition, the asymptotic directions of second order, if they exist, are self-conjugate directions of the same order. By the symmetry of the tensor $b^{i_1}_{ij}$ in the indices i and j, the notion of conjugacy of directions on the submanifold V^m possesses the symmetry property.

One can easily prove that, geometrically, the directions conjugate to a given direction ξ form a subspace of the tangent space T_x, and this subspace is a characteristic subspace for a one-parameter family of the subspaces T_x tangent to the submanifold V^m at the points of a curve $\gamma \in V^m$ passing through the point x and tangent to the vector ξ (cf. for example, [Bl 21], §54 where the case $V^2 \subset P^3$ was considered).

The conjugate directions of third order are defined in a similar manner. A triple of directions $\xi = (\xi^i), \eta = (\eta^i)$ and $\zeta = (\zeta^i)$ tangent to the submanifold V^m at the point x is said to be a *third order conjugate triple of directions* if these directions annihilate all trilinear forms associated with the third fundamental forms $\Phi^{i_2}_{(3)}$, i.e. if they satisfy the equations:

$$\Phi^{i_2}_{(3)}(\xi, \eta, \zeta) = b^{i_2}_{ijk}\xi^i \eta^j \zeta^k = 0. \tag{2.77}$$

From relations (2.55), connecting the fundamental forms of second and third orders, it follows that a pair of directions, ξ and η, that are conjugate with respect to a form $\Phi^{i_1}_{(2)}$, and any direction ζ tangent to the submanifold V^m at the point x, form a triple which is conjugate with respect to a form $\Phi^{i_2}_{(3)}$. Thus only those triples of conjugate directions of third order that do not contain pairs of conjugate directions of second order, are of interest. If a pair of directions ξ and η is not a second order conjugate pair, then the system of equations (2.77) is a linear homogeneous system with respect to the coordinates ζ^i of the vector ζ, and this system determines a vector subspace which is conjugate of third order to the pair of directions ξ and η.

The conjugate directions of fourth and higher orders in the tangent subspace $T^{(1)}_x(V^m)$ and the conjugate directions relative to the osculating hyperplanes of different orders of the submanifold V^m can be defined in a similar manner.

3. As an important example, we now consider the osculating subspaces and the fundamental forms of different orders for the Grassmannian $G(m, n)$.

As in Section **1.4**, we denote by $\Omega(m, n)$ the image of the Grassmannian $G(m, n)$ under the Grassmann mapping. This image is a submanifold of dimension $\rho = (m+1)(n-m)$ in the projective space P^N where $N = \binom{n+1}{m+1} - 1$. We found earlier (see equation (2.11)) the first differential of a point $p \in \Omega(m, n)$:

$$dp = \omega p + \omega^\alpha_i p^i_\alpha. \tag{2.78}$$

where $\omega = \omega^0_0 + \omega^1_1 + \ldots + \omega^m_m$, $i = 0, 1, \ldots, m; \alpha = m+1, \ldots, n$, and

$$p^i_\alpha = A_0 \wedge A_1 \wedge \ldots \wedge A_{i-1} \wedge A_\alpha \wedge A_{i+1} \wedge \ldots \wedge A_m.$$

Formula (2.78) proves that the forms ω^α_i are coordinates of a point in the projective space $T^{(1)}_p/p$ with respect to the moving frame $\widetilde{p}^i_\alpha = \mathcal{P}_0(p^i_\alpha)$.

To find the second differential of the point p, we first differentiate the points p^i_α and then apply projectivization with the center $T^{(1)}_p$. This gives

$$dp^i_\alpha / T^{(1)}_p = \omega^\beta_j \widetilde{p}^{ij}_{\alpha\beta}, \tag{2.79}$$

where

$$\widetilde{p}^{ij}_{\alpha\beta} = \mathcal{P}_\rho(p^{ij}_{\alpha\beta})$$

and

$$p^{ij}_{\alpha\beta} = A_0 \wedge A_1 \wedge \ldots \wedge A_{i-1} \wedge A_\alpha \wedge A_{i+1} \wedge \ldots \wedge A_{j-1} \wedge A_\beta \wedge A_{j+1} \wedge \ldots \wedge A_m.$$

Thus, the points $p^{ij}_{\alpha\beta}$ are skew-symmetric in both the upper and the lower indices. By equation (2.79), the projectivization with the center $T^{(1)}_p$ of the second differential of the point p has the form:

$$d^2p/T^{(1)}_p = \frac{1}{2} \sum_{\alpha,\beta,i,j} (\omega^\alpha_i \omega^\beta_j - \omega^\beta_i \omega^\alpha_j) \widetilde{p}^{ij}_{\alpha\beta}. \tag{2.80}$$

The right-hand side of this expression is the second fundamental form $\Phi_{(2)}$ of the image $\Omega(m, n)$ of the Grassmannian $G(m, n)$. The coordinates of this form are written as follows:

$$\omega^{\alpha\beta}_{ij} = \omega^\alpha_i \omega^\beta_j - \omega^\beta_i \omega^\alpha_j. \tag{2.81}$$

It follows that the forms $\omega^{\alpha\beta}_{ij}$ are skew-symmetric in both the upper and the lower indices. If $i < j$ and $\alpha < \beta$, the points $p^{ij}_{\alpha\beta}$ are linearly independent, and their number is equal to $\rho_1 = \binom{m+1}{2}\binom{n-m}{2}$. The number of linearly independent forms $\omega^{\alpha\beta}_{ij}$ is equal to the same number ρ_1. The points p, p^i_α and $p^{ij}_{\alpha\beta}$ determine the second osculating subspace $T^{(2)}_p$ of the manifold $\Omega(m, n)$ at the point p. Since the dimension of the tangent space $T^{(1)}_p$ of $\Omega(m, n)$ is equal to

$$\dim T^{(1)}_p = (m+1)(n-m) = \binom{m+1}{1}\binom{n-m}{1}, \tag{2.82}$$

the dimension of its second osculating subspace $T^{(2)}_p$ is given by the formula:

$$\dim T^{(2)}_p = \binom{m+1}{1}\binom{n-m}{1} + \binom{m+1}{2}\binom{n-m}{2}. \tag{2.83}$$

The equation of the asymptotic cone $C_{(2)}$ of the manifold $\Omega(m, n)$ has the form:

$$\omega^{\alpha\beta}_{ij} = \omega^\alpha_i \omega^\beta_j - \omega^\beta_i \omega^\alpha_j = 0. \tag{2.84}$$

Since the forms $\omega^{\alpha\beta}_{ij}$ are the minors of second order of the rectangular matrix

$$M = (\omega^\alpha_i), \tag{2.85}$$

the equations (2.84) are equivalent to the conditions:

$$\text{rank } M = 1. \tag{2.86}$$

But as we noted in Section **1.4**, in the projective space $T^{(1)}_p/p$ this condition defines the Segre variety $S(m-1, n-m-1)$ carrying plane generators of dimensions $m-1$ and $n-m-1$. The Segre variety $S(m-1, n-m-1)$ is the projectivization of the asymptotic cone $C_{(2)}$ which is the *Segre cone* $C(m, n-m)$. The vertex of this cone is the point p, and its director manifold is the Segre variety $S(m-1, n-m-1)$.

As follows from the results of Section **1.4**, in the space P^n, there corresponds to the asymptotic cone $C_{(2)}$ a family of m-dimensional subspaces intersecting the fixed m-dimensional subspace p in the subspaces of dimension $m-1$. In the space P^n, there corresponds to an asymptotic line of second order on the manifold $\Omega(m,n)$ a one-parameter family of m-dimensional subspaces with $(m-1)$-dimensional characteristics. The families of this kind are called the *torses*. They generalize the developable surfaces of the space P^3.

If we continue the successive differentiation of the point $p \in \Omega(m,n)$, we obtain:

$$d^k p/T_p^{(k-1)} = \sum_{\substack{i_1<i_2\cdots<i_k \\ \alpha_1<\alpha_2<\cdots<\alpha_k}} \omega_{i_1 i_2 \ldots i_k}^{\alpha_1\alpha_2\ldots\alpha_k} \tilde{p}_{\alpha_1\alpha_2\ldots\alpha_k}^{i_1 i_2\ldots i_k}, \tag{2.87}$$

where

$$\omega_{i_1 i_2\ldots i_k}^{\alpha_1\alpha_2\ldots\alpha_k} = k!\omega_{i_1}^{[\alpha_1}\ldots\omega_{i_k}^{\alpha_k]}, \quad \tilde{p}_{\alpha_1\alpha_2\ldots\alpha_k}^{i_1 i_2\ldots i_k} = \mathcal{P}_{l_{k-2}}(p_{\alpha_1\alpha_2\ldots\alpha_k}^{i_1 i_2\ldots i_k}) \tag{2.88}$$

and

$$p_{\alpha_1\alpha_2\ldots\alpha_k}^{i_1 i_2\ldots i_k} = A_0\wedge A_1\wedge\ldots\wedge A_{i_1-1}\wedge A_{\alpha_1}\wedge A_{i_1+1}\wedge\ldots\wedge A_{i_k-1}\wedge A_{\alpha_k}\wedge A_{i_k+1}\wedge\ldots\wedge A_m. \tag{2.89}$$

As in the case $k=2$, the forms $\omega_{i_1 i_2\ldots i_k}^{\alpha_1\alpha_2\ldots\alpha_k}$ and the points $p_{\alpha_1\alpha_2\ldots\alpha_k}^{i_1 i_2\ldots i_k}$ are skew-symmetric in all upper and lower indices. In formulas (2.87)–(2.89), we have the following ranges of indices: $k = 1, \ldots, m+1$; $i_\kappa = 0, 1, \ldots, m$;
$\alpha_\kappa = m+1, \ldots, n$; $\kappa = 1, \ldots, k$.

The differential forms (2.88) are of degree k, and they are the fundamental forms of order k of the manifold $\Omega(m,n)$. The points $p_{\alpha_1\alpha_2\ldots\alpha_k}^{i_1 i_2\ldots i_k}$ lying on this manifold are linearly independent. These points and the points $p, p_i^\alpha, \ldots, p_{\alpha_1\alpha_2\ldots\alpha_{k-1}}^{i_1 i_2\ldots i_{k-1}}$ form a basis of the osculating subspace $T_p^{(k)}$ of order k of the manifold $\Omega(m,n)$. The number of the points $p_{\alpha_1\alpha_2\ldots\alpha_k}^{i_1 i_2\ldots i_k}$ is equal to $\rho_k = \binom{m+1}{k}\binom{n-m}{k}$, and the dimension of the osculating subspace $T_p^{(k)}$ is given by the formula

$$\dim T_p^{(k)} = \binom{m+1}{1}\binom{n-m}{1} + \binom{m+1}{2}\binom{n-m}{2} + \ldots + \binom{m+1}{k}\binom{n-m}{k}. \tag{2.90}$$

Moreover, if $k = m+1$, we find from the last formula that

$$\begin{aligned} \dim T_p^{(m+1)} &= \tbinom{m+1}{1}\tbinom{n-m}{1} + \tbinom{m+1}{2}\tbinom{n-m}{2} + \ldots + \tbinom{m+1}{m+1}\tbinom{n-m}{m+1} \\ &= \tbinom{n+1}{m+1} - 1 = N. \end{aligned}$$

This relation shows that if $m+1 \leq n-m$, the osculating subspace $T_p^{(m+1)}$ of order $m+1$ of the manifold $\Omega(m,n)$ coincides with the ambient space P^N.

The asymptotic directions of order k at the point $p \in \Omega(m,n)$ are defined by the equation $d^k p / T_p^{(k-1)} = 0$. By (2.87), the equations of the asymptotic cone $C_{(k)}$ of order k have the form:

$$\omega_{i_1 i_2 \ldots i_k}^{\alpha_1 \alpha_2 \ldots \alpha_k} = 0. \tag{2.91}$$

This equation is equivalent to the condition

$$\operatorname{rank} \quad M \leq k-1, \tag{2.92}$$

and thus the cone $C_{(k)}$ is a determinant algebraic manifold (see Section **1.4**, p. 28). In the space P^n, there corresponds to this cone a family of m-dimensional subspaces intersecting the fixed m-dimensional subspace p in subspaces of dimension $m-k+1$. In the space P^n, there corresponds to an asymptotic line of order k on the manifold $\Omega(m,n)$ a one-parameter family of m-dimensional subspaces with $(m-k+1)$-dimensional characteristics.

As in the general case, the asymptotic cones of different orders of the manifold $\Omega(m,n)$ satisfy

$$C_{(2)} \subset C_{(3)} \subset \ldots \subset C_{(m+1)} \subset T_p^{(1)},$$

i.e. these cones form a filtration.

2.5 Some Particular Cases and Examples

1. As the first example, we consider a one-dimensional submanifold V^1—a smooth curve in the space P^n. Its tangent subspace $T_x^{(1)}$ is a straight line, its osculating subspace $T_x^{(2)}$ is a two-dimensional plane, and in general, its osculating subspace $T_x^{(k)}$ of order k is a k-dimensional plane. If we associate a family of moving frames with the curve V^1 in such a way that $T_x^{(k)} = A_0 \wedge A_1 \wedge \ldots \wedge A_k$, then the equations of infinitesimal displacements of these frames have the form:

$$\begin{aligned}
dA_0 &= \omega_0^0 A_0 + \omega^1 A_1, \\
dA_1 &= \omega_1^0 A_0 + \omega_1^1 A_1 + \omega_1^2 A_2, \\
&\ldots\ldots\ldots\ldots\ldots\ldots\ldots\ldots\ldots\ldots \\
dA_k &= \omega_k^0 A_0 + \omega_k^1 A_1 + \ldots + \omega_k^k A_k + \omega_k^{k+1} A_{k+1}, \\
&\ldots\ldots\ldots\ldots\ldots\ldots\ldots\ldots\ldots\ldots
\end{aligned}$$

The form ω^1 is the single basis form on the curve V^1, and the forms ω_k^{k+1} are proportional to this basis form:

$$\omega_k^{k+1} = b_k^{k+1} \omega^1.$$

If all quantities b_k^{k+1} are different from 0 for $k = 1, 2, \ldots, n-1$, then the curve V^1 does not belong to any subspace of the space P^n, and then we have $p = n$. On the other hand, if $b_p^{p+1} = 0$ for some $p < n$ at all points of the curve V^1, then this curve lies in its fixed osculating subspace $T_x^{(p)}$ of dimension p.

2. Consider an m-dimensional submanifold V^m belonging to a projective space P^{m+1}. Such a submanifold is called a *hypersurface*. For a hypersurface V^m, equations (2.5), (2.21) and (2.31) have the form:

$$\omega_0^{m+1} = 0, \tag{2.93}$$

$$\omega_i^{m+1} = b_{ij}\omega^j, \quad b_{ij} = b_{ji}, \tag{2.94}$$

$$\Phi_{(2)} = b_{ij}\omega^i\omega^j, \tag{2.95}$$

where $b_{ij} = b_{ij}^{m+1}$ is the second fundamental tensor of the hypersurface V^m.

If $\Phi_{(2)} \equiv 0$ at any point $x \in V^m$, then according to Theorem 2.1, the hypersurface V^m coincides with its first osculating subspace, i.e. it degenerates into a hyperplane.

If the form $\Phi_{(2)}$ does not identically vanish, then the osculating subspace $T_x^{(2)}$ coincides with the space P^{m+1}, i.e. in this case we have $p = 2$. Moreover, in this case the first normal subspace $N_x^{(1)}$ is of dimension 0 and coincides with the reduced normal subspace $\widetilde{N}_x^{(1)}$. The hypersurface V^m has a single scalar second fundamental form $\Phi^{(2)}$, which at any point x determines the cone of asymptotic directions:

$$\Phi_{(2)} = b_{ij}\omega^i\omega^j = 0. \tag{2.96}$$

The projective classification of points of a hypersurface V^m is connected with the structure of its asymptotic cone. If $\det(b_{ij}) \neq 0$ at a point x, then the point x is a *point of general type*. In a real domain, all points of general type admit classification based on the signature of the quadratic form $\Phi_{(2)}$. In a complex domain, all points of general type are equivalent.

If $\det(b_{ij}) = 0$ at a point $x \in V^m$, then this point x is called *parabolic*. The parabolic points can be classified according to the rank of the tensor b_{ij}.

If $m = 2$ and $\det(b_{ij}) \neq 0$, then the asymptotic cone of second order is reduced to a pair of tangent directions to the surface V^2 at a point x. These directions can be real or imaginary depending on the signature of the form $\Phi_{(2)}$. On the whole surface V^2, equation (2.96) determines the net of asymptotic lines. This net is real if $\det(b_{ij}) < 0$, and it is imaginary if $\det(b_{ij}) > 0$.

Equations (2.76) defining conjugate directions on a submanifold V^m, in the case of a hypersurface, are reduced to one equation:

$$b_{ij}\xi^i\eta^j = 0.$$

It follows from this equation that to each direction ξ on a hypersurface $V^m \subset P^{m+1}$ there corresponds an $(m-1)$-dimensional conjugate direction which does not contain the direction ξ if the latter direction is not asymptotic. If $\det(b_{ij}) \neq 0$, then applying a construction similar to the Gram-Schmidt orthogonalization in linear algebra (see, for example, [D 64], p. 130), at a point $x \in V^m$, we can construct a system of m linearly independent and mutually conjugate directions. If $m = 2$, such directions determine a conjugate net on the surface V^2.

3. Consider a submanifold V^m in the space P^n, and suppose that all second fundamental forms $\Phi^{\alpha}_{(2)}$, $\alpha = m = 1, \ldots, n$, of V^m are proportional. In this case, the points of the submanifold V^m are called *axial*, and the reduced normal subspaces $\widetilde{N}^{(1)}_x$ of V^m are of dimension 0, as was the case for a hypersurface.

Specializing the moving frames in the same way as we did in Section **2.3**, we obtain:

$$\Phi^{m+1}_{(2)} = b_{ij}\omega^i\omega^j, \tag{2.97}$$

$$\Phi^{\alpha_1}_{(2)} = 0, \ \alpha_1 = m+2, \ldots, n. \tag{2.98}$$

Thus, equations (2.41) and (2.42) have the form:

$$\omega^{m+1}_i = b_{ij}\omega^j, \quad \omega^{\alpha_1}_i = 0. \tag{2.99}$$

Since now the index i_1 takes on only one value, formula (2.50) can be written as follows:

$$\omega^{\alpha_1}_{m+1} = c^{\alpha_1}_k\omega^k, \tag{2.100}$$

and formula (2.52) can be written as

$$b_{ij}c^{\alpha_1}_k = b_{ik}c^{\alpha_1}_j. \tag{2.101}$$

We can now prove the following result.

Theorem 2.2 *If all points of a submanifold V^m of a projective space P^n are axial, then either the submanifold V^m belongs to its fixed osculating subspace $T^{(2)}_x$ of dimension $m+1$, or this submanifold is a torse, i.e. it is an envelope of a one-parameter family of m-dimensional subspaces.*

Proof. For the moving frame constructed above, we write the equations of infinitesimal displacements of those of its vertices that determine the second osculating subspace $T^{(2)}_x$ of the submanifold V^m:

$$\begin{aligned}
dA_0 &= \omega^0_0 A_0 + \omega^i A_i, \\
dA_i &= \omega^0_i A_0 + \omega^j_i A_j + \omega^{m+1}_i A_{m+1}, \\
dA_{m+1} &= \omega^0_{m+1} A_0 + \omega^j_{m+1} A_j + \omega^{m+1}_{m+1} A_{m+1} + \omega^{\alpha_1}_{m+1} A_{\alpha_1}.
\end{aligned}$$

One can see from these formulas that the displacement of the subspace $T_x^{(2)}$ is determined by the forms $\omega_{m+1}^{\alpha_1}$ which have the form (2.100). If $\omega_{m+1}^{\alpha_1} = 0$, i.e. if all $c_k^{\alpha_1} = 0$, then the subspace $T_x^{(2)}$ is fixed, and the submanifold V^m belongs to this subspace.

Suppose now that at least one of the components of the tensor $c_k^{\alpha_1}$ is different from 0. Let, for example, $c_1^{m+2} \neq 0$. Then it follows from relation (2.101) that

$$b_{ij} c_1^{m+2} = b_{i1} c_j^{m+2},$$

i.e. the rank of the tensor b_{ij} is equal to 1. This implies that $b_{ij} = b_i b_j$, and it follows from relation (2.99) that

$$\omega_i^{m+1} = b_i b_j \omega^j, \ \omega_i^{\alpha_1} = 0,$$

i.e. the forms ω_i^α defining the Gauss mapping $\gamma(V^m)$ of the submanifold V^m (see Section **2.2**) are expressed in terms of the single independent 1-form $b_j \omega^j$. Thus, the family of tangent subspaces $T_x^{(1)}$ of the submanifold V^m depends on one parameter, and therefore this submanifold is a torse. ■

In the case when V^m is a submanifold of an n-dimensional space of constant curvature, this theorem was proved by C. Segre (see [SegC 07], p. 571), and for this reason, it is called the *Segre theorem*. The proof given above implies that the result of this theorem does not depend on a metric but is of pure projective nature.

4. Consider a three-dimensional submanifold V^3 in a projective space P^5. For this submanifold, equations (2.5), (2.21) and (2.31) have the form:

$$\omega_0^\alpha = 0, \tag{2.102}$$

$$\omega_i^\alpha = b_{ij}^\alpha \omega^j, \ b_{ij}^\alpha = b_{ji}^\alpha, \tag{2.103}$$

$$\Phi_{(2)}^\alpha = b_{ij}^\alpha \omega^i \omega^j, \tag{2.104}$$

where $i, j = 1, 2, 3$ and $\alpha = 4, 5$.

A submanifold $V^3 \subset P^5$ has two second fundamental forms, $\Phi_{(2)}^4$ and $\Phi_{(2)}^5$. If both of these forms vanish, then V^3 is a 3-plane in P^5. If the forms $\Phi_{(2)}^4$ and $\Phi_{(2)}^5$ are proportional, then as it follows from the Segre theorem, in the general case, i.e. in the case when these two forms are not proportional to the square of the same linear form, the submanifold V^3 belongs to its fixed four-dimensional osculating subspace $T_x^{(2)}$.

The case when the forms $\Phi_{(2)}^4$ and $\Phi_{(2)}^5$ are not proportional is most interesting. Since in this case $T_x^{(2)} = P^5$, we have $p = 2$, $N_x^{(1)} = \widetilde{N}_x^{(1)}$, and $\dim N_x^{(1)} = 1$. The projectivization $\mathcal{P}_0 T_x^{(1)} = T_x^{(1)}/A_0$ is a projective plane $\widetilde{P}^2$. The cones $\Phi_{(2)}^\alpha = 0$ are represented in $\widetilde{P}^2$ by two conics whose equations are:

$$\Phi^{\alpha}_{(2)} = b^{\alpha}_{ij}\omega^i\omega^j = 0. \tag{2.105}$$

These conics themselves are not invariant but the pencil of curves defined by these two conics is invariant. The points of a submanifold V^3 can be classified according to the type of this pencil. Such a classification was done by Baimuratov in the papers [B 75a] and [B 75b]. In addition, in these papers, Baimuratov studied submanifolds $V^3 \subset P^5$ for which the pencil of second fundamental forms is of one of these types at each point $x \in V^3$.

In our considerations, we restrict ourselves to the most general case when the conics defined by equation (2.105) have four distinct common points (see Figure 2.1),

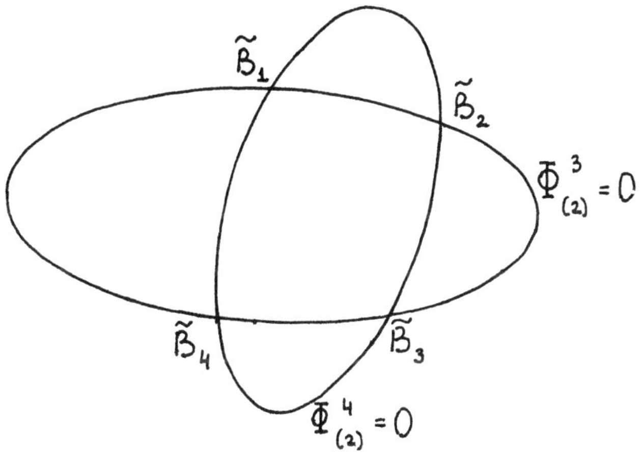

Figure 2.1

and we assume that this condition holds at any point $x \in V^3$. The common points of the conics (2.105) determine four asymptotic directions at any point $x \in V^3$. For this reason, the submanifold V^3 carries four families of asymptotic lines. Denote by $\widetilde{A}_1, \widetilde{A}_2$ and $\widetilde{A}_3$ the intersection points of the straight lines $\widetilde{B}_1\widetilde{B}_2$ and $\widetilde{B}_3\widetilde{B}_4$, $\widetilde{B}_1\widetilde{B}_3$ and $\widetilde{B}_2\widetilde{B}_4$, and $\widetilde{B}_1\widetilde{B}_4$ and $\widetilde{B}_2\widetilde{B}_3$. These points are linearly independent and can be chosen as basis points of the plane $\widetilde{P}_2$. In this basis, the fundamental forms $\Phi^4_{(2)}$ and $\Phi^5_{(2)}$ are expressed as follows:

$$\Phi^{\alpha}_{(2)} = b^{\alpha}_{11}(\omega^1)^2 + b^{\alpha}_{22}(\omega^2)^2 + b^{\alpha}_{33}(\omega^3)^2, \tag{2.106}$$

but this means that the points $\widetilde{A}_1, \widetilde{A}_2$ and $\widetilde{A}_3$ define a triple of mutually conjugate directions at a point x of the submanifold V^3. Thus, the submanifold V^3 carries a net of conjugate lines. In Chapter 3 we will study in detail the submanifolds carrying a net of conjugate lines.

5. We will now investigate a submanifold V^3 in a space P^6, and we will assume that this submanifold does not lie in any five-dimensional subspace of P^6. The basic equations of such a submanifold V^3 have the form (2.102) and (2.103) where $i, j = 1, 2, 3$ and $\alpha = 4, 5, 6$. Such a submanifold has three second fundamental forms $\Phi^{\alpha}_{(2)}$ of form (2.104). We assume that these forms are linearly independent. Then we have $T^{(2)}_x = P^6, p = 2, N^{(1)}_x = \widetilde{N}^{(1)}_x$ and $\dim N^{(1)}_x = 2$.

The equations $\Phi^{\alpha}_{(2)} = 0$ determine three conics in the plane $\widetilde{P}_2$. In general, there are no points in the plane $\widetilde{P}_2$ belonging to all these three conics. Thus, in general, the submanifold $V^3 \subset P^6$ does not have asymptotic directions.

Let us discuss the existence of pairs of conjugate directions on the submanifold V^3 in the space P^6. Let ξ be a fixed tangent direction at a point x of the submanifold V^3 under consideration. By equation (2.76), the directions conjugate to the direction ξ are determined from the following system of equations:

$$b^{\alpha}_{ij}\xi^i\eta^j = 0. \tag{2.107}$$

This system consists of three linear homogeneous equations with respect to three coordinates η^j of the vector η. In general, this system has only the trivial solution. This means that for an arbitrary direction ξ on V^3 there is no conjugate direction. However, if the coordinates ξ^i of a direction ξ satisfy the condition:

$$\det(b^{\alpha}_{ij}\xi^i) = 0, \tag{2.108}$$

then the system (2.107) has a nontrivial solution, i.e. it admits a direction η conjugate to the direction ξ. Such a direction η is unique, since by the linear independence of the forms $\Phi^{\alpha}_{(2)}$, the rank of the matrix $(b^{\alpha}_{ij}\xi^i)$ cannot be reduced by more than one.

Equation (2.108) is of third degree with respect to the coordinates ξ^i and determines a cubic cone in the space $T^{(1)}_x$, and a cubic curve in the plane

$\widetilde{P}^2 = T_x^{(1)}/A_0$ (see Figure 2.2),

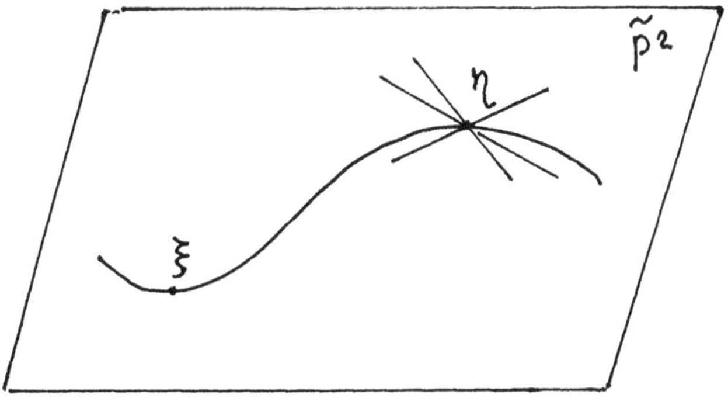

Figure 2.2

and the point $\widetilde{\xi} = \mathcal{P}\xi$ belongs to this latter curve. Since by the symmetry of the tensor b_{ij}^{α} in the indices i and j, the relation of conjugacy is symmetric, the directions ξ and η are mutually conjugate, and the point $\widetilde{\eta}$ also belongs to the cubic curve (2.108). Therefore, all conjugate pairs of directions at a point $x \in V^3$ lie on the cubic cone (2.108) of the space $T_x^{(1)}$.

6. As our final example, we consider a two-dimensional surface V^2 in the space P^n. Since $\dim T_x^{(2)} = 2 + m_1$ where $m_1 \leq \frac{1}{2}m(m+1) = 3$, we have $2 \leq \dim T_x^{(2)} \leq 5$. We consider all possible cases.

a) $m_1 = 0$, i.e. $\dim T_x^{(2)} = 2$. Then the surface V^2 is a two-dimensional plane.

b) $m_1 = 1$, i.e. $\dim T_x^{(2)} = 3$. In this case the submanifold V^2 has a single independent second fundamental form:

$$\Phi_{(2)}^3 = b_{ij}\omega^i\omega^j, \quad i, j = 1, 2.$$

By the Segre theorem, if this form is nondegenerate, i.e. if $\det(b_{ij}) \neq 0$, then the submanifold V^2 lies in its three-dimensional osculating subspace

$T_x^{(2)}$. On the other hand, if $\det(b_{ij}) = 0$, the image $\gamma(V^2)$ of the Gauss mapping of V^2 is of dimension one, and the submanifold V^2 itself is an envelope of a one-parameter family of two-dimensional planes $T_x^{(1)}$, i.e. the surface V^2 is a torse.

c) $m_1 = 2$, i.e. $\dim T_x^{(2)} = 4$. Then the submanifold V^2 has two independent second fundamental forms. These forms define two pairs of points, $\widetilde{B}_1^3, \widetilde{B}_2^3$ and $\widetilde{B}_1^4, \widetilde{B}_2^4$, on the straight line $\widetilde{P}^1 = T_x^{(1)}/A_0$. Two cases are possible:

c$_1$) These two pairs have no common points. Then there exists a unique pair of points which harmonically divides each of these pairs (see Figure 2.3)

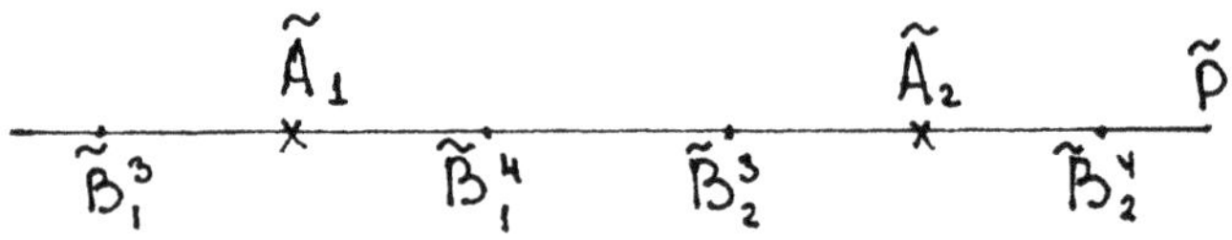

Figure 2.3

Denote the points of this unique pair by $\widetilde{A}_1$ and $\widetilde{A}_2$ and choose them as basis points of the line $\widetilde{P}^1$. Then the fundamental forms $\Phi^3_{(2)}$ and $\Phi^4_{(2)}$ are reduced to the expressions:

$$\Phi^{i_1}_{(2)} = b^{i_1}_{11}(\omega^1)^2 + b^{i_1}_{22}(\omega^2)^2,\ i_1 = 3, 4,$$

and this proves that the pair of directions on the surface V^2 defined by the points $\widetilde{A}_1$ and $\widetilde{A}_2$ is conjugate. If this situation occurs at every point of the surface V^2, then this surface carries a unique conjugate net. The submanifolds carrying conjugate nets will be studied in detail in Chapter 3.

c$_2$) The pairs of points, $\widetilde{B}_1^3, \widetilde{B}_2^3$ and $\widetilde{B}_1^4, \widetilde{B}_2^4$, have one common point, for example, let $\widetilde{B}_1^3 = \widetilde{B}_1^4 = \widetilde{A}_1$. Then we have $\widetilde{B}_2^3 \neq \widetilde{B}_2^4$ because the fundamental forms $\Phi^{i_1}_{(2)}$ are not proportional. The point $\widetilde{A}_1$ determine a unique asymptotic direction at a point $x \in V^2$. If this situation occurs at every point of the surface V^2, then this surface carries a unique family of asymptotic lines. The submanifolds carrying asymptotic lines and asymptotic submanifolds of dimension higher than one will be studied in Chapter 5.

If we choose the fourth harmonic point of the point $\widetilde{A}_1$ with respect to the points $\widetilde{B}_2^3$ and $\widetilde{B}_2^4$ as the second basis point $\widetilde{A}_2$ on the line $\widetilde{P}^1$ (see Figure 2.4),

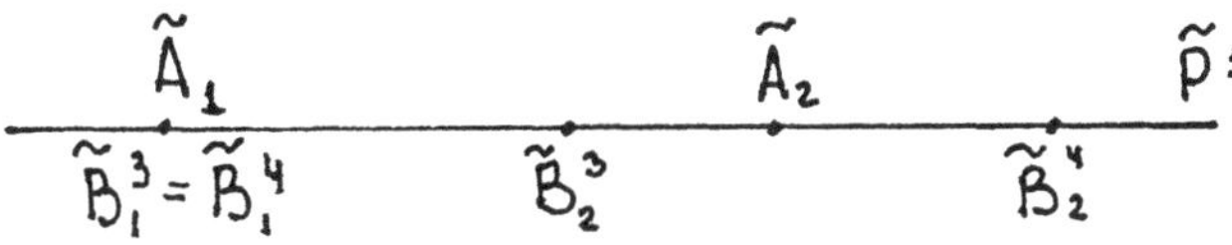

Figure 2.4

then the fundamental forms $\Phi^3_{(2)}$ and $\Phi^4_{(2)}$ are reduced to the following form:

$$\Phi^3_{(2)} = \omega^1(b_{11}\omega^1 + 2b_{12}\omega^2), \ \Phi^4_{(2)} = \omega^1(b_{11}\omega^1 - 2b_{12}\omega^2).$$

This implies that the matrices of the tensors $b^{i_1}_{ij}$ can be written as follows:

$$(b^3_{ij}) = \begin{pmatrix} b_{11} & b_{12} \\ b_{12} & 0 \end{pmatrix}, \ (b^4_{ij}) = \begin{pmatrix} b_{11} & -b_{12} \\ -b_{12} & 0 \end{pmatrix}$$

The net of lines defined on surface V^2 by the points $\widetilde{A}_1$ and $\widetilde{A}_2$ is called *semi-asymptotic*.

d) $m_1 = 3$, i.e. $\dim T_x^{(2)} = 5$. Then the submanifold V^2 has three independent second fundamental forms $\Phi^{i_1}_{(2)}$, $i_1 = 3.4, 5$. By appropriately choosing the points $\widetilde{A}_{i_1}$ in the plane $\widetilde{P}^2 = T_x^{(2)}/T_x^{(1)}$, we can reduce these three forms to the following expressions:

$$\Phi^3_{(2)} = (\omega^1)^2, \ \Phi^4_{(2)} = 2\omega^1\omega^2, \ \Phi^5_{(2)} = (\omega^2)^2$$

It follows that if $m_1 = 3$, a surface V^2 carries neither asymptotic directions nor conjugate directions.

2.6 Classification of Points of Submanifolds by Means of the Second Fundamental Form

1. The classification of the points of a surface in three-dimensional space by means of the second fundamental form is well-known: there are elliptic, hyperbolic and parabolic points. This classification is of projective nature and can be generalized for submanifolds of a projective space P^n. The main principles of such a classification were indicated by Akivis in [A 88].

Let V^m be a submanifold in a projective space P^n. In Section **2.2** we introduced its second fundamental form $\Phi_{(2)}$ (see formula (2.30)). By the formulas (2.40), the form $\Phi_{(2)}$ can be written as follows:

$$\Phi_{(2)} = b^{i_1}_{ij} \omega^i \omega^j \widetilde{A}_{i_1}, \tag{2.109}$$

where the points $\widetilde{A}_{i_1}$ form a basis of the projectivization $\widetilde{N}^1_x = T^2_x / T^1_x$ and T^1_x and T^2_x are the tangent and second osculating subspaces of the submanifold V^m at its point x, respectively.

Since the points ω^i are coordinates of a point in the projectivization $P^{m-1} = T^1_x / x$ of the tangent subspace T^1_x, the second fundamental form $\Phi_{(2)}$ defines the mapping of the space P^{m-1} into the normal subspace $\widetilde{N}^1_x$. We will denote this mapping as follows:

$$B^{(2)}: \ \mathrm{Sym}^{(2)} \, P^{m-1} \rightarrow \widetilde{N}^1_x. \tag{2.110}$$

The mapping $B^{(2)}$ can be represented as a superposition of two mappings:

$$B^{(2)} = \beta^{(2)} \circ v^{(2)}, \tag{2.111}$$

where $v^{(2)}$ is the Veronese mapping of the double symmetric transformation $\mathrm{Sym}^{(2)} \, P^{m-1}$ into a projective space P^{n_1}, where $n_1 = \frac{1}{2} m(m+1) - 1$, with the coordinates x^{ij}, $x^{ij} = x^{ji}$, defined by equations:

$$x^{ij} = \omega^i \omega^j. \tag{2.112}$$

Under this mapping, the image of the space P^{m-1} is the Veronese variety $V(m-1)$.

Besides the Veronese variety $V(m-1)$, the space P^{n_1} admits a filtration by submanifolds V_r determined by the inequalities:

$$\mathrm{rank} \ (x^{ij}) \leq r. \tag{2.113}$$

Here $V(m-1) = V_1$ and the filtration has the following form:

$$V_1 \subset V_2 \subset \ldots \subset V_{m-1} \subset V_m = P^{n_1}. \tag{2.114}$$

The submanifold V_{k-1} is the locus of singularities for the submanifold V_k, since at the points of V_{k-1} the dimension of the tangent subspaces to the submanifold V_k is reduced (see Chapter 4).

The mapping $\beta^{(2)}$ is a linear mapping of the space P^{n_1} into the subspace $\widetilde{N}^1_x$, defined by the formula:

$$x^{i_1} = b^{i_1}_{ij} x^{ij}. \tag{2.115}$$

The mapping $\beta^{(2)}$ has its kernel in the space P^{n_1}. The equations of this kernel are:

$$b_{ij}^{i_1} x^{ij} = 0, \quad i_1 = m+1, \ldots, m+m_1. \tag{2.116}$$

The kernel forms a subspace $Z^{(2)}$ of the space P^{n_1}, with $\dim Z^{(2)} = \frac{1}{2}m(m+1) - 1 - m_1$.

The points of a submanifold V^m can be classified according to the mutual location of the kernel $Z^{(2)}$ and the submanifolds in the filtration (2.114).

Before we turn to the description of this classification, we will prove the following result.

Theorem 2.3 *To the points of intersection of the kernel $Z^{(2)}$ and the Veronese variety V_1, there correspond asymptotic directions of the submanifold V^m at the point x. To the points of intersection of the kernel $Z^{(2)}$ and the submanifold V_2, there corresponds a pair of conjugate directions at the point $x \in V^m$.*

Proof. In fact, the points of intersection of the kernel $Z^{(2)}$ and the Veronese variety V_1 are defined by the equations

$$b_{ij}^{i_1} \omega^i \omega^j = 0, \tag{2.117}$$

and this system of equations is identical with the system (2.69) defining the second order asymptotic directions at a point $x \in V^m$. The second order conjugate directions at a point $x \in V^m$ are defined by system (2.76), where $\xi = (\xi^i)$ and $\eta = (\eta^i)$ are two points in the space P^{m-1}. Equation (2.76) can be written in the symmetric form:

$$b_{ij}^{i_1} (\xi^i \eta^j + \xi^j \eta^i) = 0. \tag{2.118}$$

It follows that a pair of conjugate directions ξ and η defines a point with coordinates

$$x^{ij} = \xi^i \eta^j + \xi^j \eta^i$$

in the space P^{n_1}. But for these points we have rank $(x^{ij}) = 2$, i.e. these points lie on the submanifold V_2. In fact, taking the vectors ξ and η as the basis vectors e_1 and e_2, we reduce the matrix (x^{ij}) to the form

$$(x^{ij}) = \begin{pmatrix} 0 & 1 & \ldots & 0 \\ 1 & 0 & \ldots & 0 \\ \ldots & \ldots & \ldots & \ldots \\ 0 & 0 & \ldots & 0 \end{pmatrix},$$

and clearly the rank of this matrix is equal to 2. Conversely, if rank $(x^{ij}) = 2$, then by means of a real or complex transformation of the basis, we can reduce the matrix (x^{ij}) to the form indicated above. ■

The notion of conjugate pairs of directions with respect to the second fundamental form can be generalized in the following way.

A triple (ξ, η, ζ) of directions is called *quasiconjugate* if these directions satisfy the condition:

$$\Phi_{(2)}(\xi, \eta) + \Phi_{(2)}(\eta, \zeta) + \Phi_{(2)}(\zeta, \xi) = 0, \tag{2.119}$$

or in coordinate form:

$$b_{ij}^{i_1}(\xi^i \eta^j + \eta^i \zeta^j + \zeta^i \xi^j) = 0. \tag{2.120}$$

Since

$$\text{rank } (\xi^i \eta^j + \eta^i \zeta^j + \zeta^i \xi^j) = 3,$$

the point $x \in P^{n_1}$ with coordinates

$$x^{ij} = \xi^{(i} \eta^{j)} + \eta^{(i} \zeta^{j)} + \zeta^{(i} \xi^{j)}$$

belongs to the submanifold V_3 defined by equation (2.113), where $r = 3$.

In a similar manner one can introduce the notion of quasiconjugacy for k directions $\xi_1, \ldots, \xi_k$ tangent to the submanifold V^m at a point $x \in V^m$ where $3 \leq k \leq m - 1$. To such sets of directions, there correspond points in the space P^{n_1} satisfying relation (2.113) for $r = k$ and belonging to the submanifold V_k.

If $k = 2$, then the quasiconjugate directions become usual conjugate directions on V^m, and if $k = 1$, then the quasiconjugate directions are asymptotic directions.

It is easy to see that if we have k directions $\xi_1, \ldots, \xi_k$, and any $k - 1$ of them are quasiconjugate, then all of them are quasiconjugate, but, of course, the converse is not true.

Note also that the quasiconjugate directions defined above are different from conjugate systems of directions of higher orders defined in Section **2.4** since the latter were defined by means of higher order fundamental forms of the submanifold V^m.

2. Let us apply the apparatus constructed in the previous subsection to classification of points of a two-dimensional submanifold V^2 in the space P^n. We have already considered such submanifolds in subsection **2.5.6**. Now we will look at them from a new point of view.

In this case the dimension n_1 of the space, into which the Veronese mapping $v^{(2)}$ is made, is equal to two, and this mapping can be written in the form:

$$x^{11} = (\omega^1)^2, \quad x^{12} = \omega^1 \omega^2, \quad x^{22} = (\omega^2)^2. \tag{2.121}$$

Eliminating the coordinates ω^1 and ω^2 from these equations, we find that

$$x^{11} x^{22} - (x^{12})^2 = 0, \tag{2.122}$$

i.e. in this case the Veronese variety is a plane curve of second order, and the manifold V_2 coincides with the plane P^2.

The following cases are possible:

1) If the rank m_1 of system (2.117) is maximal, i.e. $m_1 = 3$, then the mapping $\beta^{(2)}$ does not have a kernel. Because of this, in the second order differential neighborhood, all such points are equivalent.

   A basis of the system of second fundamental forms of the submanifold V^2 has the form:

$$(\omega^1)^2, \ 2\omega^1\omega^2, \ (\omega^2)^2.$$

   In this case a submanifold V^2 carries neither asymptotic nor conjugate directions.

2) Suppose now that the rank m_1 is equal to 2. Then the kernel $Z^{(2)}$ of the mapping $\beta^{(2)}$ is a point. A submanifold V^2 has two independent second fundamental forms at the point $x \in V^2$. Points of this kind are called *planar* (see [SSt 38]).

   Two cases must be distinguished:

   2i) $Z^{(2)} \notin V_1$. Since in this case the manifold V_2 coincides with the plane P^2, then *at any point x the submanifold V^2 possesses a single pair of conjugate directions.* A basis of the system of second fundamental forms of the submanifold V^2 can be reduced to the form:

$$(\omega^1)^2, \ (\omega^2)^2.$$

   The equations of a pair of conjugate directions are: $\omega^1 = 0$ and $\omega^2 = 0$.

   2ii) $Z^{(2)} \in V_1$. Then *at any point x the submanifold V^2 possesses a single asymptotic direction.* The basis of the system of second fundamental forms of the submanifold V^2 can be reduced to the form:

$$(\omega^1)^2, \ 2\omega^1\omega^2.$$

   The equation of a single asymptotic direction is: $\omega^1 = 0$.

3) Finally, suppose that the rank m_1 is equal to 1. Then the kernel $Z^{(2)}$ of the mapping $\beta^{(2)}$ is a straight line, and the submanifold V^2 has only one independent second fundamental forms at a point $x \in V^2$. Points of this kind are called *axial* (see [SSt 38]).

   From the complex point of view, we must distinguish only two cases:

   3i) The kernel $Z^{(2)}$ has two common points with the Veronese variety V_1 (see Figure 2.5), and a submanifold V^2 has two asymptotic directions corresponding to these common points and a one-parameter family of pairs of conjugate directions.

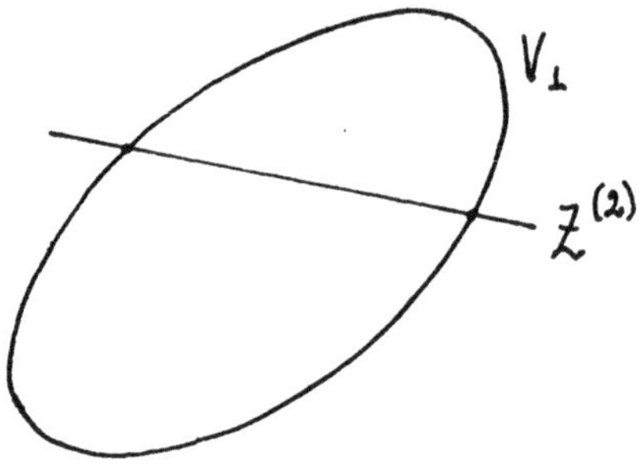

Figure 2.5

The single independent second fundamental form of the submanifold V^2 can be reduced to the form $2\omega^1\omega^2$. The equations of asymptotic directions are $\omega^1 = 0$ and $\omega^2 = 0$, and conjugate directions are determined by the equations $\omega^1 + \lambda\omega^2 = 0$ and $\omega^1 + \lambda\omega^2 = 0$, where λ is an arbitrary number.

3ii) The kernel $Z^{(2)}$ is tangent to the manifold V_1 (see Figure 2.6). A submanifold V^2 has one double asymptotic direction and a one-parameter family of conjugate directions.

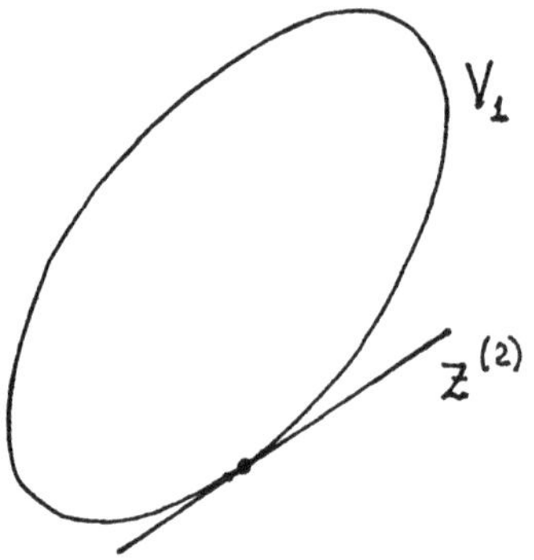

Figure 2.6

The single independent second fundamental form of the submanifold V^2 can be reduced to the form $(\omega^1)^2$. An asymptotic direction is defined by the equation $\omega^1 = 0$, and this direction is conjugate to any other direction tangent to the submanifold V^2 at a point x.

From the real point of view, the case 3i) falls into two cases: the common points of the kernel $Z^{(2)}$ and the Veronese variety are real—in which case the submanifold V^2 carries two *real* asymptotic directions, and the point x is called *hyperbolic*, or these common points are complex conjugate—in which case the submanifold V^2 does not carry real asymptotic directions, and the point x is called *elliptic*.

In the case 3ii) the point x is called *parabolic*. If all points of the submanifold V^2 are parabolic, then, as we will see in Chapter 4, this submanifold is tangentially degenerate.

3. For submanifolds V^3 of the projective space P^n, considerations similar to that in subsection **2.5.2** are much more complicated. They were completed by Polovtseva in her thesis [P 88].

We will outline below the main stages of this classification.

For a submanifold $V^3 \subset P^n$ the number m_1 of linearly independent second fundamental forms does not exceed 6. The projective plane $P^2 = PT_x^{(1)}$ is mapped by the Veronese mapping $v^{(2)}$ into the two-dimensional Veronese variety V_1 which has the order four, lies in a space P^5, and is defined in this space by equations (2.112). In the space P^5 the filtration (2.114) has the form:

$$V_1 \subset V_2 \subset V_3 = P^5, \tag{2.123}$$

where V_2 is a determinant submanifold in P^5 defined by the equation:

$$\det(x^{ij}) = 0, \tag{2.124}$$

and is a cubic hypersurface which is called the *cubic symmetroid* (see subsection **1.4.2** and also [SR 85]).

For classification of points of submanifolds $V^3 \subset P^n$ we must consider all possible values of the number m_1: $0 \leq m_1 \leq 6$. We will consider the cases when the rank m_1 is equal to 6, 5, 4 and 3. The cases when this rank is equal to 2, 1, and 0 are similar to the cases when it is equal to 4, 5 and 6, respectively (for details see subsection **1.4.2** and also [P 88]).

1) $m_1 = 6$. The six independent second fundamental forms of the submanifold have a general form. The mapping $\beta^{(2)}$ does not have a kernel. In the second order differential neighborhood, all points of the $V^3 \subset P^n$ are equivalent, and a submanifold V^2 carries neither asymptotic nor conjugate directions.

2) $m_1 = 5$. Then the kernel $Z^{(2)}$ of the mapping $\beta^{(2)}$ is a point. The submanifold V^3 has five independent second fundamental forms at $x \in V^3$.

The following three cases must be distinguished:

2i) $Z^{(2)} \notin V_2$. A submanifold V^2 carries neither asymptotic nor conjugate directions. In this case a basis of the system of second fundamental forms of the submanifold V^2 can be reduced to the form:

$$(\omega^1)^2 - (\omega^3)^2, \;\; (\omega^2)^2 - (\omega^3)^2, \;\; 2\omega^1\omega^2, \;\; 2\omega^2\omega^3, \;\; 2\omega^3\omega^1.$$

2ii) $Z^{(2)} \in V_2, Z^{(2)} \notin V_1$. Then *at any point x the submanifold V^2 has a single pair of conjugate directions.* A basis of the system of second fundamental forms of the submanifold V^2 can be reduced to the form:

$$(\omega^1)^2, \ (\omega^2)^2, \ (\omega^3)^2, \ 2\omega^1\omega^2, \ 2\omega^1\omega^3.$$

The directions $\omega^1 = \omega^2 = 0$ and $\omega^1 = \omega^3 = 0$ are the conjugate directions at the point $x \in V^3$.

2iii) $Z^{(2)} \in V_1$. Then *at any point x the submanifold V^2 has a single asymptotic direction.* A basis of the system of second fundamental forms of the submanifold V^3 can be reduced to the form:

$$(\omega^2)^2, \ (\omega^3)^2, \ 2\omega^1\omega^2, \ 2\omega^2\omega^3, \ 2\omega^3\omega^1,$$

and the direction $\omega^2 = \omega^3 = 0$ is the single asymptotic direction at a point $x \in V^3$.

In the last three cases the systems of second fundamental forms are of different projective types.

3) $m_1 = 4$. The submanifold V^3 has four independent second fundamental forms at the point $x \in V^3$, and the kernel $Z^{(2)}$ of the mapping $\beta^{(2)}$ is a straight line in a space P^5.

The following 8 cases must be distinguished:

3i) The straight line $Z^{(2)}$ intersects the cubic symmetroid V_2 at three distinct points. Then *at any point $x \in V^3$ there are three distinct pairs of conjugate directions.* A basis of the system of second fundamental forms of the submanifold V^3 can be reduced to the form:

$$(\omega^1)^2, \ (\omega^2)^2, \ (\omega^3)^2, \ 2(\omega^1\omega^2 + \omega^2\omega^3 + \omega^3\omega^1).$$

Three pairs of conjugate directions are defined by the equations: $\omega^1 = \omega^2 = 0$ and $\omega^3 = \omega^1 + \omega^2 = 0$; $\omega^2 = \omega^3 = 0$ and $\omega^1 = \omega^2 + \omega^3 = 0$; and $\omega^3 = \omega^1 = 0$ and $\omega^2 = \omega^1 + \omega^3 = 0$.

3ii) The straight line $Z^{(2)}$ is tangent to the cubic symmetroid V_2 at one nonsingular point, and intersects V_2 at another point, but does not have common points with the Veronese variety V_1. Then *at any point $x \in V^3$ there are one double and one simple pair of conjugate directions.* A basis of the system of second fundamental forms of the submanifold V^3 can be reduced to the form:

$$(\omega^1)^2, \ (\omega^2)^2, \ (\omega^3)^2, \ 2(\omega^1\omega^2 + \omega^3\omega^1).$$

The double pair of conjugate directions is defined by the equations: $\omega^1 = \omega^2 = 0$ and $\omega^1 = \omega^3 = 0$, and the equations of the simple pair of conjugate directions are: $\omega^2 = \omega^3 = 0$ and $\omega^1 = \omega^2 + \omega^3 = 0$.

3iii) The straight line $Z^{(2)}$ has a double tangency with the cubic symmetroid V_2 at a nonsingular point (i.e. three common points of $Z^{(2)}$ and V_2 coincide), but does not have common points with the Veronese variety V_1. Then *at any point $x \in V^3$ there is one threefold pair of conjugate directions.* A basis of the system of second fundamental forms of the submanifold V^3 can be reduced to the form:

$$(\omega^1)^2 + 2\omega^2\omega^3, \ (\omega^2)^2, \ (\omega^3)^2, \ 2\omega^1\omega^2,$$

and the threefold pair of conjugate directions is defined by the equations: $\omega^1 = \omega^2 = 0$ and $\omega^2 = \omega^3 = 0$.

3iv) The straight line $Z^{(2)}$ intersects the cubic symmetroid V_2 at a double point, which belongs to the Veronese variety V_1, and a simple point. Then *at any point $x \in V^3$ there is an asymptotic direction and a pair of conjugate directions.* A basis of the system of second fundamental forms of the submanifold V^3 can be reduced to the form:

$$(\omega^2)^2, \ (\omega^3)^2, \ 2\omega^1\omega^2, \ 2\omega^1\omega^3,$$

and the asymptotic direction is defined by the equations $\omega^2 = \omega^3 = 0$, and the single pair of conjugate directions is defined by the equations: $\omega^1 = \omega^2 = 0$ and $\omega^1 = \omega^3 = 0$.

3v) The straight line $Z^{(2)}$ has double tangency with the cubic symmetroid V_2 at its singular point, which belongs to the Veronese variety V_1, but $Z^{(2)}$ is not tangent to the Veronese variety V_1. Then *at any point $x \in V^3$ there is a threefold asymptotic direction.* A basis of the system of second fundamental forms of the submanifold V^3 can be reduced to the form:

$$(\omega^1)^2 + 2\omega^2\omega^3, \ (\omega^3)^2, \ 2\omega^1\omega^2, \ 2\omega^1\omega^3,$$

and the threefold asymptotic direction is defined by the equations $\omega^1 = \omega^3 = 0$.

3vi) The straight line $Z^{(2)}$ belongs to the cubic symmetroid V_2 but does not have common points with the Veronese variety V_1. Then the straight line $Z^{(2)}$ belongs to the tangent subspace of V_1, and *at any point $x \in V^3$ there is a one-parameter family of conjugate directions but there are no asymptotic directions.* A basis of the system of second fundamental forms of the submanifold V^3 can be reduced to the form:

$$(\omega^1)^2, (\omega^2)^2, \ (\omega^3)^2, \ 2\omega^2\omega^3,$$

and the direction $\omega^2 = \omega^3 = 0$ is conjugate to the two-dimensional direction $\omega^1 = 0$.

3vii) The straight line $Z^{(2)}$ intersects the Veronese variety V_1 at two distinct points. Then this subspace belongs to the cubic symmetroid V_2, and *at any point $x \in V^3$ there are two asymptotic directions and a one-parameter family of mutually conjugate directions.* A basis of the system of second fundamental forms of the submanifold V^3 can be reduced to the form:

$$(\omega^3)^2, \ 2\omega^1\omega^2, \ 2\omega^2\omega^3, \ 2\omega^3\omega^1,$$

and the directions $\omega^1 = \omega^3 = 0$ and $\omega^2 = \omega^3 = 0$ are asymptotic directions and the directions $\omega^3 = \omega^1 + \lambda\omega^2 = 0$ and $\omega^3 = \omega^1 - \lambda\omega^2 = 0$ are conjugate for any λ.

3viii) The straight line $Z^{(2)}$ is tangent to the Veronese variety V_1. Then *at any point $x \in V^3$ there is a double asymptotic direction, and this direction is conjugate to a one-parameter family of directions.* A basis of the system of second fundamental forms of the submanifold V^3 can be reduced to the form:

$$(\omega^2)^2, \ (\omega^3)^2, \ 2\omega^1\omega^2, \ 2\omega^2\omega^3,$$

and the double asymptotic direction is defined by the equations $\omega^2 = \omega^3 = 0$, and this direction is conjugate to any direction in the subspace $\omega^2 = 0$.

4) $m_1 = 3$. A submanifold V^3 has three independent second fundamental forms at a point $x \in V^3$, and the kernel $Z^{(2)}$ of the mapping $\beta^{(2)}$ is a 2-plane in a space P^5. In the general case the plane $Z^{(2)}$ intersects the cubic symmetroid V_2 along a plane cubic curve C_3 which does not have common points with the Veronese variety V_1. *To any point of this cubic curve C_3 there corresponds a pair of conjugate directions at a point $x \in V^3$.*

If $m_1 = 3$, classification of points of the three-dimensional submanifold $V^3 \subset P^n$ can be connected with the projective classification of plane cubic curves over the field of complex numbers. This work has been done by Polovtseva in [P 88]. She found 15 different types of points on the submanifolds in question.

NOTES

2.1. Our presentation of the projectivization of the tangent and osculating subspaces of different orders of a submanifold V^m is close to that in the paper [GH 79] by Griffiths and Harris.

The differential geometry of the Grassmannian was considered by Akivis in [A 82b] and [A 82c].

Some submanifolds of the Grassmannian were studied by Bubyakin in [Bu 91], Kruglyakov in [Kr 67a], [Kr 67b], [Kr 68], [Kr 69], [Kr 71a], [Kr 71b], [Kr 73], [KR

73], and [ShK 73] (see also his book [Kr 80]), Ivanova-Karatopraklieva (see [IK 67] and (IK 68]),Khasin (see [Kh 74]) and Ruscior (see [Ru 59] and [Ru 63]).

2.2–2.3. The osculating spaces and fundamental forms of different orders of a submanifold V^m and the relations similar to relation (2.57) were considered by É. Cartan in [Ca 19].

2.4. The asymptotic and conjugate directions and lines of different orders on a submanifold V^m were also considered by É. Cartan in his paper [Ca 19].

Trushin considered nets on a submanifold $V^m \subset P^n$ for which the tangents to any three net lines form a third order conjugate triple of directions (see [Tr 59], [Tr 60], [Tr 61a], [Tr 61b], [Tr 61c], [Tr 62a] and [Tr 62b]). In particular, he studied different cases of holonomicity of such a net on V^3 and semi-conjugate (or semi-asymptotic) nets on V^m for which the tangent directions to any two lines satisfy the equations

$$\Phi^{i_2}_{(3)}(\xi_1, \xi_1, \xi_2) = 0, \quad \Phi^{i_2}_{(3)}(\xi_1, \xi_2, \xi_2) = 0.$$

The notion of asymptotic lines on a submanifold V^m in a projective space was generalized in different directions. We will indicate some of these generalizations.

The quasiasymptotic curves were introduced in a Riemannian space V^m imbedded into V^n by Bompiani (see [Bom 21]) who used them to characterize the Veronese varieties. Bompiani results are reviewed in Appendix III of the monograph [FČ 26] (pp. 729–769) written by Terrachini.

A curve $\gamma \subset V^m \subset P^n$ is called *quasiasymptotic of type* $\gamma_{r,s}$ if the osculating s-plane of γ and the osculating space $T^{(r)}_x(V^m)$ at any point $x \in \gamma$ belong to a subspace having abnormally small dimension. For example, if $r = 1$ and $s = 2$, the dimension must be $m + 1$ or less.

A submanifold $V^h \subset V^m$, then V^h is called *quasiasymptotic of the type* $\sigma^t_{r,s}$ if the osculating spaces $T^{(r)}_x(V^h)$ and $T^{(s)}_x(V^m)$ lie in a subspace having abnormally small dimension where t indicates the difference between its dimension and the maximal possible dimension.

Villa in [Vi 37a], [Vi 38], [Vi 39a], [Vi 39b], [Vi 39c], [Vi 39f] and [Vi 39g] considered quasiasymptotic lines of type $\gamma_{r,s}$ (in particular, of type $\gamma_{1,3}$) and submanifolds V^4 with ∞^7 and ∞^8 of such lines, and established their connection with Veronese varieties. In [Vi 37b] he generalized these results for quasiasymptotic lines $\gamma_{k,3}$, and in [Vi 39d] and [Vi 39e] he considered the curves of type $\gamma_{s-1,s}$ and $\gamma_{s-2,s}$ and found conditions for V^m to belong to $T^{(s-1)}_x(V^m)$ (Segre's type theorem).

In [Vi 40] Villa considered quasiasymptotic submanifolds of type $\sigma_{1,2}$ and proved that the dimension of the spanning space of $T^{(2)}_x(V^h)$ and $T^{(1)}_x(V^M)$ can be $m+2$, $m+1$ or m. He also proved that the submanifold of type $\sigma_{1,2}$ on a Segre variety (pairs of points of two planes) can be of all three kinds. (The submanifold V^h in this case is a submanifold representing degenerate correspondence between planes.)

In [Vi 52] he considered 1-to-1 correspondence of two vector spaces and the properties of corresponding quasiasymptotic lines on the corresponding Segre variety.

In [VV 50] Villa and Vaona introduced $\gamma_{p,q,r}$-curves on $V^h \subset V^m$ (for these, the spanning space of the the osculating subspaces $T^{(p)}_x(V^h), T^{(q)}_x(V^m)$ and $T^{(r)}_x(\gamma)$ has abnormally small dimension) and their generalizations and apply them to study of Segre varieties.

In [M 51b] Muracchini studied the submanifolds V^3 with ∞^1 and ∞^2 of submanifolds of type $\sigma^3_{1,2}$. In [M 51c] he considered the case when t is maximal and proved Segre type theorems (gave conditions under which V^m is of class k). In [M 53b] he considered quasiasymptotic submanifolds $\sigma^t_{s-1,s+1}$, where $t = \binom{s=h}{s+1}$, $1 \leq h \leq m+1$, and generalized the Bompiani result on characterization of Veronese varieties.

2.5. C. Segre proved the theorem named after him for a submanifold of an n-dimensional space of constant curvature (see [SegC 07], p. 571).

Vangeldère considered a submanifold $V^3 \subset P^5$ whose two asymptotic cones at any point $x \in V^3$ are tangent one another along two distinct generators (see [V 61]). He also gave a simple proof of the Togliatti theorem (see [To 28]): a submanifold V^3 belongs to the class indicated above if and only if it is foliated by one-parameter family of two-dimensional submanifolds V^2, and each V^2 belongs to a three-dimensional generator of some tangentially degenerate submanifold V^4_1 of rank one. He gave a classification of such submanifolds V^3 according to the focal properties of the corresponding submanifold V^4_1.

In [B 75a] and [B 75b] Baimuratov gave a classification of points of a submanifold V^3 according to the type of the pencil of curves defined by two conics (2.105). He also studied submanifolds $V^3 \subset P^5$ for which the pencil of second fundamental forms is of one of the types indicated in the text at any point $x \in V^3$. A submanifold $V^3 \subset P^5$ was also considered by Akivis and Baimuratov in [AB 75].

In [M 51a], [M 51d] and [M 52], Muracchini studied and classified the submanifolds $V^5 \subset P^n$ with some special restrictions for n and the dimension of the osculating space $T^{(2)}_x(V^5)$. In [M 53a] he used the second fundamental forms of different orders to study μ-ruled submanifolds $V^3 \subset P^n$, i.e. submanifolds containing ∞^2 straight lines such that μ of these lines pass through each point of V^3 (the tangents to any three of them are not coplanar). He proved that if $n > 10$ and V^3 contains two families of ∞^2 asymptotic curves, then V^3 is foliated by ∞^1 of submanifolds containing curves of both families. He noted that it is possible to prove a similar result for a submanifold $V^3 \subset P^7$: if $V^3 \subset P^7$ has three families of ∞^2 asymptotic curves (with noncoplanar tangents), then V^3 is foliated by three families of ∞^1 of submanifolds containing curves of two of the asymptotic families. In [M 54] he considered 2-ruled submanifolds $V^3 \subset P^n, n > 10$, and proved that if V^3 carries ∞^2 straight lines, then it is foliated by ∞^1 of quadrics. In [M 58] he studied submanifolds $V^m \subset P^n$, which are foliated by projective spaces, found the minimal value for n and established conditions for these submanifolds to be Segre varieties or their projections.

The μ-ruled submanifolds $V^3 \subset P^n$ were also studied by Severi in [Sev 01] ($\mu = \infty$), Sisam in [Si 30] ($\mu = 6, n = 4$), Blaschke and Bol in the monograph [BB 38], p. 138 and p. 204 ($\mu = 3, n = 7$), and by Baldasari in [Bal 50], who studied algebraic μ-ruled V^3.

2.6. For classification of points of two-dimensional submanifolds V^2 in Riemannian spaces V^n, see the book [SS 38] by Schouten and Struik. The results on classification of points of three-dimensional submanifolds V^3 in the projective space P^n are due to Polovtseva (see [P 88]). The case $m_1 = 3$ was considered in detail in her three recent papers [P 90], [P 91a], and [P 91b]. In these papers she studied the submanifolds V^3 with a six-dimensional osculating subspace and with 0, 1 or 2 families of two-dimensional asymptotic directions.

Chapter 3

Submanifolds Carrying a Net of Conjugate Lines

3.1 Basic Equations and General Properties

In this chapter we will study special classes of tangentially nondegenerate submanifolds V^m of a projective space P^n which have m linearly independent mutually conjugate directions at each point. For any pair of vectors ξ_i and ξ_j determined by these directions, the following relation holds:

$$\Phi^\alpha_{(2)}(\xi_i, \xi_j) = 0.$$

We now consider a subbundle of moving frames whose vertices A_i lie on the straight lines determined by the vectors indicated above. Then the vectors ξ_i have coordinates: $\xi_i = (\delta^k_i)$, and the condition of their conjugacy takes the form

$$\Phi^\alpha_{(2)}(\xi_i, \xi_j) = b^\alpha_{kl}\delta^k_i\delta^l_j = b^\alpha_{ij} = 0, \quad i \neq j. \tag{3.1}$$

Thus, the second fundamental forms of the submanifold V^m are reduced to sums of squares:

$$\Phi^\alpha_{(2)} = \sum_i b^\alpha_{ii}(\omega^i)^2. \tag{3.2}$$

Each field of directions ξ_i considered above determines a family of integral curves, and all m fields ξ_i determine a net Σ_2 of curves on the submanifold V^m, which is called a *conjugate net of second order*.

The result which we obtained above can be formulated as the following theorem:

Theorem 3.1 *If a submanifold V^m carries a net Σ_2 of conjugate lines, then its second fundamental forms can be simultaneously reduced to the form* (3.2), *i.e. they are sums of squares.* ■

Since the points A_i lie on the tangents to the lines of the conjugate net, the group $\widetilde{G}_1$ of admissible transformations of first order frames associated with a point of the submanifold V^m is smaller than in the general case. This group is a subgroup of the group G of centroprojective transformations of first order frames of the submanifold V^m. The transformations of this subgroup keep the straight lines A_0A_i fixed. A subbundle $\widetilde{\mathcal{R}}^1(V^m)$ of the bundle $\mathcal{R}^1(V^m)$ arises on the submanifold V^m carrying a conjugate net. The fiber $\widetilde{\mathcal{R}}^1_x$ of this subbundle is diffeomorphic to the group $\widetilde{G}_1$ and is an orbit of a faithful representation of this group.

If we denote by δ the value of the differential d on the fiber $\widetilde{\mathcal{R}}^1_x$ of the subbundle $\widetilde{\mathcal{R}}^1(V^m)$ (i.e. the value of d when $\omega^i = 0$), then the equations of the infinitesimal displacement of moving frames associated with a point x of the submanifold V^m carrying a conjugate net have the form:

$$\delta A_0 = \pi_0^0 A_0, \quad \delta A_i = \pi_i^0 A_0 + \pi_i^i A_i, \tag{3.3}$$

where there is no summation over the index i. It follows from equations (3.3) that

$$\pi_j^i = \omega_j^i(\delta) = 0, \quad i \neq j, \tag{3.4}$$

and the forms $\pi_0^0 = \omega_0^0(\delta)$, $\pi_i^0 = \omega_i^0(\delta)$, $\pi_i^i = \omega_i^i(\delta)$ are invariant forms of the group $\widetilde{G}_1$.

To simplify our notation, we set:

$$b_{ii}^\alpha = b_i^\alpha.$$

Then formulas (2.21) and (3.2) take the form:

$$\omega_i^\alpha = b_i^\alpha \omega^i \tag{3.5}$$

and

$$\Phi_{(2)}^\alpha = \sum_i b_i^\alpha (\omega^i)^2. \tag{3.6}$$

Note that in formulas (3.5) there is no summation over the index i. In general, in this chapter summation over the indices $i, j, k = 1, \ldots, m$, is assumed only if there is the summation sign $\sum$.

Let us write the matrix of the coefficients of form (3.6):

$$B = (b_i^\alpha) = \begin{pmatrix} b_1^{m+1} & b_2^{m+1} & \ldots & b_m^{m+1} \\ \multicolumn{4}{c}{\dotfill} \\ b_1^{m+n} & b_2^{m+n} & \ldots & b_m^{m+n} \end{pmatrix}. \tag{3.7}$$

Since rank $B \leq m$, the matrix B does not have more than m linearly independent rows. Hence, the number m_1 of linearly independent forms (3.6) also does not exceed m, and *the dimension $m + m_1$ of the osculating space $T_x^{(2)}$ does not exceed $2m$.*

Since we suppose that the submanifold V^m is tangentially nondegenerate, the matrix (3.7) has no columns with all zero entries. This allows us to consider the points

$$B_i = b_i^\alpha A_\alpha \tag{3.8}$$

(cf. (2.26)). These points and the points A_0 and A_i define the subspace $T_x^{(2)}$. If we place the points A_{i_1}, where $i_1 = m+1, \ldots, m+m_1$, into the subspace $T_x^{(2)}$, then decompositions (3.8) take the form:

$$B_i = b_i^{i_1} A_{i_1}, \tag{3.9}$$

and the matrix (3.7) will have only the first m_1 rows as nonzero rows.

Discarding rows with all zero entries, we write this matrix in the form:

$$B = (b_i^{i_1}) = \begin{pmatrix} b_1^{m+1} & b_2^{m+1} & \ldots & b_m^{m+1} \\ \multicolumn{4}{c}{\dotfill} \\ b_1^{m+m_1} & b_2^{m+m_1} & \ldots & b_m^{m+m_1} \end{pmatrix}. \tag{3.10}$$

Then equations (3.5) take the form:

$$\omega_i^{i_1} = b_i^{i_1} \omega^i, \quad \omega_i^{\alpha_1} = 0, \quad \alpha_1 = m + m_1 + 1, \ldots, n. \tag{3.11}$$

If we factorize formulas (3.9) with respect to the space $T_x^{(1)}$, then these formulas take the form:

$$\widetilde{B}_i = b_i^{i_1} \widetilde{A}_{i_1}. \tag{3.12}$$

Note also one more general property of submanifolds V^m carrying a net Σ_2 of conjugate lines. If a submanifold V^m is referred to its conjugate net Σ_2, then $b_{ij}^{i_1} = 0$ for $i \neq j$ (cf. (3.1)). Because of this and the symmetry of the tensor $b_{ijk}^{\alpha_1}$, it follows from equations (2.26) that the tensor $b_{ijk}^{\alpha_1}$ also has a diagonal form, i.e. $b_{ijk}^{\alpha_1} = 0$ if at least one pair of the indices i, j and k consists of different indices. Hence the fundamental forms of third order can be reduced to the form:

$$\Phi_{(3)}^{\alpha_1} = \sum_i b_{iii}^{\alpha_1} (\omega^i)^3. \tag{3.13}$$

It is obvious that this consideration can be applied to the fundamental forms of any order k.

This implies the following theorem:

Theorem 3.2 *If a submanifold V^m carrying a net Σ_2 is referred to this net, then its fundamental forms of any order k can be reduced to the canonical form of type (3.13), i.e. they are sums of kth powers of the basis forms ω^i.* ■

In a similar manner we can consider conjugate nets $\Sigma_k, k \geq 3$, of higher orders k. For example, an independent system of m directions ξ_i is called *conjugate of third order* if any three of these directions, two of which can coincide, are conjugate with respect to all the forms $\Phi^{\alpha_1}_{(3)}$, but each pair of these directions is not conjugate with respect to the forms $\Phi^{i_1}_{(2)}$. Analytically, this means that

$$\Phi^{\alpha_1}_{(3)}(\xi_i, \xi_j, \xi_k) = 0,$$

where at least one pair of indices i, j and k consists of different indices, but

$$\Phi^{i_1}_{(2)}(\xi_i, \xi_j) \neq 0$$

for at least one pair (i, j), $i \neq j$.

For a net Σ_3 the cubic forms $\Phi^{\alpha_1}_{(3)}$ can be reduced to the form (3.13), but the quadratic forms $\Phi^{i_1}_{(2)}$ have general form.

Applying the same method used for the nets Σ_2, we can prove that *for nets Σ_k all fundamental forms of order greater than or equal to k can be reduced to the sums of lth powers of the basis forms ω^i, where $l \geq k$.*

We will consider below only conjugate nets Σ_2 of second order.

3.2 The Holonomicity of the Conjugate Net Σ_2

Let a net Σ (not necessarily conjugate) of curves be given on a submanifold V^m. Suppose that the points A_i of moving frames associated with a point $x \in V^m$ are placed on the tangents to the net curves passing through the point x. If the point $x = A_0$ is fixed, then the straight lines A_0A_i are also fixed, and if $\omega^i = 0$, then equations (3.3) and (3.4) hold. It follows from these equations that the 1-forms ω^i_j, $i \neq j$, can be expressed linearly in terms of the basis forms ω^k of a submanifold V^m, i.e.

$$\omega^i_j = \sum_k l^i_{jk}\omega^k, \quad i \neq j. \tag{3.14}$$

In the tangent subspace $T^{(1)}_x$ of a submanifold V^m carrying a net of curves, the points $A_0, A_1, \ldots, A_{i-1}, A_{i+1}, \ldots, A_m$ define an $(m-1)$-dimensional subspace $\Delta_i(x)$. The field of these subspaces forms a *distribution* Δ_i on V^m, which is given by the equation $\omega^i = 0$, where i is a fixed index.

Definition 3.3 A net Σ on V^m is said to be *holonomic* if each of the m distributions Δ_i is involutive.

This means that the subspaces $\Delta_i(x)$ are tangent to $(m-1)$-dimensional submanifolds V_i^{m-1}, and the lines of the net Σ belong to these submanifolds.

Each pair of families of lines of a net Σ defines a two-dimensional distribution Δ^{ij}. If a net Σ is holonomic, all these distributions are also holonomic, since the systems of equations

$$\omega^k = 0, \quad k \neq i, j, \tag{3.15}$$

defining these distributions are completely integrable.

Each pair of families of lines of a net Σ also defines a curvilinear two-web in the sense of Blaschke (see [Bl 55], pp. 96–98). If the distributions Δ^{ij} are holonomic, then this two-web is quadrilateral (see [Bl 55], pp. 99–100), i.e. any quadruple of curves of this web forms a closed quadrangle (see Figure 3.1).

Figure 3.1

As was indicated above, a distribution Δ_i on V^m is defined by the equation:

$$\omega^i = 0, \quad i \text{ is fixed.} \tag{3.16}$$

If the net Σ is holonomic, each of equations (3.16) must be completely integrable.

Applying exterior differentiation to the forms ω^i and using equations (3.14), we obtain

$$d\omega^i = \sum_j \omega^j \wedge \omega_j^i = \omega^i \wedge (\omega_i^i - l_{ji}^i \omega^j) + \sum_{j,k \neq i} l_{jk}^i \omega^j \wedge \omega^k. \tag{3.17}$$

By the Frobenius theorem (see Theorem 1.2), equation (3.16) is completely integrable if and only if the last term in equation (3.17) vanishes, i.e. if

$$l_{[jk]}^i = 0, \; j, k \neq i. \tag{3.18}$$

For fixed i, this condition is the same as the involutivity of the distribution Δ_i. If condition (3.18) holds for all values of i, then all distributions Δ_i are involutive, and the net Σ is holonomic. Thus, *condition* (3.18) *is necessary and sufficient for an arbitrary net* Σ *on a submanifold* V^m *to be holonomic.*

Suppose now that Σ is a conjugate net Σ_2. Then each distribution $\Delta_i(x)$ is conjugate to the direction A_0A_i. The curves of a conjugate net Σ_2 form nets on the submanifolds V_i^{m-1}. In general, these nets are not conjugate.

Definition 3.4 If a conjugate net Σ_2 is holonomic and each of the nets determined on the submanifolds V_i^{m-1} by the net Σ_2 is conjugate itself, then the conjugate net Σ_2 is called an *m-conjugate system.*

Let us find necessary and sufficient conditions for a conjugate net Σ_2 on V^m to be an m-conjugate system. Take, for example, a submanifold V_1^{m-1} which is defined by the equations $\omega^1 = 0, \ \omega^\alpha = 0, \ \alpha = m+1, \ldots, n$, in the space P^n. On this submanifold we have:

$$dA_0 = \omega_0^0 A_0 + \omega^a A_a, \quad a = 2, \ldots, m,$$

and

$$d^2 A_0 / T_x^{(1)}(V_1^{m-1}) = l_{ab}^1 \omega^a \omega^b A_1 + b_{ab}^\alpha \omega^a \omega^b \widetilde{A}_\alpha.$$

Since a net Σ_2 is conjugate on the submanifold V^m, we have $b_{ab}^\alpha = 0$ for $a \neq b$. Hence, the net defined by the net Σ_2 on the submanifold V_1^{m-1} is conjugate if and only if the conditions $l_{ab}^1 = 0, \ a \neq b$, hold. It follows that the conjugate net Σ_2 is an m-conjugate system on the submanifold V^m if and only if the following conditions hold:

$$l_{jk}^i = 0, \quad i \neq j, k, \quad j \neq k. \tag{3.19}$$

Therefore, we proved the following result.

Theorem 3.5 *A conjugate net Σ_2 on a submanifold V^m is holonomic if and only if conditions* (3.18) *are satisfied. This net is an m-conjugate system if and only if conditions* (3.19) *hold.* ■

By (3.19), if a conjugate net Σ_2 is an m-conjugate system, the forms ω_j^i, defined by equations (3.14) for an arbitrary conjugate net Σ_2, take the form:

$$\omega_j^i = l_{ji}^i \omega^i + l_{jj}^i \omega^j. \tag{3.20}$$

We will now prove the following property of m-conjugate systems.

Theorem 3.6 *If a submanifold V^m carries an m-conjugate system, then all its submanifolds V_i^{m-1} also carry $(m-1)$-conjugate systems.*

Proof. Consider, for example, the submanifold V_1^{m-1} which is defined by the equations

$$\omega^1 = 0, \ \omega^\alpha = 0, \ \alpha = m+1, \ldots, n.$$

On this submanifold, equations (3.5) take the form:

$$\omega_1^\alpha = 0, \quad \omega_a^\alpha = b_a^\alpha \omega^a, \quad a = 2, \ldots, m.$$

If we take $i = 1$ and $j = a$, then from equations (3.20) we find that

$$\omega_a^1 = l_{aa}^1 \omega^a.$$

This implies that the submanifold V_1^{m-1} has the following second fundamental forms: the second fundamental forms (3.6) of V^m, which on the submanifold V_1^{m-1} are expressed by the formula

$$\Phi^\alpha_{(2)} = \sum_{a=2}^m b^\alpha_a (\omega^a)^2,$$

and one additional form

$$\Phi^1_{(2)} = \sum_{a=2}^m l^1_{aa} (\omega^a)^2,$$

which is also a sum of squares. Thus, the submanifold V_1^{m-1} carries a conjugate net.

On the submanifold V_1^{m-1}, equations (3.20) take the form:

$$\omega^a_b = l^a_{ba}\omega^a + l^a_{bb}\omega^b,$$

which proves that this conjugate net is an $(m-1)$-conjugate system. ■

It follows from Theorem 3.6 that if on V^m we consider the submanifolds $V^{m-q}_{i_1 \ldots i_q}$ defined by the equations

$$\omega^{i_1} = 0, \ldots, \omega^{i_q} = 0,$$

then the net consisting of $m-q$ families of curves defined on each of the submanifolds $V^{m-q}_{i_1 \ldots i_q}$ by the m-conjugate net Σ_2 is also an $(m-q)$-conjugate system.

The properties of an m-conjugate net Σ_2 indicated above are connected with the structure of the matrix B. To establish this connection, we find the exterior differentials of equation (3.5). This gives the following quadratic equations:

$$\Delta b^\alpha_i \wedge \omega^i - \sum_{j \neq i} (b^\alpha_i \omega^i_j + b^\alpha_j \omega^j_i) \wedge \omega^j = 0, \tag{3.21}$$

where

$$\Delta b^\alpha_i = db^\alpha_i + b^\beta_i \omega^\alpha_\beta - b^\alpha_i (2\omega^i_i - \omega^0_0),$$

and as we indicated above, there is no summation over the index i. Substituting the expressions of the forms ω^i_j from equations (3.14) into equations (3.21), we obtain

$$\widetilde{\Delta} b^\alpha_i \wedge \omega^i - \sum_{k \neq i} \sum_{j \neq i} (b^\alpha_i l^i_{jk} + b^\alpha_j l^j_{ik}) \omega^k \wedge \omega^j = 0, \tag{3.22}$$

where

$$\widetilde{\Delta} b^\alpha_i = \nabla b^\alpha_i + \sum_{j \neq i} (b^\alpha_i l^i_{ji} + b^\alpha_j l^j_{ii}) \omega^j.$$

Since this relation must be satisfied identically, the alternated coefficients in $\omega^k \wedge \omega^j$ must vanish, i.e. we have

$$b_i^\alpha(l^i_{jk} - l^i_{kj}) + b_j^\alpha l^j_{ik} - b_k^\alpha l^k_{ij} = 0, \quad k, j \neq i. \tag{3.23}$$

Since the quantities b_i^α define the points B_i (see (3.8)) in the subspace $T_x^{(2)}$, by contracting equations (3.23) with the points $\widetilde{A}_\alpha$, we find that

$$\widetilde{B}_i(l^i_{jk} - l^i_{kj}) + \widetilde{B}_j l^j_{ik} - \widetilde{B}_k l^k_{ij} = 0. \tag{3.24}$$

We can now prove the following result.

Theorem 3.7 *Suppose that a submanifold V^m carries a conjugate net Σ_2 and that one of the following two conditions is satisfied:*

a) *Any three of the points $\widetilde{B}_i$ are linearly independent.*

b) *The net Σ_2 is holonomic on V^m and any two points $\widetilde{B}_i$ and $\widetilde{B}_j$, $j \neq i$, are linearly independent.*

Then the net Σ_2 is an m-conjugate system.

Proof. a) Since the points $\widetilde{B}_i$, $\widetilde{B}_j$ and $\widetilde{B}_k$ for distinct i, j and k are linearly independent, equation (3.24) implies equation (3.19).

b) The holonomicity of the net Σ_2 implies the equations $l^i_{jk} = l^i_{kj}$. Hence equation (3.24) takes the form:

$$\widetilde{B}_j l^j_{ik} - \widetilde{B}_k l^k_{ij} = 0.$$

Since the points $\widetilde{B}_j$ and $\widetilde{B}_k$ are linear independent, equation (3.25) again yields (3.19). ■

It is interesting to find submanifolds V^m that carry a nonholonomic conjugate net Σ_2. To this end we give the following definition.

Definition 3.8 A conjugate net Σ_2 is said to be *irreducible* if each distribution Δ^{ij} is not involutive.

Since the distribution Δ^{ij} is defined by equations (3.15) and

$$d\omega^k \equiv 2l^k_{[ij]}\omega^i \wedge \omega^j \pmod{\omega^k}, \quad k \neq i, j,$$

the conditions for a net Σ_2 to be irreducible are given by the following inequalities:

$$l^k_{ij} \neq l^k_{ji} \tag{3.25}$$

which must hold for any three different values i, j and k.

Theorem 3.9 *Only submanifolds V^m, whose osculating subspaces $T_x^{(2)}$ are of dimension $m+1$ or $m+2$, can carry an irreducible net Σ_2.*

Proof. In fact, the subspace $T_x^{(2)}$ is defined by the points A_0, A_i and B_i. It follows from equations (3.24) that condition (3.25) can hold only if the dimension of the normal subspace $\widetilde{N}_x^{(2)}$, to which the points $\widetilde{B}_i = B_i/T_x^{(2)}$ belong, is equal to 0 or 1, i.e. if $m_1 = 1$ or 2. ■

As we saw in Section **2.5**, if $m_1 = 1$, a submanifold V^m carries a set of conjugate nets.

On the other hand, if $m_1 = 2$, then the matrix B (see (3.10)) has the form:

$$B = (b_i^{i_1}) = \begin{pmatrix} b_1^{m+1} & b_2^{m+1} & \dots & b_m^{m+1} \\ b_1^{m+2} & b_2^{m+2} & \dots & b_m^{m+2} \end{pmatrix}. \tag{3.26}$$

Thus the points $\widetilde{B}_i$ lie on a straight line. If any two pairs of these points are distinct, then matrix (3.26) does not have proportional columns, and the submanifold V^m carries a unique conjugate net which is in general nonholonomic.

3.3 Classification of the Conjugate Nets Σ_2

1. Consider the points

$$\widetilde{B}_i = B_i/T_x^{(1)} = b_i^\alpha \widetilde{A}_\alpha. \tag{3.27}$$

Since the coordinates of the points $\widetilde{B}_i$ are the columns of the matrix B, the number of linearly independent points is equal to the rank of this matrix B, i.e. this number is equal to m_1. These independent points generate the reduced normal subspace $\widetilde{N}_x^{(1)}$, whose dimension is equal to $m_1 - 1$.

We can now prove the result generalizing the Segre theorem which was proved in Section **2.5**.

Theorem 3.10 *Suppose that a submanifold V^m carries a net of conjugate lines and that at any point of V^m the following two conditions hold:*

a) *$m_1 < m$, and*

b) *Any subset of the points $\widetilde{B}_i$ consisting of $m-1$ points has rank m_1 and generates the subspace $\widetilde{N}_x^{(1)}$.*

Then the submanifold V^m belongs to its osculating subspace $T_x^{(2)}$ of dimension $m + m_1$.

Proof. Consider equation (2.27) for $j = i \neq k$. Since $b_{ik}^{i_1} = 0$ if $i \neq k$ and $b_{ii}^{i_1} = b_i^{i_1}$, equation (2.27) takes the form

$$b_i^{i_1} c_{i_1 k}^{\alpha_1} = 0, \; i \neq k. \tag{3.28}$$

By the condition b) of the theorem, there are m_1 linearly independent points among the points B_i, $i \neq k$. Thus, if $i \neq k$, then $\operatorname{rank}(b_i^{i_1}) = m_1$. This implies that the system (3.28) has only the trivial solution $c_{i_1 k}^{\alpha_1} = 0$. It follows that

$$\omega_{i_1}^{\alpha_1} = 0. \tag{3.29}$$

Taking into account equations (3.29), we can write the equations of infinitesimal displacement of the moving frame in the following form:

$$\begin{aligned} dA_0 &= \omega_0^0 A_0 + \omega^i A_i, \\ dA_i &= \omega_i^0 A_0 + \omega_i^j A_j + \omega_i^{i_1} A_{i_1}, \\ dA_{i_1} &= \omega_{i_1}^0 A_0 + \omega_{i_1}^j A_j + \omega_{i_1}^{j_1} A_{j_1}. \end{aligned}$$

These equations show that the second osculating subspace $T_x^{(2)}$ of the submanifold V^m remains constant. Thus, the submanifold V^m belongs to this osculating subspace. ■

The generalized Segre theorem was first proved by Bazylev (see [Ba 55a] and [Ba 55b]) and Akivis (see [A 61a]).

2. Suppose now that $m_1 = m$. Then any subset of the points $\widetilde{B}_i$ consisting of $m-1$ points has rank $m-1$. In this case $\dim T_x^{(2)} = 2m$, and the submanifold V^m has precisely m independent second fundamental forms $\Phi_{(2)}^{m+i}$. Consider the subbundle of moving frames whose vertices A_{m+i} belong to the osculating 2-plane of the curve $C_i \subset \Sigma_2$ defined by the equations $\omega^j = 0,\ j \neq i$. The matrix (3.10) becomes a square matrix, and for the frames from the subbundle indicated above, this matrix is diagonal. If we appropriately normalize the points A_{m+i}, this matrix will be the identity matrix:

$$B = \begin{pmatrix} 1 & 0 & \dots & 0 \\ 0 & 1 & \dots & 0 \\ \dots & \dots & \dots & \dots \\ 0 & 0 & \dots & 1 \end{pmatrix}. \tag{3.30}$$

This implies that equations (3.11) take the following form:

$$\omega_i^{m+i} = \omega^i, \quad \omega_i^{m+j} = 0, \ j \neq i, \tag{3.31}$$

$$\omega_i^{\alpha_1} = 0, \quad \alpha_1 = 2m+1, \dots, n. \tag{3.32}$$

The second fundamental forms of V^m take the form:

$$\Phi_{(2)}^{m+i} = (\omega^i)^2, \quad \Phi_{(2)}^{\alpha_1} = 0, \quad \alpha_1 = 2m+1, \dots, n. \tag{3.33}$$

Submanifolds of this type were first considered by É. Cartan in his paper [Ca 19]. This is the reason that these submanifolds are called *Cartan varieties.*

Proposition 3.11 *A Cartan variety is an m-conjugate system.*

Proof. In fact, exterior differentiation of equations (3.31) leads to the following exterior quadratic equations:

$$(\omega_{m+i}^{m+i} - 2\omega_i^i + \omega_0^0) \wedge \omega^i - \sum_{k \neq i} \omega_k^i \wedge \omega^k = 0, \quad -\omega_j^i \wedge \omega^i + \omega_{m+j}^{m+i} \wedge \omega^j = 0.$$

If we apply Cartan's lemma to these equations, we get equations (3.20), which imply conditions (3.19). By Theorem 3.5, the submanifold V^m under consideration is an m-conjugate system. ■ Note that the converse is not true, since for an m-conjugate system the number m_1 is not necessarily equal to m: $m_1 \leq m$.

Since the forms $\Phi_{(3)}^{\alpha_1}$ can be reduced to the form (3.13), the number m_2 of linearly independent forms $\Phi_{(3)}^{\alpha_1}$ does not exceed m: $m_2 \leq m$. This implies that $\dim T_x^{(3)} \leq 3m$.

If $m_2 = m$, then we consider the subbundle of moving frames whose vertices A_{2m+i} belong to the osculating three-plane of the curve $C_i \subset \Sigma_2$. By appropriately normalizing the points A_{2m+i}, we can reduce the forms $\Phi_{(3)}^{\alpha_1}$ to the form:

$$\Phi_{(3)}^{2m+i} = (\omega^i)^3, \quad \Phi_{(3)}^{\alpha_2} = 0, \quad \alpha_2 = 3m+1, \ldots, n. \tag{3.34}$$

3. Suppose further that $m_1 < m$, but the set of points $\widetilde{B}_i$ has the property that among its subsets consisting of $m-1$ points there are both subsets of rank m_1 and subsets of rank less than m_1.

We will now prove the following result for submanifolds V^m carrying a conjugate net Σ_2 satisfying the above condition.

Theorem 3.12 *Suppose that a submanifold V^m carries a conjugate net Σ_2 and that at any point of V^m the following four conditions hold:*

a) $m_1 < m$.

b) *The subsets $\widetilde{B}_1, \ldots, \widetilde{B}_{l-1}, \widetilde{B}_{l+1}, \ldots, \widetilde{B}_q, \widetilde{B}_{q+1}, \ldots, \widetilde{B}_m$ are of rank $m_1 - 1$.*

c) *The subsets $\widetilde{B}_1, \ldots, \widetilde{B}_q, \widetilde{B}_{q+1}, \ldots, \widetilde{B}_{l-1}, \widetilde{B}_{l+1}, \ldots, \widetilde{B}_m$ are of rank m_1.*

d) *No two points $\widetilde{B}_u$ and $\widetilde{B}_v$, $u \neq v$, coincide.*

Then the submanifold V^m is doubly foliated: V^m is an $(m-q)$-parameter family of the Cartan varieties V^q of dimension q, and V^m is a q-parameter family of submanifolds V^{m-q} of dimension $m-q$ which belong to their second osculating subspaces $T_x^{(2)}$ of dimension $m + m_1 - 2q$.

Proof. By the condition b) of the theorem, the points $\widetilde{B}_1, \ldots, \widetilde{B}_q$ are linearly independent, and they are not linearly dependent on the points $\widetilde{B}_{q+1}, \ldots, \widetilde{B}_m$

which determine a subspace L of dimension $m_1 - q - 1$. Therefore, we can include the points $\widetilde{B}_1, \ldots, \widetilde{B}_q$ into a basis of the normal subspace $\widetilde{N}_x^{(1)}$ by setting

$$\widetilde{A}_{m+1} = \widetilde{B}_1, \ldots, \widetilde{A}_{m+q} = \widetilde{B}_q,$$

and place the points $\widetilde{A}_{m+q+1}, \ldots, \widetilde{A}_{m+m_1}$ into the subspace L. As a result of this specialization of moving frames, the matrix (3.10) is reduced to the form:

$$B = \begin{pmatrix} 1 & \ldots & 0 & 0 & \ldots & 0 \\ \ldots & \ldots & \ldots & \ldots & \ldots & \ldots \\ 0 & \ldots & 1 & 0 & \ldots & 0 \\ 0 & \ldots & 0 & b_{q+1}^{m+q+1} & \ldots & b_m^{m+q+1} \\ \ldots & \ldots & \ldots & \ldots & \ldots & \ldots \\ 0 & \ldots & 0 & b_{q+1}^{m+m_1} & \ldots & b_m^{m+m_1} \end{pmatrix}. \tag{3.35}$$

Moreover, the matrix

$$\begin{pmatrix} b_{q+1}^{m+q+1} & \ldots & b_m^{m+q+1} \\ \ldots & \ldots & \ldots \\ b_{q+1}^{m+m_1} & \ldots & b_m^{m+m_1} \end{pmatrix}$$

has rank $m_1 - q$, and this rank is not decreased if we delete any column of this matrix. Suppose that $a, b = 1, \ldots, q$ and $u = q+1, \ldots, m$. Then (3.24) for $i = a, j = b$ and $k = u$ takes the form:

$$\widetilde{B}_a (l_{bu}^a - l_{ub}^a) + \widetilde{B}_b l_{au}^b - \widetilde{B}_u l_{ab}^u = 0. \tag{3.36}$$

From the form of matrix (3.35) it follows that the points $\widetilde{B}_a, \widetilde{B}_b, a \neq b$, and $\widetilde{B}_u$ are linearly independent. This implies

$$l_{ab}^u = 0. \tag{3.37}$$

Since

$$d\omega^u \equiv l_{ab}^u \omega^a \wedge \omega^b \pmod{\omega^v},$$

it follows from equation (3.37) that the system $\omega^u = 0$ is completely integrable. This proves that the submanifold V^m under consideration is foliated by an $(m-q)$-parameter family of submanifolds V^q of dimension q. By (3.35) and (3.37), the second fundamental forms of the submanifolds V^q have the form:

$$\Phi_{(2)}^{m+a} = (\omega^a)^2, \quad \Phi_{(2)}^{m+u} = 0, \quad \Phi_{(2)}^{u_1} = 0, \quad \Phi_{(2)}^{\alpha_1} = 0,$$

where $a = 1, \ldots, q$; $u = q+1, \ldots, m$; $u_1 = m+q+1, \ldots, m+m_1$ and $\alpha_1 = m+m_1+1, \ldots, n$. Thus the dimension of the osculating space of each of the submanifolds V^q is equal to $2q$, and these submanifolds are Cartan varieties.

Next, we write system (3.24) for $i = u$, $j = v$, $u \neq v$ and $k = a$:

$$\widetilde{B}_u(l^u_{va} - l^u_{av}) + \widetilde{B}_v l^v_{ua} - \widetilde{B}_a l^a_{uv} = 0. \tag{3.38}$$

From the form of the matrix (3.35) and the condition d) of the theorem, it follows that the points $\widetilde{B}_u, \widetilde{B}_v, u \neq v$, and $\widetilde{B}_a$ are linearly independent. This and equation (3.38) yield

$$l^b_{uv} = 0. \tag{3.39}$$

Since

$$d\omega^a \equiv l^a_{uv}\omega^u \wedge \omega^v \pmod{\omega^b},$$

it follows from equation (3.38) that the system $\omega^a = 0$ is completely integrable. This proves that the submanifold V^m under consideration is foliated by a q-parameter family of submanifolds V^{m-q} of dimension $m - q$. By (3.35) and (3.38), only the following second fundamental forms of submanifolds V^{m-q} do not vanish:

$$\Phi^{u_1}_{(2)} = \sum_u b^{u_1}_u(\omega^u)^2, \quad u = q+1, \ldots, m; u_1 = m+q+1, \ldots, m+m_1.$$

Since all these forms are sums of squares, each of the submanifolds V^{m-q} carries a conjugate net. But by condition c) of the theorem, the hypotheses of the generalized Segre theorem (Theorem 3.10) are satisfied for each of the submanifolds V^{m-q}. Thus, each of these submanifolds belongs to its second osculating subspaces $T^{(2)}_x$ of dimension $m + m_1 - 2q$. ■

3.4 Some Existence Theorems

1. First of all, we prove an existence theorem for submanifolds V^m carrying a conjugate system Σ_2.

Theorem 3.13 *Submanifolds V^m carrying a conjugate system Σ_2 exist, and the solution of the system defining such submanifolds depends on $m(m-1)$ arbitrary functions of two variables.*

Proof. As we saw earlier, the system of Pfaffian equations defining an m-conjugate system Σ_2 consists of equations (2.5), (3.5) and (3.20):

$$\omega^\alpha = 0, \tag{3.40}$$

$$\omega^\alpha_i = b^\alpha_i \omega^i, \tag{3.41}$$

$$\omega^j_i = l^j_{ii}\omega^i + l^j_{ij}\omega^j. \tag{3.42}$$

By (3.41), exterior differentiation of equations (3.40) leads to identities. By (3.22) and (3.42), exterior differentiation of equations (3.41) gives the following exterior quadratic equations:

$$\Delta b_i^\alpha \wedge \omega^i = 0, \tag{3.43}$$

where

$$\Delta b_i^\alpha = db_i^\alpha + b_i^\alpha(\omega_0^0 - 2\omega_i^i) + \sum_{k\neq i}(l_{ki}^i b_i^\alpha + l_{ii}^k b_k^\alpha)\omega^k + b_i^\beta \omega_\beta^\alpha.$$

Finally, applying exterior differentiation to equation (3.42), we obtain:

$$\Delta l_{ii}^j \wedge \omega^i + \Delta l_{ij}^j \wedge \omega^j = 0, \tag{3.44}$$

where

$$\Delta l_{ii}^j = dl_{ii}^j + l_{ii}^j(\omega_0^0 - 2\omega_i^i + \omega_j^j) + b_i^\alpha \omega_\alpha^j + l_{ii}^j \sum_{k\neq j} l_{ki}^i \omega^k + \sum_{k\neq i,j} l_{ii}^k \omega_k^j, \tag{3.45}$$

and

$$\Delta l_{ij}^j = dl_{ij}^j + l_{ij}^j(\omega_0^0 - \omega_i^i) - \omega_i^0 + l_{ij}^j \sum_{k\neq j} l_{kj}^j \omega^k - \sum_{k\neq i,j} l_{kj}^j l_{ik}^k \omega^k, \tag{3.46}$$

The system of equations (3.40)–(3.44) is closed with respect to the operation of exterior differentiation. We will apply the Cartan test (see Theorem 1.3 in Section **1.2**) to investigate the consistency of this system. The number q of unknown functions Δb_i^α, Δl_{ii}^j and Δl_{ij}^j in the exterior quadratic equations is: $q = (n-m)m + 2m(m-1) = m(n+m-2)$. The first character s_1 of the system under consideration is equal to the number of independent exterior quadratic equations, i.e. $s_1 = (n-m)m + m(m-1) = m(n-1)$. Its second character $s_2 = q - s_1 = m(m-1)$, and the third and all subsequent characters are equal to 0: $s_3 = \ldots = s_m = 0$. This implies that the Cartan number $Q = s_1 + 2s_2 = (n-m)m + 3m(m-1)$.

Let us find the number N of parameters on which the most general integral element of the system of equations (3.40)–(3.44) depends. To find N, we apply the Cartan lemma to equations (3.43) and (3.44):

$$\begin{aligned} \Delta b_i^\alpha &= b_{iii}^\alpha \omega^i, \\ \Delta l_{ji}^i &= l_{jii}^i \omega^i + l_{jij}^i \omega^j, \\ \Delta l_{jj}^i &= l_{jij}^i \omega^i + l_{jjj}^i \omega^j. \end{aligned}$$

The number N is equal to the number of independent coefficients in these equations, i.e.

$$N = m(n-m) + 3m(m-1).$$

Since $N = Q$, the system of equations (3.40)–(3.44) is in involution, and an m-dimensional integral manifold V^m, defined by this system, depends on $s_2 = m(m-1)$ arbitrary functions of two variables. ■

A submanifold V^m, carrying a conjugate net Σ_2 which is not an m-conjugate system, can be more arbitrary. For example, a submanifold V^m of general type in a projective space P^{m+2} carries a conjugate net Σ_2 which is not an m-conjugate system (see Theorem 3.10). It is easy to see that this submanifold is defined by two arbitrary functions of m variables.

2. Consider now a submanifold V^m carrying a conjugate net Σ_3 of third order and suppose that the osculating subspace $T_x^{(2)}(V^m)$ has the maximal possible dimension $m + \frac{1}{2}m(m+1) = \frac{1}{2}m(m+3)$ and $m_2 = m$. For such a submanifold V^m we have: $m_1 = \frac{1}{2}m(m+1)$. In this case $i_1 = m+1, \ldots, \frac{1}{2}m(m+3)$, and the fundamental forms $\Phi^{i_1}_{(2)}$ can be reduced to the form:

$$\Phi^{(ii)}_{(2)} = (\omega^i)^2, \quad \Phi^{(ij)}_{(2)} = 2\omega^i\omega^j, \quad i \neq j, \; i, j = 1, \ldots, m. \tag{3.47}$$

We use here a double index notation for numbering the forms $\Phi^{i_1}_{(2)}$. The fundamental forms $\Phi^{i_2}_{(3)}$ can be reduced to the form:

$$\Phi^{(m+m_1+i)}_{(3)} = (\omega^i)^3. \tag{3.48}$$

To achieve this, the vertices A_α of a moving frame must be chosen in such a way that $A_{i_1} \in T_x^{(2)}$, $A_{i_2} \in T_x^{(3)}$ and $A_{\alpha_2} \notin T_x^{(3)}$.

Note that the quadratic forms (3.47) on the submanifold under consideration have the same structure which they had on the Veronese varieties (see Section **2.2**). However, in contrast to the Veronese varieties, we now have $m_2 > 0$.

Theorem 3.14 *The submanifolds V^m described above exist and depend on $\frac{1}{2}m(m-1)$ arbitrary functions of m variables.*

Proof. For the submanifolds V^m under consideration the system (2.5) takes the form:

$$\omega^{(ij)} = 0, \quad \omega^{i_2} = 0, \quad \omega^{\alpha_2} = 0, \tag{3.49}$$

where $i_2 = m + m_1 + 1, \ldots, 2m + m_1$; $\alpha_2 = 2m + m_1 + 1, \ldots, n$. By (3.47), equations (2.18) become:

$$\omega_i^{(ii)} = \omega^i, \quad \omega_j^{(ii)} = 0, \quad \omega_i^{(ij)} = \omega^j, \quad \omega_k^{(ij)} = 0, \tag{3.50}$$

where the indices denoted by different letters take distinct values. By (3.48), equations (2.25) take the form:

$$\omega^{i_2}_{(ii)} = \omega^i, \quad \omega^{i_2}_{(ij)} = 0, \quad \omega^{i_2}_{(jk)} = 0, \tag{3.51}$$

where $i_2 = m + m_1 + i$. Moreover, the specialization of the moving frame indicated above implies the following Pfaffian equations:

$$\omega_i^{j_2} = 0, \quad \omega_i^{\alpha_2} = 0, \quad \omega_{i_1}^{\alpha_2} = 0. \tag{3.52}$$

Exterior differentiation of all Pfaffian equations (3.49)–(3.52) leads to the following exterior quadratic equations:

$$\theta_i^{(ii)} := \omega^i \wedge (\omega_{(ii)}^{(ik)} + \omega_0^0 - 2\omega_i^i) + \sum_{k \neq i} \omega^k \wedge (\omega_{(ii)}^{(ii)} - \omega_k^i) = 0, \tag{3.53}$$

$$\theta_j^{(ii)} := \omega^i \wedge (\omega_{(ij)}^{(ii)} - \omega_j^i) + \sum_{k \neq i} \omega^k \wedge \omega_{(jk)}^{(ii)} = 0, \tag{3.54}$$

$$\begin{aligned} \theta_i^{(ij)} := \; & \omega^i \wedge (\omega_{(ii)}^{(ij)} - 2\omega_i^j) + \omega^j \wedge (\omega_{(ij)}^{(ij)} + \omega_0^0 - \omega_i^i - \omega_j^j) \\ & + \sum_{k \neq i,j} \omega^k \wedge (\omega_{(ik)}^{(ij)} - \omega_k^j) = 0, \end{aligned} \tag{3.55}$$

$$\theta_k^{(ij)} := \omega^i \wedge (\omega_{(ik)}^{(ij)} - \omega_k^j) + \omega^j \wedge (\omega_{(jk)}^{(ij)} - \omega_k^i) + \sum_{h \neq i,j} \omega^h \wedge \omega_{(kh)}^{(ij)} = 0, \tag{3.56}$$

$$\theta_{(ii)}^{i_2} := \omega^i \wedge (\omega_{i_2}^{i_2} + \omega_0^0 - \omega_{(ii)}^{(ii)} - \omega_i^i) - \sum_{k \neq i} \omega^k \wedge \omega_k^i = 0, \tag{3.57}$$

$$\theta_{(ij)}^{i_2} := \omega^i \wedge \omega_{(ij)}^{(ii)} = 0, \tag{3.58}$$

$$\theta_{(jk)}^{i_2} := \omega^i \wedge \omega_{(jk)}^{(ii)} = 0, \tag{3.59}$$

$$\theta_{(ii)}^{\alpha_2} := \omega^i \wedge \omega_{i_2}^{\alpha_2} = 0. \tag{3.60}$$

In equations (3.53)–(3.60) the forms θ denote the exterior differentials of the corresponding equations of system (3.49)–(3.52). Note that the exterior differentiation of all equations (3.49) and the first two equations of (3.52) lead to identities.

Each of $(n - 2m - m_1)m$ exterior quadratic equations (3.60) contains one form $\omega_{i_2}^{\alpha_2}$ which does not enter into the other exterior quadratic equations. Thus we have for each of the quadratic equations (3.60):

$$\sigma_1 = 1, \quad \sigma_2 = \ldots = \sigma_m = 0.$$

Next, consider the subsystem consisting of $\frac{1}{2}m^2(m+3)$ exterior quadratic equations (3.53), (3.54), (3.57), (3.58) and (3.59). This subsystem contains five groups of forms:

$$\omega^{(ii)}_{(ii)} + \omega^0_0 - 2\omega^i_i, \ \omega^{(ii)}_{(ij)}, \ \omega^i_j, \ \omega^{i_2}_{i_2} + \omega^0_0 - \omega^{(ii)}_{(ii)} - \omega^i_i, \ \omega^{(ii)}_{(jk)}$$

which do not enter into the other exterior quadratic equations. The number of these forms coincides with the number of exterior quadratic equations. Therefore, the characters of this subsystem are:

$$\sigma_1 = \frac{1}{2}m(m+3), \quad \sigma_2 = \ldots = \sigma_m = 0.$$

Finally, consider the subsystem consisting of m exterior quadratic equations (3.55)–(3.56) where the indices i and j, $i \neq j$, are fixed. The number of such subsystems is $\frac{1}{2}m(m+1)$, and each of these subsystems contains six group of forms:

$$\omega^{(ij)}_{(ii)} - 2\omega^j_i, \ \omega^{(ij)}_{(jj)} - 2\omega^i_j, \ \omega^{(ij)}_{(ij)} + \omega^0_0 - \omega^i_i - \omega^j_j, \omega^{(ij)}_{(ik)} - \omega^j_k, \ \omega^{(ij)}_{(jk)} - \omega^i_k, \ \omega^{(ij)}_{(kh)},$$

which do not enter into the other exterior quadratic equations. Thus, we have the following values of characters for each of these subsystems:

$$\sigma_1 = m, \quad \sigma_2 = m-1, \ldots, \sigma_m = 1.$$

Since these subsystems are independent, the characters of the entire system are the sums of the corresponding characters of the subsystems:

$$\begin{aligned}
s_1 &= (n - 2m - m_1) + \tfrac{1}{2}m^2(m+3) + \tfrac{1}{2}m^2(m-1),\\
s_2 &= \tfrac{1}{2}m(m-1)^2,\\
s_3 &= \tfrac{1}{2}m(m-1)(m-2),\\
&\ldots\ldots\ldots\ldots\ldots\ldots\ldots\ldots\ldots\ldots\\
s_m &= \tfrac{1}{2}m(m-1).
\end{aligned}$$

Therefore, the Cartan number for the system (3.55)–(3.56) is:

$$Q = s_1 + 2s_2 + \ldots + ms_m = (n - 2m - m_1)m + \frac{1}{2}m^2(m+3) + \frac{1}{6}m(m+1)(m+2).$$

One can easily show that the number N of parameters, on which the most general integral element of the system under consideration depends, is equal to Q: $N = Q$, i.e. the Cartan criterion is satisfied.

Thus, the system is in involution, and an m-dimensional integral manifold V^m, defined by this system, depends on $s_m = \frac{1}{2}m(m-1)$ arbitrary functions of m variables. ■

This example was considered by É. Cartan in Chapter 4 of his paper [Ca 19].

3.5 Laplace Transforms of Conjugate Nets and Their Generalizations

1. First we consider a two-dimensional surface V^2 in a three-dimensional projective space P^3 and a one-parameter family λ of its lines. none of which is tangent to any of the asymptotic lines of V^2. We associate a family of frames with the surface V^2 in such a way that the vertex A_0 of a frame coincides with a point $x \in V^2$, the vertex A_1 lies on the line of λ passing through the point x, and the vertex A_2 belongs to the tangent plane $T_x(V^2)$. Then we have

$$\begin{aligned} dA_0 &= \omega_0^0 A_0 + \omega_0^1 A_1 + \omega_0^2 A_2, \\ dA_1 &= \omega_1^0 A_0 + \omega_1^1 A_1 + \omega_1^2 A_2 + \omega_1^3 A_3, \end{aligned} \tag{3.61}$$

where

$$\begin{aligned} \omega_1^3 &= b_{11}\omega^1 + b_{12}\omega^2, \\ \omega_1^2 &= l_{11}^2\omega^1 + l_{12}^2\omega^2, \end{aligned} \tag{3.62}$$

$\omega^1 = \omega_0^1$, $\omega^2 = \omega_0^2$ and $b_{11} \neq 0$ since the line of λ is not tangent to an asymptotic direction.

Let us find a point F on the tangent A_0A_1 to the line of λ which describes a two-dimensional surface, as the point A_0 does. Since $F = A_1 + zA_0$, then it follows from equations (3.61) and (3.62) that

$$\begin{aligned} dF = \; & (\omega_1^1 + z\omega^1)F + (dz + \omega_1^0 + z(\omega_0^0 - \omega_1^1) - z^2\omega^1)A_0 \\ & + (l_{11}^2\omega^1 + (l_{12}^2 + z)\omega^2)A_2 + (b_{11}\omega^1 + b_{12}\omega^2)A_3. \end{aligned}$$

The point F will describe a two-dimensional surface if and only if the 1-forms

$$\theta^2 = l_{11}^2\omega^1 + (l_{12}^2 + z)\omega^2 \quad \text{and} \quad \omega_1^3 = b_{11}\omega^1 + b_{12}\omega^2 \tag{3.63}$$

are linearly dependent, i.e. the following condition holds:

$$\det \begin{pmatrix} l_{11}^2 & l_{12}^2 + z \\ b_{11} & b_{12} \end{pmatrix} = 0.$$

It follows that

$$z = -l_{12}^2 + \frac{b_{12}}{b_{11}} l_{11}^2. \tag{3.64}$$

Thus, on the straight line A_0A_1, there exists a unique point F which is different from A_0 and which describes a two-dimensional surface $\widetilde{V}^2$ when the point A_0 describes the surface V^2. The point F (as well as the point A_0) is called a *focus* of the straight line A_0A_1. The lines A_0A_1 form a two-parameter family in the space P^3. Usually such families are called *congruences of straight lines* (see [Fi 50]). The surface $\widetilde{V}^2$ described by the point F is called the *Laplace transform* of the surface V^2.

The equation $\omega_1^3 = 0$ defines a family $\widetilde{\lambda}$ of lines on the surface V^2, and the lines of this family are conjugate to the lines of the family λ. In fact, the lines of λ on V^2 are defined by the equation $\omega^2 = 0$, and thus, by (2.76), the directions which are conjugate to the directions tangent to the lines of λ are defined by the equation

$$\varphi(\omega, \widetilde{\omega}) = b_{11}\omega^1\widetilde{\omega}^1 + b_{12}\omega^1\widetilde{\omega}^2 = 0.$$

Factoring out $\omega^1 \neq 0$, we arrive at the equation $\omega_1^3 = 0$. If $\omega_1^3 = 0$, the point A_0 describes a line of $\widetilde{\lambda}$ on V^2, and this line is conjugate to a line of the family λ. Moreover, the point F describes a line which is tangent to the straight line A_0F and which lies on the surface $\widetilde{V}^2$. We say that the families λ and $\widetilde{\lambda}$ form a *conjugate net* on the surface V^2.

We further specialize the moving frame associated with a point $A_0 \in V^2$ by placing the vertex A_2 on the tangent to the line of $\widetilde{\lambda}$ passing through the point A_0. Then the form ω_1^3 becomes proportional to the form ω^1, the coefficient b_{12} becomes 0, and equation (3.64) takes the form:

$$z = -l_{12}^2.$$

If we place the vertex A_1 at the focus F of the straight line A_0A_1, then we obtain $z = 0$ and subsequently $l_{12}^2 = 0$. This implies that equations (3.61) take the form:

$$\omega_1^3 = b_{11}\omega^1, \quad \omega_1^2 = l_{11}^2\omega^1. \tag{3.65}$$

The fundamental form $\Phi = b_{ij}\omega^i\omega^j$ of the surface V^2 becomes the sum of squares:

$$\Phi = b_{11}(\omega^1)^2 + b_{22}(\omega^2)^2,$$

and the form ω_2^3 has the following expression:

$$\omega_2^3 = b_{22}\omega^2. \tag{3.66}$$

In the same way, the straight line A_0A_2 tangent to the line of $\widetilde{\lambda}$ has a unique focus F' describing a two-dimensional surface $'\widetilde{V}^2$. If we place the vertex A_2 into the focus F', then the form ω_2^1 has the following expression:

$$\omega_2^1 = l_{22}^1\omega^2. \tag{3.67}$$

The surfaces $\widetilde{V}^2$ and $'\widetilde{V}^2$ are two Laplace transforms of the initial surface V^2.

Let us prove that the lines $\omega^2 = 0$ and $\omega^1 = 0$ are conjugate on the surfaces $\widetilde{V}^2$ and $'\widetilde{V}^2$. We will prove this, for example, for the surface $\widetilde{V}^2$ described by the vertex A_1. By (3.65), we have for this surface

$$dA_1 = \omega_1^0 A_0 + \omega_1^1 A_1 + (l_{11}^2 A_2 + b_{11}A_3)\omega^1.$$

Thus, the equation of the tangent plane ξ to the surface $\widetilde{V}^2$ has the form

$$b_{11}x^2 - l_{11}^2 x^3 = 0, \tag{3.68}$$

where x^0, x^1, x^2 and x^3 are the coordinates of a point in a frame $\{A_0A_1A_2A_3\}$.

Let us find the second fundamental form of the surface $\widetilde{V}^2$. For this, we first apply exterior differentiation to equations (3.65). A simple calculation leads to the following equations:

$$\begin{aligned} &\{db_{11} + b_{11}(\omega_0^0 - 2\omega_1^1 + \omega_3^3) + l_{11}^2\omega_2^3\} \wedge \omega^1 = 0, \\ &\{dl_{11} + l_{11}^2(\omega_0^0 - 2\omega_1^1 + \omega_2^2) + b_{11}\omega_3^2\} \wedge \omega^1 - \omega_1^0 \wedge \omega^2 = 0. \end{aligned}$$

The first of these equations can also be obtained from formulas (3.44) taken with $m = 2$ and $n = 3$. Applying Cartan's lemma, we find from the above equations that

$$\begin{aligned} db_{11} + b_{11}(\omega_0^0 - 2\omega_1^1 + \omega_3^3) + l_{11}^2\omega_2^3 &= r\omega^1, \\ dl_{11}^2 + l_{11}^2(\omega_0^0 - 2\omega_1^1 + \omega_2^2) + b_{11}\omega_3^2 &= p_{11}\omega^1 + p_{12}\omega^2, \\ -\omega_1^0 &= p_{21}\omega^1 + p_{22}\omega^2, \end{aligned} \tag{3.69}$$

where the quantities r and p_{ij} are related to the third order differential neighborhood of the surface V^2 and $p_{12} = p_{21}$.

Further, let us calculate the second differential of the point A_1:

$$\begin{aligned} d^2A_1 \equiv\ & (\omega_1^0\omega^2 + \omega^1(dl_{11}^2 + l_{11}^2\omega_2^2 + b_{11}\omega_3^2))A_2 \\ & + \omega^1(db_{11} + b_{11}\omega_3^3 + l_{11}^2\omega_2^3)A_3 \pmod{T_{A_1}(\widetilde{V}^2)}. \end{aligned}$$

Using equations (3.69), we can write the above relation as follows:

$$\begin{aligned} d^2A_1 \equiv\ & (-p_{22}(\omega^2)^2 - l_{11}^2\omega^1(\omega_0^0 - 2\omega_1^1) + p_{11}(\omega^1)^2)A_2 \\ & (-b_{11}\omega^1(\omega_0^0 - 2\omega_1^1) + r(\omega^1)^2)A_3 \pmod{T_{A_1}(\widetilde{V}^2)}. \end{aligned}$$

Next, we find the the second fundamental form of the surface $\widetilde{V}^2$. By (3.68), we obtain

$$\widetilde{\Phi} = <\xi, d^2A_1> = (b_{11}p_{11} - l_{11}^2 r)(\omega^1)^2 - b_{11}p_{22}(\omega^2)^2.$$

Since this form does not contain the term with the product $\omega^1\omega^2$ of the basis forms, the lines $\omega^2 = 0$ and $\omega^1 = 0$ on the surface $\widetilde{V}^2$, which correspond to the lines of the families λ and $\widetilde{\lambda}$ of the surface V^2, compose a conjugate net. The same property holds for the surface $'\widetilde{V}^2$.

Thus, *under Laplace transform, a conjugate net of the surface V^2 corresponds to a conjugate net of the surfaces $\widetilde{V}^2$ and $'\widetilde{V}^2$*. This is the main property of the Laplace transform.

2. We now turn to the multidimensional case. Let V^m be an m-dimensional submanifold V^m of a projective space P^n, and let λ be a one-dimensional foliation on V^m such that the lines of λ are not tangent to asymptotic directions at any of its points. As we did in the two-dimensional case, we associate with a submanifold V^m the bundle of frames of first order in such a way that the point A_0 coincides with the point x : $A_0 = x$, the points A_0, A_i, $i = 1, \ldots, m$, lie in the tangent subspace $T_x(V^m)$, and the line A_0A_1 is tangent to the line of the foliation λ. Then this line is defined by the equations

$$\omega^2 = \ldots = \omega^m = 0.$$

After this specialization the forms ω_1^a, $a, b = 2, \ldots, m$, become principal forms, i.e. they are expressed in terms of the basis forms ω^1 and ω^b:

$$\omega_1^a = l_{11}^a \omega^1 + l_{1b}^a \omega^b. \tag{3.70}$$

We write equations (2.21) on the submanifold in question in the form

$$\omega_1^\alpha = b_{11}^\alpha \omega^1 + b_{1b}^\alpha \omega^b, \quad \omega_a^\alpha = b_{a1}^\alpha \omega^1 + b_{ab}^\alpha \omega^b. \tag{3.71}$$

In this equation and in the rest of this subsection, we have the following ranges of indices: $a, b, c = 2, \ldots, m$ and $\alpha = m + 1, \ldots, n$. Since the direction A_0A_1 is not asymptotic, at least one of the coefficients b_{11}^α is different from 0.

Let us find under what condition the straight line A_0A_1 has points that describe submanifolds of dimension m, as the point A_0 does[1]. Such points are called the *foci* of the straight line A_0A_1. Since

$$\begin{aligned} dA_0 &= \omega_0^0 A_0 + \omega^1 A_1 + \omega^a A_a, \\ dA_1 &= \omega_1^0 A_0 + \omega_1^1 A_1 + \omega_1^a A_a + \omega_1^\alpha A_\alpha, \end{aligned}$$

then, by (3.70) and (3.71), for a point $F = A_1 + zA_0$ lying on the straight line A_0A_1, we find that

$$\begin{aligned} dF = \; & (\omega_1^1 + z\omega^1)F + (dz + \omega_1^0 + z(\omega_0^0 - \omega_1^1) - z^2\omega^1)A_0 \\ & + (l_{11}^a \omega^1 + (l_{1b}^a + z\delta_b^a)\omega^b)A_a + (b_{11}^\alpha \omega^1 + b_{1a}^\alpha \omega^a)A_\alpha. \end{aligned}$$

Since the point F must describe a submanifold $\widetilde{V}^m$ of dimension m, the forms

$$\theta^a = l_{11}^a \omega^1 + (l_{1b}^a + z\delta_b^a)\omega^b, \quad \omega_1^\alpha = b_{11}^\alpha \omega^1 + b_{1a}^\alpha \omega^a \tag{3.72}$$

must be expressed in terms of $m - 1$ linearly independent forms. Note that the form ω^1 is one of those independent forms since at least one of the coefficients b_{11}^α is different from 0.

Placing the vertex A_1 at the focus F describing the m-dimensional manifold $\widetilde{V}^m$, we obtain $z = 0$. Assuming that the forms $\omega^1, \omega^3, \ldots, \omega^m$ are linearly independent on $\widetilde{V}^m$, we find from (3.72) that

$$l_{12}^a = 0, \quad b_{12}^\alpha = 0. \tag{3.73}$$

The second from these equations means that on the submanifold V^m, there exists a one-parameter distribution $\{A_0A_2\}$ which is conjugate to the distribution

[1]Note that it is possible to consider points on the straight lines A_0A_1 that describe submanifolds of dimension less than m. For now, we do not consider this case.

$\{A_0A_1\}$ tangent to the given foliation λ. The integral lines of the distribution $\{A_0A_2\}$ give rise to a foliation $\widetilde{\lambda}$ on V^m, and this foliation is conjugate to the foliation λ.

Now it is clear that *if the straight line* A_0A_1 *carries* $m-1$ *foci, then at the point* $A_0 \in V^m$, *there are* $m-1$ *directions conjugate to the direction* A_0A_1, *and these* $m-1$ *directions define an* $(m-1)$*-dimensional direction which is conjugate to the direction* A_0A_1.

We will now prove the converse statement.

Theorem 3.15 *Suppose that a submanifold* V^m *of the space* P^n *carries a one-dimensional foliation* λ *whose lines are not tangent to asymptotic directions and which has a complementary* $(m-1)$*-dimensional conjugate distribution. Then each straight line* l *which is tangent to a line of* λ, *carries precisely* $m-1$ *foci, if we count each of them as many times as its multiplicity.*

Proof. In fact, let us associate with a point x of the submanifold V^m a family of frames of first order in such a way that the vertex A_1 lies on the straight line l, and the points $A_2, \ldots, A_m$ span the $(m-1)$-dimensional direction which is conjugate to A_0A_1. Then the coefficients b^{α}_{1a} in formulas (3.71) must vanish, and equations (3.72) take the form:

$$\begin{aligned} \theta^a &= l^a_{11}\omega^1 + (l^a_{1b} + z\delta^a_b)\omega^b, \\ \omega^{\alpha}_1 &= b^{\alpha}_{11}\omega^1. \end{aligned} \tag{3.74}$$

The condition defining the foci on the straight line A_0A_1 can be written as

$$\det(l^a_{1b} + z\delta^a_b) = 0. \tag{3.75}$$

This equation is an algebraic equation of degree $m-1$, and hence it has precisely $m-1$ roots, if we count each root as many times as its multiplicity. Each of these roots defines a focus $F = A_1 + zA_0$ on the straight line A_0A_1. ■

Thus, each straight line l tangent to a line of the foliation $\lambda \subset V^m$ carries m foci, since, besides the foci defined by equation (3.75), the point A_0 also describes an m-dimensional submanifold. Such a family of straight lines l is called *focal.*

In the general case, i.e. in the case when all the foci on the straight lines l are mutually distinct, a focal family places in correspondence to each of its focal submanifolds V^m $m-1$ other submanifolds of the same kind as V^m. This correspondence is called a *transformation of a submanifold* V^m *by means of a focal family of rays.*

3. The straight lines l of a focal family form a ruled submanifold V^{m+1} of dimension $m+1$ in the space P^n. This submanifold is characterized by the following theorem.

Theorem 3.16 *An* m*-parameter family of straight lines of a projective space* P^n, $n > m+1$, *is focal if and only if the tangent subspace* $T_x(V^{m+1})$ *to the submanifold* V^{m+1} *formed by the straight lines of this family depends on* m *parameters and is constant along a straight line of the family.*

Proof. In fact, let us place the vertices A_0 and A_1 on the straight line l of the given family of straight lines. Then

$$\begin{aligned} dA_0 &= \omega_0^0 A_0 + \omega_0^1 A_1 + \omega_0^u A_u, \\ dA_1 &= \omega_1^0 A_0 + \omega_1^1 A_1 + \omega_1^u A_u, \end{aligned} \tag{3.76}$$

where $u = 2, \ldots, n$. The forms ω_0^u and ω_1^u in formulas (3.74) are expressed in terms of m independent forms $\theta^p, p = 2, \ldots, m+1$, determining a displacement of the straight line A_0A_1:

$$\omega_0^u = c_{0p}^u \theta^p, \quad \omega_1^u = c_{1p}^u \theta^p. \tag{3.77}$$

The foci $F = A_1 + zA_0$ on the straight line A_0A_1 can be found from the condition $dF \in A_0A_1$. Since

$$dF \equiv (\omega_1^u + z\omega_0^u)A_u \pmod{A_0, A_1},$$

they are determined by the condition

$$\omega_1^u + z\omega_0^u = 0,$$

which, by (3.77), takes the form

$$(c_{1a}^u + zc_{0a}^u)\theta^a = 0. \tag{3.78}$$

This equation must determine those displacements of the straight line A_0A_1 for which it describes one-parameter torses. Thus, equation (3.78) must have nontrivial solutions with respect to the forms θ^a. The existence of such nontrivial solutions is guaranteed by the condition

$$\operatorname{rank}\,(c_{1a}^u + zc_{0a}^u) \le m - 1. \tag{3.79}$$

Suppose now that the family of straight lines A_0A_1 is focal, i.e. each straight line A_0A_1 has m foci (taking account of their multiplicity). This will be the case if and only if the system (3.79) has exactly m solutions, i.e. this system is reduced to one algebraic equation of degree m. Therefore, among the rank determinants of order m of the matrix $(c_{1a}^u + zc_{0a}^u)$, only one is essential. Assuming that this essential determinant is

$$\det(c_{1p}^q + zc_{0p}^q), \quad p, q = 2, \ldots, m+1,$$

and making an appropriate transformation of moving frames associated with the family of straight lines considered above, we easily arrive at the following equations:

$$c_{0p}^\alpha = 0, \quad c_{1p}^\alpha = 0, \quad \alpha = m+2, \ldots, n.$$

It follows that the tangent subspace $T_x(V^{m+1})$ to the submanifold V^{m+1} at the points $x \in A_0A_1$, spanned by the points A_0, A_1, A_p, is constant along the straight line A_0A_1 and depends on m parameters, in terms of the differentials of which the forms θ^a are expressed.

As was noted in Section **2.2**, submanifolds of a projective space P^n whose tangent subspace depends on a smaller number of parameters than a point, are called *tangentially degenerate*, and the number of parameters on which this tangent subspace actually depends is called the *rank* of this submanifold. Hence, the focal family of straight lines in P^n considered above is a tangentially degenerate submanifold V^{m+1} of rank m. The general tangentially degenerate submanifolds will be considered in Chapter 4.

Conversely, suppose that a submanifold V^{m+1} described by straight lines A_0A_1 is tangentially degenerate of rank m. If we place the vertices $A_2, \ldots, A_{m+1}$ in the tangent subspace common to all points of the straight line A_0A_1, then in equations (3.76), the index u takes the values $2, \ldots, m+1$. Thus, the foci will be defined from the single equation

$$\det(c^q_{1p} + zc^q_{0p}) = 0, \qquad (3.80)$$

which defines precisely m foci on the straight line A_0A_1. This implies that the family of straight lines A_0A_1 is focal. ■

We will denote a tangentially degenerate submanifold V^{m+1} of rank m by V^{m+1}_m (cf. Section **4.1**).

4. Suppose that a focal family V^{m+1}_m of straight lines has m mutually distinct foci. We may choose the vertex A_0 in such a way that it does not coincide with any focus. Then the forms ω^p_0, $p = 2, \ldots, m+1$, are linearly independent and can be taken as basis forms on the family V^{m+1}_m. In this basis, equations (3.77) can be written as

$$\omega^p_1 = c^p_q \omega^q_0, \;\; \omega^\alpha_0 = \omega^\alpha_1 = 0, \;\; p, q = 2, \ldots, m+1; \; \alpha = m+2, \ldots, n. \qquad (3.81)$$

(As above, we assume that the points A_p are located in the $(m+1)$-dimensional tangent subspace of the submanifold V^{m+1}_m common to all points of the straight line A_0A_1.) This implies that equation (3.80) takes the form

$$\det(c^q_p + zc^q_p) = 0. \qquad (3.82)$$

This equation has precisely m mutually distinct roots. Thus the operator $\mathbf{C} = (c^q_p)$ is diagonalizable, i.e we have

$$c^p_q = \delta^p_q c^p, \qquad (3.83)$$

where $c^p \neq c^q$ if $p \neq q$. As a result, the first group of equations (3.81) can be written as

$$\omega^p_1 = c^p \omega^p_0. \qquad (3.84)$$

The forms ω_0^p define a net of torses on the focal family V_m^{m+1}, and this net is called the *focal net*. In general, the focal net is not holonomic since in general, the forms ω_0^p are not multiples of total differentials.

By (3.83), equation (3.82) takes the form

$$\prod_p (c^p + z) = 0,$$

and the foci of the straight line A_0A_1 can be written as

$$F^p = A_1 - c^p A_0.$$

Consider a second order neighborhood of a focal submanifold $V_{(p)}^m$ described by the focus F^p. First, we have

$$dF^p \equiv \sum_{q \neq p} (c^q - c^p)\omega^q A_q \pmod{A_0, A_1}, \tag{3.85}$$

and thus the tangent subspace T_{F^p} to the submanifold $V_{(p)}^m$ is spanned by the points A_0, A_1, A_q where $q \neq p$. Equation (3.84) implies that the corresponding tangent subspaces to the submanifolds described by the points F^p and F^q have in common an $(m-1)$-dimensional subspace spanned by the points $A_0, A_1, A_r, r \neq p, q$.

To find the second differentials d^2F^p of the points F^p, we need some additional formulas. However, we will not derive them. Instead, we take them from Section **4.3**. According to that section, the following relations hold on a focal family V_m^{m+1}:

$$\omega_p^\alpha = b_p^\alpha \omega^p, \; b_p^\alpha = b_{pp}^\alpha, \tag{3.86}$$

$$\omega_q^p = \sum_{s=2}^{m+1} l_{qs}\omega^s, \tag{3.87}$$

$$\Delta c^p = \sum_{q=2}^{m+1} c_q^p \omega^q. \tag{3.88}$$

(The first two of the above equations correspond to equations (4.28) and (4.37), and the third follows from (4.40).)

Differentiation of equation (3.85) gives

$$\begin{aligned} d^2F^p \equiv \; & \{-c_p^p(\omega^p)^2 + \sum_{q\neq p}(c^q - c^p)l_{qq}^p(\omega^q)^2 + \sum_{\substack{q,s\neq p \\ q\neq s}}(c^q - c^p)l_{qs}^p\omega^q\omega^s\}A_p \\ & + \sum_{q\neq p}(c^q - c^p)b_q^\alpha(\omega^q)^2 A_\alpha \pmod{T_{F_p}(V_{(p)}^m)}. \end{aligned}$$

The coefficients in the points A_p and A_α in the above expression are the second fundamental forms of the submanifold $V^m_{(p)}$ described by the point F^p:

$$\Phi^p_{(p)} = -c^p_p(\omega^p)^2 + \sum_{q \neq p}(c^q - c^p)l^p_{qq}(\omega^q)^2 + \sum_{\substack{q,s \neq p \\ q \neq s}}(c^q - c^p)l^p_{qs}\omega^q\omega^s,$$
$$\Phi^\alpha_{(p)} = (c^q - c^p)b^\alpha_q(\omega^q)^2.$$

In the above equations, the lower index (p) denotes, in contrast to Chapter 2, not the order of the corresponding form—all these forms are defined in a second order neighborhood—but that the corresponding form is a second fundamental form of the submanifold $V^m_{(p)}$.

An analysis of these quadratic fundamental forms allows us to make the following conclusions:

1. On the submanifold $V^m_{(p)}$, the directions F^pA_0 (focal directions), tangent to the straight lines A_0A_1 and defined by the system of equations
$$\omega^q = 0, \quad q \neq p, \tag{3.89}$$
are conjugate to the $(m-1)$-dimensional distribution defined by the equation $\omega^p = 0$. This result matches Theorem 3.15 proved above.

2. *If at least one of the coefficients $l^p_{qs}, q, s \neq p, q \neq s$, on a focal family V^{m+1}_m is different from 0, then on the submanifold $V^m_{(p)}$, the net of lines corresponding to the focal net of the family V^{m+1}_m is not conjugate.* We will still call this new net on $V^m_{(p)}$ *focal.*

3. *The focal net on the submanifold $V^m_{(p)}$ is conjugate if and only if the following condition holds on V^{m+1}_m:*
$$l^p_{qs} = 0, \quad q, s \neq p, \ q \neq s,$$
where the index p is fixed. Then the focal submanifolds $V^m_{(q)}, q \neq p$, are called the *quasi-Laplace transforms* of the conjugate net Σ_2 on the submanifold $V^m_{(p)}$. In general, the focal nets on the submanifolds $V^m_{(q)}$ are not conjugate since some of the coefficients l^q_{st} may be different from 0. Of course, a conjugate net Σ_2 admits quasi-Laplace transforms along any of one-dimensional foliations composing this net. Focal submanifolds of all focal families tangent to these foliations are quasi-Laplace transforms of the original submanifold $V^m_{(p)}$. In general, the total number of all quasi-Laplace transforms of $V^m_{(p)}$ is equal to $m(m-1)$.

4. *The focal nets on all submanifolds $V^m_{(p)}$, $p = 2, \ldots, m+1$, of a focal family V^{m+1}_m are conjugate if and only if all coefficients $l^p_{qs} = 0, \ q, s \neq p, \ q \neq s$.* Then a focal net of the family V^{m+1}_m is *holonomic*, and the conjugate nets

corresponding to this holonomic net on all focal submanifolds of V_m^{m+1} are *m-conjugate systems.* In this case the focal nets on the submanifolds $V_{(q)}^m$, $q \neq p$, are called the *Laplace transforms* of the conjugate net Σ_2 on the submanifold $V_{(p)}^m$. Since an m-conjugate system Σ_2 on the submanifold $V_{(p)}^m$ consists of m one-dimensional foliations, the total number of its Laplace transforms is equal to $m(m-1)$.

5. Suppose now that a submanifold V^m carries a conjugate net Σ_2. If the vertices A_i are placed on the tangents to the lines of this net, then, as we saw in Sections **3.1** and **3.2** (see equations (3.5) and (3.14)), the following equations hold on V^m:

$$\omega^\alpha = 0, \quad \omega_i^\alpha = b_i^\alpha \omega^i, \quad \omega_j^i = \sum_k l_{jk}^i \omega^k.$$

Let us find the foci of the straight line $A_0 A_i$ when this straight line is displaced along the lines which are conjugate to the direction $A_0 A_i$. The latter lines satisfy the equation

$$\omega^i = 0,$$

which implies

$$\begin{aligned} dA_0 &= \omega_0^0 A_0 + \sum_{j \neq i} \omega^j A_j, \\ dA_i &= \omega_i^0 A_0 + \omega_i^i A_i + \sum_{j,k \neq i} l_{ik}^j \omega^k A_j. \end{aligned} \tag{3.90}$$

If a point $F = A_i + zA_0$ is a focus, then

$$dF \wedge A_0 \wedge A_i = 0.$$

But by relations (3.90), we have

$$dF \equiv \sum_{h,k \neq i} (l_{ik}^h + z\delta_k^h)\omega^k A_h \pmod{A_0, A_1}.$$

This implies that the foci of of the straight line $A_0 A_i$ are determined by the following system of equations:

$$(l_{ik}^h + z\delta_k^h)\omega^k = 0. \tag{3.91}$$

This system has nontrivial solutions if and only if the quantity z satisfies the following equation:

$$\det(l_{ik}^h + z\delta_k^h) = 0. \tag{3.92}$$

Since in this equation $h, k \neq i$, it is an algebraic equation of degree $m-1$ and has $m-1$ roots z_j, $j \neq i$, if we count each root as many times as its multiplicity. Each of these roots defines a focus

$$A_i^j = A_i + z_j A_0$$

on the straight line A_0A_i. This focus describes an m-dimensional submanifold (A_i^j) which is a quasi-Laplace transform of the submanifold V^m.

If we substitute a root z_j of equation (3.92) into the system (3.91), then the determinant of this system vanishes. This implies that the system has at least a one-dimensional solution

$$\underset{j}{\omega}^k = \underset{j}{\lambda}^k \theta. \tag{3.93}$$

This solution defines a one-parameter family of straight lines A_0A_i forming a torse whose edge of regression is described by the focus A_i^j.

Next consider the case when the torse defined by equation (3.93) corresponds to the jth line of the conjugate net Σ_2 passing through a point x. This will be the case if the system (3.93) reduces to the form:

$$\omega^k = 0, \quad k \neq j.$$

Substituting these values of ω^k into (3.91), we obtain

$$l_{ij}^h + z_j \delta_i^h = 0.$$

This implies

$$l_{ij}^h = 0, \quad h, j \neq i, \ h \neq j; \quad z_j = -l_{ij}^j. \tag{3.94}$$

Thus, the focus A_i^j is expressed by

$$A_i^j = A_i - l_{ij}^j A_0. \tag{3.95}$$

If the correspondence described above between torses of the families described by the straight lines A_0A_i and the lines of the conjugate net of the submanifold V^m holds for any i and j, then condition (3.94) is valid for all mutually distinct values i, j and h. This means (see Theorem 3.5) that the conjugate net on V^m is an m-conjugate system.

Since the converse is obvious, we have proved the following result.

Theorem 3.17 *Torses of of the families, described by the tangents to the lines of the conjugate net on the submanifold V^m, correspond to the lines of this net if and only if this net is an m-conjugate system. The focal submanifolds of these congruences are Laplace transforms of the submanifold V^m.* ■

6. As an example, we consider a submanifold V^m of general type in a projective space P^{m+2}. As was indicated at the end of Section **3.2**, in general such a submanifold carries a unique conjugate m-net which is not holonomic. Thus, such a submanifold admits quasi-Laplace transforms along any of the one-dimensional foliations composing the conjugate net, and these transforms are not Laplace transforms. Moreover, the focal nets on focal families of straight lines realizing quasi-Laplace transforms do not correspond to the lines of the conjugate net of V^m. Submanifolds $\widetilde{V}^m$, which are quasi-Laplace transforms of V^m, also carry conjugate nets but they do not correspond to the conjugate net of V^m.

A similar situation occurs for a hypersurface $V^m \subset P^{m+1}$ which carries a family of nonholonomic conjugate nets.

As another example, we consider a tangentially degenerate submanifold $V_m^{m+1} \subset P^{m+3}$. Focal submanifolds of such V_m^{m+1} are of codimension three, and hence in general they do not carry nets of conjugate lines. This implies that a transformation by means of a focal family of rays is not reduced either to a quasi-Laplace transform or to a Laplace transform. The existence theorem for such focal families for $m = 3$ and $m = 4$ was proved by Akivis in his paper [A 62a].

7. Consider a submanifold V^m carrying an m-conjugate system Σ_2. With a point $A_0 \in V^m$, we associate a family of frames whose edges A_0A_i are tangent to the lines of the conjugate system. Denote by A_i^j, $j \neq i$, the focus of the straight line A_0A_i corresponding to the displacement of the point A_0 along the jth line of the m-conjugate system Σ_2. As was indicated above, the points A_j^i describe m-dimensional submanifolds carrying m-conjugate systems. Each of those new m-conjugate systems admits Laplace transforms which are also m-conjugate systems. Denote the points describing these new Laplace transforms by A_{ik}^{jl} where the index k indicates a direction emanating from the point A_j^i, and the index l indicates the number of the focus on the straight line emanating from the point A_j^i in this direction. The submanifolds described by the points A_{ik}^{jl}, are called the *second Laplace transforms* of the conjugate system Σ_2.

It is not difficult to prove the following relations:

$$A_{ij}^{ji} = A_0, \quad A_{ij}^{jk} = A_i^k, \; k \neq i. \tag{3.96}$$

—see Figure 3.2.

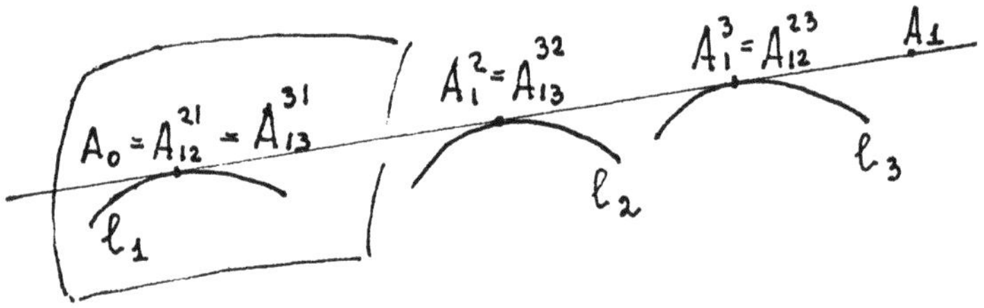

Figure 3.2

This figure represents the case $m = 3$, and in this figure the line defined by the equations $\omega^j = 0, \ \ j \neq i$, is denoted by l_i.

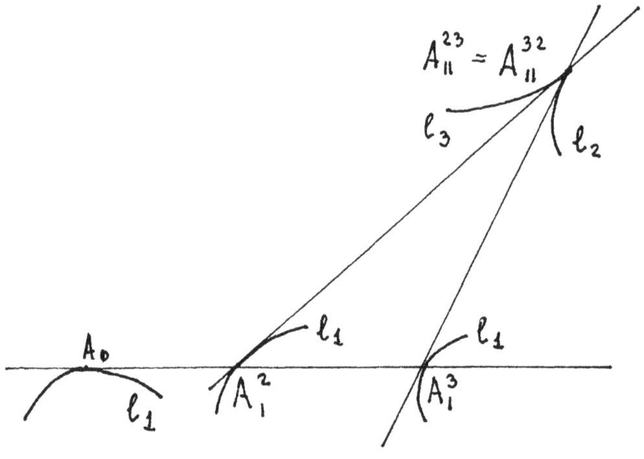

Figure 3.3

The proof of the relations

$$A_{ik}^{ji} = A_k^j, \quad A_{ii}^{jk} = A_{ii}^{kj}, \tag{3.97}$$

where all indices i, j and k take mutually distinct values, is more complicated. These relations were proved by Smirnov in his paper [Sm 50]. If $m = 3$, relations (3.97) are illustrated in Figures 3.3 and 3.4.

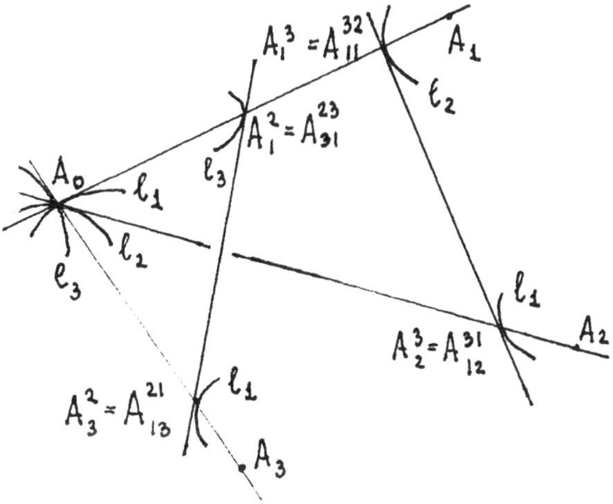

Figure 3.4

3.6 Conic m-Conjugate Systems

1. We now consider a tangentially nondegenerate submanifold $V^m \subset P^n$ carrying a conjugate net $\Sigma_2 \subset V^m$.

With a point $x \in V^m$ we associate a family of moving frames $\{A_u\}$, $u = 0, 1, \ldots, n$, whose vertex A_0 coincides with the point x: $A_0 = x$, and the edges $A_0A_i, i = 1, \ldots, m$, are tangent to the lines of the conjugate net Σ_2 passing through the point x.

The equations of the infinitesimal displacement of these moving frames can be written as

$$dA_u = \omega_u^v A_v, \quad u, v = 0, 1, \ldots, n. \tag{3.98}$$

The specialization of the moving frames described above gives (see Section 3.1):

$$\omega^\alpha = 0, \quad \alpha = m+1, \ldots, n, \tag{3.99}$$

$$\omega_i^\alpha = b_i^\alpha \omega^i, \quad i = 1, \ldots, m, \tag{3.100}$$

$$\omega_i^j = \sum_k l_{ik}^j \omega^k \tag{3.101}$$

(cf. equations (3.5) and (3.14)). In equations (3.99) and (3.100) and in other equations of this section there is no summation over the indices $i, j, k, \ldots$ unless the summation sign Σ is shown.

Definition 3.18 A submanifold $V^m \subset P^n$ carrying a holonomic conjugate net Σ_2 is said to be *conic* or a *Peterson submanifold* if the tangents to any line of each family (C_i) of Σ_2 taken at the points of intersection of this line with the $(m-1)$-dimensional submanifold conjugate to (C_i) form a cone.

This definition implies that all foci of each of the straight lines A_0A_i, except A_0, coincide, and each of these foci describes a curve.

If we place the point A_i at the vertex of the cone associated with the family (C_i), we must have

$$dA_i \wedge A_i \equiv 0 \pmod{\omega^i}. \tag{3.102}$$

Equations (3.98) and (3.102) imply that in equations (3.101) we have $l_{ik}^j = 0, \ k \neq i$, i.e.

$$\omega_i^j = l_{ii}^j \omega^i, \tag{3.103}$$

and

$$\omega_i^0 = l_{ii}\omega^i. \tag{3.104}$$

Since equations (3.104) imply $l_{ik}^j = 0, \ k \neq i, j, \ i \neq j$, a conjugate net on V^m is an m-conjugate system.

We will now prove the following result.

Theorem 3.19 *If a submanifold $V^m \subset P^n$ carrying a holonomic conjugate net Σ_2 is conic, then*

(i) *The system defining a conic m-conjugate system is in involution, and an m-dimensional integral manifold V^m, defined by this system, depends on mn arbitrary functions of one variable.*

(ii) *In homogeneous projective coordinates the equations of a conic submanifold can be reduced to the form:*

$$\rho A_0 = L_1(u^1) + L_2(u^2) + \ldots + L_m(u^m), \tag{3.105}$$

where $L_i(u^i)$ equations of the curves described by the foci A_i.

Proof. (i) To prove the existence of conic m-conjugate systems, note that such a system is defined by Pfaffian equations (3.99), (3.100), (3.103) and (3.104). Exterior differentiation of these Pfaffian equations leads to the following exterior quadratic equations:

$$\Delta b_i^\alpha \wedge \omega^i = 0, \quad \Delta l_{ii}^j \wedge \omega^i = 0, \quad \Delta l_{ii}^j \wedge \omega^i = 0, \qquad (3.106)$$

where

$$\Delta b_i^\alpha = db_i^\alpha + b_i^\alpha(\omega_0^0 - 2\omega_i^i) + \sum_{k\neq i} l_{ii}^k b_k^\alpha \omega^k + b_i^\beta \omega_\beta^\alpha,$$
$$\Delta l_{ii}^j = dl_{ii}^j + l_{ii}^j(\omega_0^0 - 2\omega_i^i + \omega_j^j) + b_i^\alpha \omega_\alpha^j + \sum_{k\neq i,j} l_{ii}^k l_{kk}^j \omega^k,$$
$$\Delta l_{ii} = dl_{ii} + 2l_{ii}(\omega_0^0 - \omega_i^i) + \sum_{k\neq i} l_{ii}^k \omega_k^0 + b_i^\alpha \omega_\alpha^0.$$

The system of equations (3.99), (3.100), (3.103), (3.104) and (3.106) is closed with respect to the operation of exterior differentiation. We will apply the Cartan test (see Theorem 1.3 in Section **1.2**) to investigate the consistency of this system. The number q of unknown functions $\Delta b_i^\alpha, \Delta l_{ii}^j$ and Δl_{ii}, and the number of the exterior quadratic equations are equal: $q = s_1 = m(m-1) + m(n-m) + m = mn$. Thus $s_2 = 0$ and $Q = s_1 + 2s_2 = mn$. It is easy to check that the number N of parameters, on which the most general integral element of the system under consideration depends, is also equal to mn. Since $N = Q$, the system of equations (3.99), (3.100), (3.103), (3.104) and (3.106) is in involution, and an m-dimensional integral manifold V^m, defined by this system, depends on mn arbitrary functions of one variable.

(ii) Note that equations (3.104) imply that $d\omega_0^0 = 0$. Thus, the form ω_0^0 is a total differential. It is convenient to write this form as

$$\omega_0^0 = -d\ln|\rho|. \qquad (3.107)$$

We will integrate the system (3.98) where the components of infinitesimal displacement are defined by equations (3.99), (3.100), (3.103), (3.104) and (3.107).

Since the point A_i describes a curve, this point can be written as

$$A_i = \rho_i B_i,$$

where ρ_i is a normalizing factor. The structure equations and equation (3.107) imply that

$$d\omega^i = (-d\ln|\rho| - \omega_i^i) \wedge \omega^i,$$

and this means that each of the forms ω^i can be represented as

$$\omega^i = \sigma^i du^i.$$

Then

$$dA_0 = -d\ln|\rho|A_0 + \sum_i \mu_i B_i du^i, \tag{3.108}$$

where $\mu_i = \sigma^i \rho_i$ (no summation over i). Exterior differentiation of the last equation gives

$$\sum_i (d\mu_i + \mu_i d\ln|\rho|) \wedge du^i B_i = 0.$$

Since the points B_i are linearly independent, we find from the last equations that

$$(d\mu_i + \mu_i d\ln|\rho|) \wedge du^i = 0 \quad \text{(no summation over } i\text{)}.$$

Multiplying this relation by ρ, we obtain

$$d(\mu_i \rho) \wedge du^i = 0 \quad \text{(no summation over } i\text{)}.$$

This implies

$$\mu_i \rho = \varphi_i(u^i),$$

or

$$\mu_i \rho = \frac{1}{\rho}\varphi_i(u^i).$$

Substituting this expression for μ_i into equation (3.108), we arrive at

$$dA_0 = -\frac{d\rho}{\rho}A_0 + \frac{1}{\rho}\sum_i \varphi_i(u^i)B_i(u^i)du^i,$$

or

$$d(\rho A_0) = \sum_i C_i(u^i)du^i, \tag{3.109}$$

where

$$C_i(u^i) = \varphi_i(u^i)B_i(u^i).$$

Integrating (3.109), we find that

$$\rho A_0 = L_1(u^1) + L_2(u^2) + \ldots + L_m(u^m),$$

where

$$L_i = \int C_i(u^i)du^i. \blacksquare$$

2. We will now define generalized conic systems. First, we suppose from the beginning that a net Σ_2 on a submanifold V^m is an m-conjugate system.

With a point $x \in V^m$ we associate a bundle of moving frames $\{A_u\}$, $u = 0, 1, \dots, n$, as described in the beginning of this section. Thus, we have equations (3.99), (3.100) and (3.101). In Section **3.4** we proved that for a submanifold V^m carrying an m-conjugate system, we have equations (3.19):

$$\omega_i^j = l_{ii}^j \omega^i + l_{ij}^j \omega^j. \tag{3.110}$$

We showed in Section **3.4** that the exterior differentiation of equations (3.99) and (3.110) gives the following exterior quadratic equations:

$$\Delta b_i^\alpha \wedge \omega^i = 0, \tag{3.111}$$

where

$$\Delta b_i^\alpha = db_i^\alpha + b_i^\alpha(\omega_0^0 - 2\omega_i^i) + \sum_{k \neq i} (l_{ki}^i b_i^\alpha + l_{ii}^k b_k^\alpha)\omega^k + b_i^\beta \omega_\beta^\alpha,$$

and

$$\Delta l_{ii}^j \wedge \omega^i + \Delta l_{ij}^j \wedge \omega^j = 0, \tag{3.112}$$

where

$$\Delta l_{ii}^j = dl_{ii}^j + l_{ii}^j(\omega_0^0 - 2\omega_i^i + \omega_j^j) + b_i^\alpha \omega_\alpha^j + l_{ii}^j \sum_{k \neq j} l_{ki}^i \omega^k + \sum_{k \neq i,j} l_{ii}^k \omega_k^j,$$
$$\Delta l_{ij}^j = dl_{ij}^j + l_{ij}^j(\omega_0^0 - \omega_i^i) - \omega_i^0 + l_{ij}^j \sum_{k \neq j} l_{kj}^j \omega^k - \sum_{k \neq i,j} l_{kj}^j l_{ik}^k \omega^k,$$

and that by the Cartan lemma we have from equations (3.111) and (3.112):

$$\Delta b_i^\alpha = b_{iii}^\alpha \omega^i, \tag{3.113}$$

and

$$\Delta l_{ii}^j = l_{iii}^j \omega^i + l_{iij}^j \omega^j, \Delta l_{ij}^j = l_{iij}^j \omega^i + l_{ijj}^j \omega^j, \tag{3.114}$$

respectively.

Definition 3.20 We will say that a submanifold V^m carrying an m-conjugate system is a *generalized conic submanifold* or a *generalized Peterson submanifold* if any two-dimensional submanifold V_{ij}^2, defined by equations $\omega^k = 0, \ k \neq i, j$, is a conic system.

The submanifolds defined by equations (3.105) are particular cases of generalized conic submanifolds.

Theorem 3.21 *A generalized conic submanifold has degenerate Laplace transforms of dimension at most $m - 1$.*

Proof. In fact, the Laplace transform (A_i^j) of V^m in the direction ω^i corresponding to the direction ω^j can be obtained as the Laplace transform of the surface V_{ij}^2 defined by the system $\omega^k = 0,\ k \neq i, j$. Then, since the surface V_{ij}^2 is conic, it follows that the submanifold (A_i^j) does not depend on u^i but can depend on all other $u^s,\ \ s \neq j$, i.e. the dimension of (A_i^j) is at most $m-1$. ■

We will now prove an existence theorem for generalized conic submanifolds.

Theorem 3.22 *The system defining generalized conic submanifolds is in involution, and an m-dimensional integral manifold V^m, defined by this system, depends on $m(n+m-2)$ arbitrary functions of one variable.*

Proof. As we saw in Section **3.5**, the focus A_i^j of the tangent A_0A_i corresponding to the line ω^j has the form $A_i^j = -l_{ij}^j A_0 + A_i$ (see equations (3.95)). By (3.98), (3.100), (3.110) and (3.114), its differential has the following form:

$$\begin{aligned} dA_i^j = \ & (\omega_i^i - l_{ij}^j\omega^i)A_i^j + (-l_{iji}^j\omega^i - l_{ijj}^j\omega^j + l_{ij}^j\sum_{k\neq j} l_{kj}^j\omega^k - \sum_{k\neq i,j} l_{kj}^j l_{ik}^k\omega^k)A_0 \\ & +\omega^i(\sum_{k\neq j} l_{ii}^k A_k + b_i^\alpha A_\alpha). \end{aligned} \tag{3.115}$$

Since $\omega^k = 0,\ k \neq i, j$, on the submanifold V_{ij}^2, and since V_{ij}^2 is conic, the point A_i^j describes the curve l_i with equations $\omega^k = \omega^i = 0$. Expression (3.115) implies that the necessary and sufficient condition for this is

$$l_{ijj}^j = 0. \tag{3.116}$$

These conditions and the equations (3.114) imply that each term of the exterior quadratic equations (3.112) vanishes, i.e. we have

$$\Delta l_{ii}^j \wedge \omega^i = 0, \quad \Delta l_{ij}^j \wedge \omega^j = 0. \tag{3.117}$$

Thus, generalized conic submanifolds are defined by Pfaffian equations (3.99), (3.100) and (3.110) and the exterior quadratic equations (3.111) and (3.117).

The system is closed with respect to the operation of exterior differentiation. The number q of unknown functions $\Delta b_i^\alpha, \Delta l_{ii}^j$ and Δl_{ii} and the number of the exterior quadratic equations are equal: $q = s_1 = 2m(m-1) + m(n-m) = m(n+m-2)$. Thus $s_2 = 0$ and $Q = s_1 + 2s_2 = m(n+m-2)$. It follows from equations (3.113), (3.114) and (3.116) that the number N of parameters, on which the most general integral element of the system depends, is also equal to $m(n+m-2)$. Since $N = Q$, the system of equations (3.99), (3.100), (3.110), (3.111) and (3.117) is in involution, and an m-dimensional integral manifold V^m, defined by this system, depends on $m(n+m-2)$ arbitrary functions of one variable. ■

3. We will now describe a procedure which can be used to construct a generalized conic submanifold.

First, note that if in the space P^{n+1}, we fix two curves l_1 and l_2 and a hyperplane P^n, then the locus of the common points of P^n with the straight lines joining arbitrary points of the curves l_1 and l_2 is a two-dimensional conic submanifold of general type.

In fact, let N be a fixed point on the curve l_2 (see Figure 3.5).

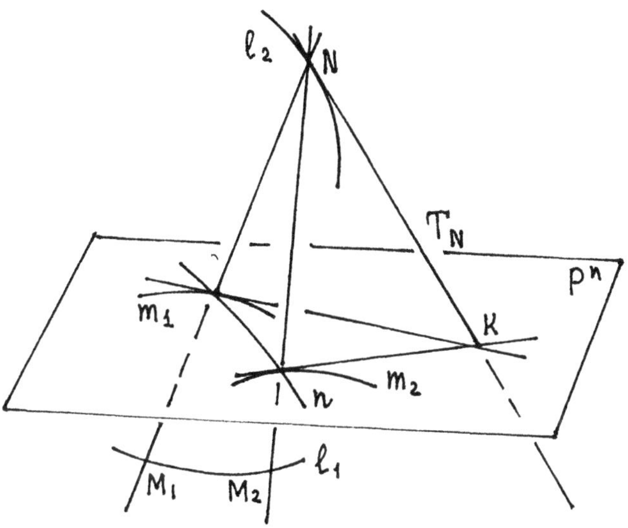

Figure 3.5

Projecting the line l_1 from the point N onto the hyperplane P^n, we obtain a line n. Next take two points $M_1, M_2 \in l_1$. Projecting the curve l_1 from these two points onto the hyperplane P^n, we obtain the curves m_1 and m_2. Moreover, the tangents to the curves m_1 and m_2 at the points of their intersection with the curve n are the projections of the tangent T_N to the curve l_2 at the point N from the points M_1 and M_2. Therefore, the tangents to the curves m_1 and m_2 have in common the point K at which the tangent T_N intersects the hyperplane P^n. Since the point K does not depend on the position of the points M_1 and M_2 on the curve l_1, this point is the vertex of the cone passing through all curves m at the points of their intersection with a fixed curve n.

In the general case, in a space P^{n+m-1}, we consider m arbitrary curves l_i, $i = 1, \ldots, m$, and a fixed subspace P^n. Suppose that the curves l_i and the subspace P^n are in general position. An arbitrary subspace P^{m-1} passing through m points $t_i \in l_i$ intersects P^n at a point x which depends on m parameters. This point x describes a submanifold V^m.

Let us prove that the submanifold V^m is a generalized conic submanifold. In fact, consider an arbitrary two-dimensional submanifold obtained by fixing points on some $m-2$ of m curves l_i. Let P^{m-3} be a subspace defined by the $m-2$ fixed points indicated above and P^{n+m-2} be a subspace defined by P^n and P^{m-3}. Then the straight lines joining arbitrary points of the two remaining curves intersect P^{n+m-2} at the points of a conic two-dimensional submanifold W^2 located in P^{n+m-2}. If we project the submanifold W^2 onto P^n from P^{m-3} as the center of projection, then we obtain a two-dimensional submanifold V^2. Since the property of a submanifold to be conic is invariant under such a projection, the submanifold V^2 is conic.

4. Ryzhkov in [Ry 58] found the general form of parametric equations of a generalized conic submanifold. We will describe how these equations can be obtained and indicate their connection with the construction of the kind of submanifolds given above.

For simplicity, we assume that the subspace P^n is the coordinate subspace defined by vanishing of the first $m-1$ homogeneous coordinates.

Let m curves l_i be given in the space P^{n+m-1} by quantities $a_{i1}, \ldots, a_{i,m-1}$, U_{m-1} where $a_{i1}, \ldots, a_{i,m-1}$ are scalar functions of a single parameter u^i and $U_i = U_i(u^i)$ is a point in the space P^{n+m-1}. Then it is possible to prove that an equation of a generalized conic submanifold V^m can be written as follows:

$$A_0 = \det \begin{pmatrix} a_{11} & a_{12} & \cdots & a_{1,m-1} & U_1 \\ a_{21} & a_{22} & \cdots & a_{2,m-1} & U_2 \\ \cdots & \cdots & \cdots & \cdots & \cdots \\ a_{m1} & a_{m2} & \cdots & a_{m,m-1} & U_m \end{pmatrix} = \alpha_1 U_1 + \alpha_2 U_2 + \ldots + \alpha_m U_m, \tag{3.118}$$

where α_i are the cofactors of the entries U_i in this determinant.

We now state Ryzhkov's result.

Theorem 3.23 *The submanifold $A_0 = A_0(u^1, \ldots, u^m)$ defined by equations* (3.118) *is a generalized conic submanifold whose conjugate system is formed by the coordinate lines, and the equations of any generalized conic submanifold can be reduced to the form* (3.118). ■

To establish the relationship between this result and the above construction of generalized conic submanifolds, we consider $a_{i1}, a_{i2}, \ldots, a_{i,m-1}, U_i$ as coordinates of the point U_i describing the curve in the space P^{n+m-1}. Then A_0 is the point of intersection of the coordinate subspace P^n and the subspace P^{m-1} defined by m points taken on the curves U_i. Therefore, $A_0 = A_0(u^1, \ldots, u^m)$ is a submanifold which, according to the construction described above, is a generalized conic submanifold.

NOTES

3.1. Submanifolds carrying conjugate nets were studied by É. Cartan in [Ca 19], by Akivis in [A 61a], [A 62c], [A 63] and by Bazylev in [Ba 53], [Ba 55a], [Ba 55b] and [Ba 67] (see also the papers [A 70a], [Ry 56] and [Ry 58] devoted to conjugate systems). The papers [AR 64], [Ba 65a] and [Lu 75] give surveys of the results on this subject. Akivis in [A 62c] investigated $V^m \subset P^n$ carrying a conjugate net and admitting a field of axial $(n-m)$-planes (he called them the Voss normals) such that the osculating 2-planes to the lines of the net at a point $x \in V^m$ meet the axial $(n-m)$-plane passing through the point x along straight lines. He classified such submanifolds and proved existence theorems for some of them.

3.2. Theorem 3.7 was proved by Akivis in [A 61a].

A particular case of m-conjugate systems V^m in P^n, $n \geq 2m$, for which $\dim\ T_x^{(2)}(V^m) = 2m$, was introduced by É. Cartan in [Ca 19].

3.3. The generalized Segre theorem was first proved by Bazylev in [Ba 55a] and [Ba 55b]. Akivis gave a simpler proof of this theorem in [A 61a] and [A 63] and stated it in the form which we used in Theorem 3.10.

Segre type theorems (giving conditions under which a submanifold V^m is of class k) were proved by Villa and Muracchini when they studied quasiasymptotic curves and submanifolds (see [Vi 39e] and [M 51c]).

Theorem 3.11 was proved by Akivis in [A 63].

Submanifolds with a conjugate net, for which $m_1 = m$, were first considered by É. Cartan in [Ca 19]. Chern studied these submanifolds in detail in [C 47] and named them the Cartan varieties. In [C 47], Chern also considered Laplace transforms of Cartan varieties.

3.4. The example which we presented in this section was considered by É. Cartan in Chapter 4 of his paper [Ca 19].

3.5. The theory of Laplace transforms is a very well developed chapter of differential geometry of conjugate nets on surfaces in a three-dimensional projective space P^3. For the first time the geometric theory of Laplace transforms of two-dimensional surfaces in P^3 was constructed by Darboux in [Da 72], vol. 2. This theory was also presented in the books [Tz 24], [Fi 37], [Fi 50], [Bo 50a], [Mi 58] and [Go 64]. Note that in the book [Fi 50] the Laplace transforms of two-dimensional conjugate nets in an n-dimensional projective space were also studied. B. Segre in his book [SegB 71] presented Laplace transforms of two-dimensional conjugate nets in P^n from the point of view of algebraic geometry.

In the 1940's–1950's this theory was extended to the m-conjugate systems $(m > 2)$ in P^n. Koz′mina studied Laplace transforms of three-conjugate systems in P^3 (see [Koz 47]). Chern in [C 44] (see also [C 47]) generalized Laplace transforms to m-conjugate systems. However, he considered only Laplace transforms of m-conjugate systems that are Cartan varieties, i.e. the case when $n \geq 2m$. Smirnov in [Sm 49] and [Sm 50] generalized Koz′mina's and Chern's results by constructing Laplace transforms of m-conjugate systems in P^n where $n \geq m$. Four years later, Bell (who did not know the paper [Sm 50]) in [Bel 54] independently arrived to the results similar to those in [Sm 50].

Korovin in [Kor 55a] constructed a closed Laplace network consisting of 16 three-conjugate systems in P^3. In [Kor 55b] Korovin extended the theory of Darboux invariants to three-conjugate systems $V^3 \subset P^4$ and introduced three-conjugate systems R (for them their asymptotic directions and asymptotic directions of their Laplace transforms correspond to each other). Geidelman in [Ge 57] generalized these results of Korovin and introduced m-conjugate systems R in the set of m-conjugate systems belonging to its second osculating space P^n and possessing $m_1 = n - m < m$ linearly independent forms.

Focal families of tangents to $V^3 \subset P^5$ were first introduced by Akivis in [A 49] and [A 50] when he studied the T-pairs of complexes.

Later Akivis gave the general theory of transformations of multidimensional submanifolds $V^m \subset P^n$ by means of a focal family of rays (see [A 61b] and [A 62a]).

Bazylev in [Ba 53], [Ba 55a], [Ba 55b], [Ba 61] and [Ba 64] constructed quasi-Laplace transforms of the general and some special submanifolds V^m carrying conjugate nets.

In the papers [V 64a] and [V 64b] Vangeldère, who did not know the Akivis papers [A 49], [A 50], [A 61b], [A 62a] and the Bazylev papers [Ba 53], [Ba 55a], [Ba 55b], considered the submanifolds $V^3 \subset P^5$, whose two asymptotic cones at any point $x \in V^3$ have four distinct asymptotic directions, and therefore such submanifolds V^3 carry a conjugate net. He generalized the notion of conjugacy of two families of lines on V^3 (one can find this generalized notion in [Ba 53]) and constructed a particular case of quasi-Laplace transforms of V^3 introduced earlier by Bazylev for general $V^m \subset P^{m+2}$ (see [Ba 53] and [Ba 55b]).

3.6. The results on conic (Peterson) submanifolds for $n = 3$ can be found in the paper [Bla 71] by Blank and for $n \geq 3$ in the paper [Ry 58] by Ryzhkov (see also [Ry 56]). However, in these papers the conic submanifolds were considered in an Euclidean and affine spaces, respectively. The presentation of conic submanifolds in a projective space P^n appears in this book for the first time.

Degen in [De 67] considered two-component conjugate systems on $V^{n-1} \subset P^n$ with involutive distributions Δ_1 and Δ_2. In particular, in [De 67] he introduced the Darboux tensors of Δ_1 and Δ_2, proved that if Δ_1 defines the shadow submanifolds (along which V^{n-1} and a hypercone of the same dimension as Δ_2 are tangent), then one of the Darboux tensors vanishes, and considered the case when the shadow surfaces belong to some $P^s \subset P^n$.

Blank and Gormasheva in [BG 70] proved that a Peterson hypersurface V^3 in P^4 can carry at most a six-parameter family of conic three-nets, and that if this is the case, then V^3 is a hyperquadric.

Chapter 4

Tangentially Degenerate Submanifolds

4.1 Basic Notions and Equations

1. We consider the Gauss mapping γ of a submanifold V^m which maps a point $x \in V^m$ into its tangent subspace $T_x^{(1)}$ (see Section **2.2**):

$$\gamma : V^m \rightarrow G(m, n).$$

The basis forms on the Grassmann image $\gamma(V^m)$ of the submanifold V^m are the forms ω_i^α. These forms are expressed in terms of the basis forms of the submanifold V^m according to formulas (2.21):

$$\omega_i^\alpha = b_{ij}^\alpha \omega^j, \;\; b_{ij}^\alpha = b_{ji}^\alpha. \tag{4.1}$$

As we noted in Section **2.2**, if the submanifold V^m is tangentially nondegenerate, its tangent subspace depends on m parameters, and hence the forms ω_i^α cannot be expressed in terms of less than m linearly independent forms.

Suppose now that the Grassmann mapping γ of the submanifold V^m is degenerate, i.e. its Grassmann image $\gamma(V^m)$ depends on r parameters, where $0 \leq r < m$. Then we say that the submanifold V^m is *tangentially degenerate and has rank* r. We denote such submanifolds by V_r^m.

For a tangentially degenerate submanifold V_r^m of rank r, the forms ω_i^α can be expressed in terms of precisely r linearly independent forms. Let us take the forms $\omega^{l+1}, \ldots, \omega^m$, where $l = m - r$, as these independent forms, and write the expressions of the forms ω_i^α in terms of the forms ω^q, $q = l+1, \ldots, m$, as follows:

$$\omega_i^\alpha = b_{iq}^\alpha \omega^q, \;\; q = l+1, \ldots, m. \tag{4.2}$$

Since the matrix (b_{ij}^α) is symmetric, this matrix takes the form:

$$(b^\alpha_{ij}) = \begin{pmatrix} 0 & 0 \\ 0 & b^\alpha_{pq} \end{pmatrix}, \tag{4.3}$$

and, by (4.2) and (4.3), the forms ω^α_i can be written as follows:

$$\omega^\alpha_a = 0, \ a = 1, \ldots, l, \tag{4.4}$$

$$\omega^\alpha_p = b^\alpha_{pq}\omega^q, \ p, q = l+1, \ldots, m, \tag{4.5}$$

where, as earlier, $b^\alpha_{pq} = b^\alpha_{qp}$. The 1-forms ω^q are basis forms of the Grassmann image $\gamma(V^m_r)$ of the submanifold V^m_r, and the quantities b^α_{pq} form a tensor.

By (4.4), the equations of infinitesimal displacement of the moving frame associated with the tangentially degenerate submanifold V^m_r have the form:

$$\begin{aligned} dA_0 &= \omega^0_0 A_0 + \omega^a A_a + \omega^p A_p, \\ dA_a &= \omega^0_a A_0 + \omega^b_a A_b + \omega^p_a A_p, \\ dA_p &= \omega^0_p A_0 + \omega^a_p A_a + \omega^q_p A_p + \omega^\alpha_p A_\alpha, \\ dA_\alpha &= \omega^0_\alpha A_0 + \omega^\alpha_a A_a + \omega^\alpha_q A_q + \omega^\beta_\alpha A_\beta. \end{aligned} \tag{4.6}$$

Taking exterior derivatives of equations (4.4), we obtain the following exterior quadratic equations:

$$\omega^p_a \wedge \omega^\alpha_p = 0. \tag{4.7}$$

Substituting expressions (4.5) into equations (4.7), we find that

$$b^\alpha_{pq}\omega^p_a \wedge \omega^q = 0. \tag{4.8}$$

Let us prove that, as was the case for the forms ω^α_p, the forms ω^p_a can be expressed in terms of the basis forms ω^q alone. Suppose that decompositions of the forms ω^p_a have the following form:

$$\omega^p_a = c^p_{aq}\omega^q + c^p_{a\xi}\omega^\xi, \tag{4.9}$$

where the forms ω^ξ are some linearly independent forms. Substituting these expressions into equations (4.8), we find that

$$b^\alpha_{sq}c^s_{ap}\omega^p \wedge \omega^q + b^\alpha_{pq}c^p_{a\xi}\omega^\xi \wedge \omega^q = 0.$$

Since the exterior products $\omega^p \wedge \omega^q$ and $\omega^\xi \wedge \omega^q$ are independent, it follows from these relations that

$$b^\alpha_{sq}c^s_{ap} = b^\alpha_{sp}c^s_{aq} \tag{4.10}$$

and

$$b^\alpha_{pq}c^p_{a\xi} = 0. \tag{4.11}$$

Since the forms ω_p^α cannot be expressed in terms of less than r linearly independent forms, the rank of system (4.11) is equal to r, and the system has only the trivial solution: $c_{a\xi}^p = 0$. Hence equations (4.9) take the form:

$$\omega_a^p = c_{aq}^p \omega^q, \tag{4.12}$$

where the coefficients c_{aq}^p are related to the coefficients b_{pq}^α by (4.10).

Note that under transformations of the points A_p, the quantities c_{aq}^p are transformed as tensors. As to the index a, the quantities c_{aq}^p do not form a tensor with respect to this index. Nevertheless, under transformations of the points A_0 and A_a, the quantities c_{aq}^p along with the identity tensor δ_q^p do transform as tensors. For this reason, the system of quantities c_{aq}^p is called a *quasitensor*.

2. We will now prove the following result.

Theorem 4.1 *A tangentially degenerate submanifold V_r^m of rank r is an r-parameter family of l-dimensional planes (where $l = m - r$), and the tangent subspaces $T_x^{(1)}(V^m)$ remain constant along these l-planes.*

Proof. On a submanifold V_r^m consider the system of equations:

$$\omega^p = 0. \tag{4.13}$$

By (4.12), we have

$$d\omega^p = \omega^q \wedge (\omega_q^p - \delta_q^p \omega_0^0 - c_{aq}^p \omega^a). \tag{4.14}$$

By the Frobenius theorem, equations (4.14) imply that the system of equations (4.13) is completely integrable and defines a foliation of the submanifold V_r^m into $(m-r)$-dimensional submanifolds.

Let us prove that all these submanifolds are planes. In fact, if $\omega^p = 0$, then the first two of equations (4.6) take the form:

$$\begin{aligned} dA_0 &= \omega_0^0 A_0 + \omega^a A_a, \\ dA_a &= \omega_a^0 A_0 + \omega_a^b A_b. \end{aligned} \tag{4.15}$$

This means that if $\omega^p = 0$, then the l-plane defined by the points A_0 and A_q remains constant. Thus the submanifolds defined on V_r^m by the system of equations (4.13) are planes of dimension l.

The same system of equations (4.13) implies that if $\omega^p = 0$, then in addition to equations (4.15), we have the equation:

$$dA_p = \omega_p^0 A_0 + \omega_p^a A_a + \omega_p^q A_q. \tag{4.16}$$

Equations (4.15) and (4.16) mean that the tangent subspace $T_x^{(1)}(V_r^m)$ along the fixed l-plane $A_0 \wedge A_1 \wedge \ldots \wedge A_l$ remains constant. ■

In what follows, we assume that if a submanifold V_r^m contains a plane l-dimensional submanifold, then the latter is extended to a complete l-dimensional plane. We will call this l-plane an *l-dimensional generator* of a submanifold V_r^m and will denote it by E^l.

We denote by u^p the first integrals of the completely integrable system of equations (4.13). These first integrals are parameters on which the generator E^l and the tangent subspace $T_x^{(1)}$ of a tangentially degenerate submanifold V_r^m depend.

The simplest particular case of a tangentially degenerate submanifold is a two-dimensional surface V_1^2 in the space P^n formed by the tangents to a nonplanar curve. Such a surface is called a *torse*. This torse is a particular case of a tangentially degenerate submanifold V_1^m of rank one in P^n which is also called a *torse*. Its generator is of dimension $m-1$, and its second osculating subspace is of dimension $m+1$. The latter torse is a family of osculating planes of the mth order of a curve which does not belong to a projective space of dimension m.

Note also the extreme values of the rank r of a submanifold V^m. If $r = 0$, then a submanifold V^m is an m-dimensional projective subspace, since in this case the matrix (b_{ij}^α) is the zero matrix. On the other hand, if $r = m$, then it is obvious that the submanifold V^m is nondegenerate.

3. Consider a correlative transformation $\mathcal{K}$ in the space P^n which maps a point $x \in P^n$ into a hyperplane $\xi \in P^n$, $\xi = \mathcal{K}(x)$, and preserves the incidence of points and hyperplanes. A correlation $\mathcal{K}$ maps a k-dimensional subspace $P^k \subset P^n$ into an $(n-k-1)$-dimensional subspace $P^{n-k-1} \subset P^n$.

Consider a smooth curve C in the space P^n and suppose that this curve does not belong to a hyperplane. A correlation $\mathcal{K}$ maps points of C into hyperplanes forming a one-parameter family. The hyperplanes of this family envelope a tangentially degenerate hypersurface of rank one with $(n-2)$-dimensional generators.

If the curve C lies in a subspace $P^s \subset P^n$, then a correlation $\mathcal{K}$ maps points of C into hyperplanes which envelop a hypercone with an $(n-s-1)$-dimensional vertex.

Further, let V^m be an arbitrary tangentially nondegenerate submanifold in the space P^n. A correlation $\mathcal{K}$ maps points of such V^m into hyperplanes forming an m-parameter family. The hyperplanes of this family envelop a tangentially degenerate hypersurface V_m^{n-1}. The generators of this hypersurface are of dimension $n-m-1$ and correspond to the tangent subspaces $T_x^{(1)}(V^m)$.

If the tangentially degenerate submanifold V^m belongs to a subspace $P^s \subset P^n$, $s > m$, then the hypersurface V_m^{n-1} corresponding to V^m under a correlation $\mathcal{K}$ is a hypercone with an $(n-s-1)$-dimensional vertex.

Now let V^m be a tangentially degenerate submanifold of rank r. Then we can prove the following result.

Theorem 4.2 *A correlation $\mathcal{K}$ maps a tangentially degenerate submanifold V_r^m of rank r into a tangentially degenerate submanifold V_r^{n-l-1}, $l = m - r$, of the same rank r with $(n-m-1)$-dimensional plane generators.*

Proof. Under a correlation $\mathcal{K}$, to an l-dimensional plane generator $E^l \subset V_r^m$ there corresponds an $(n-l-1)$-dimensional plane E^{n-l-1}, and to a tangent subspace $T_x^{(1)}(V_r^m) = E^m$ there corresponds an $(n-m-1)$-dimensional plane E^{n-m-1}, where $E^{n-m-1} \subset E^{n-l-1}$. Since both of these planes depend on r parameters, the planes E^{n-m-1} are generators of the submanifold $\mathcal{K}(V_r^m)$, and the planes E^{n-l-1} are its tangent subspaces. Thus, the submanifold $\mathcal{K}(V_r^m)$ is a tangentially degenerate submanifold V^{n-l-1} of dimension $n-l-1$ and of rank r. ■

In particular, we note that, among tangentially degenerate submanifolds of rank r with l-dimensional generators, where $l = m - r$, there are submanifolds for which $l + m = n - 1$. A correlation $\mathcal{K}$ maps these tangentially degenerate submanifolds into submanifolds of the same dimension and with l-dimensional plane generators. These tangentially degenerate submanifolds are called *self-dual* or *autodual*.

4.2 Focal Images

Let V_r^m be a tangentially degenerate submanifold of rank r. By Theorem 4.1, such a submanifold carries an r-parameter family of l-dimensional plane generators E^l, where $l = m - r$. Let $x = x^0 A_0 + x^a A_a$ be an arbitrary point of this generator. We will call this point *focal* if $dx \in E^l$. Since

$$dx/E^l = (x^0 \omega_0^p + x^a \omega_a^p) A_p,$$

the following conditions must hold for focal points x:

$$x^0 \omega_0^p + x^a \omega_a^p = 0. \tag{4.17}$$

Using relations (4.12), we can write these equations in the form:

$$(x^0 \delta_q^p + x^a c_{aq}^p)\omega^q = 0. \tag{4.18}$$

These equations form a homogeneous system with respect to ω^q. This system has a nontrivial solution if and only if its determinant vanishes:

$$\det(x^0 \delta_q^p + x^a c_{aq}^p) = 0. \tag{4.19}$$

The latter equation defines an algebraic variety of order r in the plane E^l. To every point of this variety there corresponds a certain displacement of the plane E^l which can be found from equations (4.18). In this displacement the family of generators E^l has an envelope to which the point x belongs. The point x is called a *focus* of the plane E^l, and the algebraic variety defined by equations (4.19) is called the *focus variety*. Let us denote this variety by F.

Note that the point A_0 does not belong to the variety F, since if $x^a = 0$, the left-hand side of equation (4.19) takes the form:

$$\det(x^0 \delta^p_q) = (x^0)^r \neq 0.$$

Theorem 4.3 *The focus variety F is a set of singular points of a tangentially degenerate submanifold V^m, which are located on its generator E^l.*

Proof. In fact, at these points the rank of the system of forms on the left-hand side of equation (4.17) is reduced. This implies a reduction of the dimension of the tangent subspace $T^{(1)}_x(V^m_r)$. The points that do not belong to the focus variety F are *regular points* of the submanifold V^m_r. The dimension of the tangent subspace $T^{(1)}_x(V^m_r)$ at these regular points is not reduced. ■

Since each generator E^l of a tangentially degenerate submanifold V^m_r contains an $(l-1)$-dimensional variety F, the set of all these varieties is an $(m-1)$-dimensional *submanifold of singular points* of the tangentially degenerate submanifold V^m_r.

For the two-dimensional tangentially degenerate submanifold V^2_1 of rank one (a torse) considered above, each of its generators E^1 has one singular point (a focus), and the tangent subspace of V^2 at this point degenerates into a straight line. The set of all singular points of such V^2_1 form the *edge of regression* of this tangentially degenerate submanifold.

Consider one more example of a tangentially degenerate submanifold. In the space P^n, $n \geq 4$, we take two smooth space curves C^1 and C^2, which do not belong to the same three-dimensional space, and the set of all straight lines intersecting these two curves. These straight lines form a three-dimensional submanifold V^3. It is easy to see that the submanifold V^3 is tangentially degenerate. In fact, the three-dimensional tangent subspace $T^{(1)}_x(V^3)$ to V^3 at a point x lying on a rectilinear generator E^1 is defined by this generator E^1 and two straight lines tangent to the curves C^1 and C^2 at the points F^1 and F^2 of their intersection with the line E^1. Since this tangent subspace does not depend on the location of the point x on the generator E^1, the submanifold under consideration is a tangentially degenerate submanifold V^3_2 of rank two. The points F^1 and F^2 are foci of the generator E^1, and the curves C^1 and C^2 are degenerate focus varieties. There are two cones through every generator E^1. These cones are described by generators passing through the focus F_1 or the focus F_2. On the submanifold V^3_2 these cones form two one-parameter families comprising a focal net of the submanifold V^3_2.

This example can be generalized by taking k space curves in the space P^n, where $n \geq 2k$ and $k > 2$, and considering a k-parameter family of $(k-1)$-planes intersecting all these k curves.

The question arises: do there exist in the space P^n tangentially degenerate submanifolds without singularities ? Theorem 4.3 implies that from the complex point of view *a tangentially degenerate submanifold V^m_r does not have singularities if and only if it is an m-dimensional plane P^m, i.e. if $r = 0$.* From the real point of view a tangentially degenerate submanifold V^m_r does not have real singularities if and only if its focal images in the plane generators are pure imaginary, and this situation can occur only if the rank r is even. An example of such a submanifold will be considered in subsection **4.7.5**.

One can consider dual focal images of a submanifold V_r^m. The tangent subspace $T_x^{(1)}(V_r^m) = E^m$ of a tangentially degenerate submanifold V_r^m of rank r depends on r parameters u^p, $p = l+1, \ldots, m$. Consider two tangent subspaces E^m and $'E^m$ corresponding to the values u^p and $u^p + du^p$ of these parameters. Let $\xi = \xi^\alpha x_\alpha = 0$, $\alpha = m+1, \ldots, n$, be a hyperplane passing through the tangent subspace E^m. We say that this hyperplane is a *focus hyperplane* if it contains not only the tangent subspace E^m but also the tangent subspace $'E^m$.

Since the tangent subspace $'E^m$ is determined by the points $'A_0 = A_0 + dA_0, 'A_a = A_a + dA_a$ and $'A_p = A_p + dA_p$, by formulas (4.6), the subspace $'E^m$ belongs to the hyperplane ξ if an only if the following conditions hold:

$$\xi_\alpha \omega_p^\alpha = \xi_\alpha b_{pq}^\alpha \omega^q = 0. \tag{4.20}$$

This is a system of homogeneous equations with respect to ω^q. It has a non-trivial solution if and only if its determinant vanishes:

$$\det(\xi_\alpha b_{pq}^\alpha) = 0. \tag{4.21}$$

This equation defines an algebraic family of hyperplanes, which is an algebraic hypersurface of degree r in the dual projective space P^{n*}. The equation (4.21) defines a hypercone in the space P^n. The vertex of this hypercone is the subspace E^m tangent to the submanifold V_r^m.

To each hyperplane ξ of this family, there corresponds a displacement of the subspace E^m defined by equations (4.20). Under this displacement, the $(m+1)$-dimensional submanifold described by the subspace E^m is tangent to a hyperplane ξ. The hyperplanes ξ are the focus hyperplanes, and the cone (4.21) with the vertex E^m is called the *focus cone* Φ of the tangentially degenerate submanifold V^m.

The cone Φ is the set of *singular hyperplanes* tangent to the tangentially degenerate submanifold V_r^m along its generator E^l. A correlative transformation $\mathcal{K}$ maps the focus variety F of a generator E^l of a submanifold V_r^m into the focus cone $'\Phi$ whose vertex is the $(n-l-1)$-dimensional tangent subspace to the submanifold $\mathcal{K}(V_r^m)$, and maps the focus cone Φ with the vertex E^m into the focus variety $'F$ belonging to a $(n-m-1)$-dimensional generator of the submanifold $\mathcal{K}(V_r^m)$.

Consider the second osculating subspace $T_x^{(2)}(V_r^m)$ of a tangentially degenerate submanifold V_r^m of rank r. This subspace is defined by the points A_0, A_a, A_p and A_{pq}, where $A_{pq} = b_{pq}^\alpha A_\alpha$. Since the indices p and q take r values, and $A_{pq} = A_{qp}$, then the dimension of the subspace $T_x^{(2)}(V^m)$ is equal to $m + m_1$, where $m_1 \leq \frac{1}{2}r(r+1)$. If we place the points A_{i_1}, where $i_1 = m+1, \ldots, m+m_1$, into the subspace $T_x^{(2)}(V_r^m)$, then the equation (4.21) of the focus cone Φ takes the form:

$$\det(\xi_{i_1} b^{i_1}_{pq}) = 0. \tag{4.22}$$

This equation describes the essential part of the cone which belongs to the subspace $T^{(2)}_x(V^m_r)$. Precisely this part will figure in our future considerations.

Consider now the focus variety $F \subset E^l$. This variety can also be a cone with an $(l - l_1)$-dimensional vertex E^{l-l_1}. If this is the case, we place the points A_{a_2}, where $a_2 = 1, \ldots, l_2 + 1$ and $l_2 = l - l_1$, into the vertex of this cone. Then equation (4.19) of the cone will not contain the coordinates x^{a_2} and will take the form:

$$\det(x^0 \delta^p_q + x^{a_1} c^p_{a_1 q}) = 0, \tag{4.23}$$

where $a_1 = l_2 + 2, \ldots, l$, i.e. the index a_1 takes $l_1 - 1$ values. Thus, equation (4.18) takes the form:

$$(x^0 \delta^p_q + x^{a_1} c^p_{a_1 q})\omega^q = 0. \tag{4.24}$$

If a point belongs to the vertex E^{l-l_1} of the cone F, its coordinates x^0 and x^{a_1} vanish: $x^0 = x^{a_1} = 0$, and equation (4.24) is identically satisfied. This means that the subspace E^{l-l_1} is a characteristic and belongs to the envelope of the r-parameter family of plane generators E^l of a tangentially degenerate submanifold V^m_r of rank r.

A correlation $\mathcal{K}$ maps this characteristic subspace E^{l-l_1} into the osculating subspace $T^{(2)}_x$ of the submanifold $\mathcal{K}(V^m_r)$ and maps the osculating subspace $T^{(2)}_x(V^m_r)$ into the characteristic plane belonging to the generator E^{n-m-1} of the submanifold $\mathcal{K}(V^m_r)$.

If the focus variety F is not a cone, then the osculating subspace $T^{(2)}_x$ of the submanifold $\mathcal{K}(V^m_r)$ coincides with the entire space P^n. Conversely, if the subspace $T^{(2)}_x(V^m_r)$ coincides with the entire space P^n, then the focus variety $'F$ of an $(n - m - 1)$-dimensional generator of the submanifold $\mathcal{K}(V^m_r)$ is not a cone.

4.3 Decomposition of Focal Images

Suppose that $l \geq 1$ and that the focus variety F of a tangentially degenerate submanifold V^m_r of rank r does not have multiple components. This means that a general straight line belonging to a generator E^l intersects the variety F precisely in r distinct points. We place the points A_0 and A_1 of our moving frame on such a general straight line. Then the coordinates of the intersection points of the line $A_0 A_1$ and the variety F are defined by the equation:

$$\det(x^0 \delta^p_q + x^1 c^p_{1q}) = 0. \tag{4.25}$$

This equation differs from the characteristic equation of the matrix (c^p_{1q}) only in one factor. Since the intersection points of the line $A_0 A_1$ and the variety

F are distinct, the matrix (c^p_{1q}) has distinct eigenvalues. Therefore, by an appropriate choice of the points A_p, this tensor can be diagonalized. We write the diagonal form of this matrix as follows:

$$(c^p_{1q}) = \begin{pmatrix} c^{l+1} & \dots & 0 \\ \dots\dots & \dots & \dots\dots \\ 0 & \dots & c^m \end{pmatrix}, c^p \neq c^q \text{ if } p \neq q. \tag{4.26}$$

Consider now equations (4.10) for $a = 1$:

$$b^\alpha_{sq} c^s_{1p} = b^\alpha_{sp} c^s_{1q}.$$

By (4.26), these equations take the form

$$b^\alpha_{pq} c^p = b^\alpha_{qp} c^q,$$

where there is no summation over the indices p and q. Since $b^\alpha_{pq} = b^\alpha_{qp}$, it follows that

$$b^\alpha_{pq}(c^p - c^q) = 0,$$

and since $c^p \neq c^q$, we have $b^\alpha_{pq} = 0$ for $p \neq q$. This means that together with the tensor c^p_{1q} all the tensors b^α_{pq} are simultaneously reduced to diagonal form. Thus, equation (4.21) of the focus cone Φ takes the form:

$$\prod_{p=l+1}^{m} (\xi_\alpha b^\alpha_p) = 0, \quad b^\alpha_p = b^\alpha_{pp}, \tag{4.27}$$

i.e. the cone Φ decomposes into r bundles of hyperplanes whose centers are the $(m+1)$-planes determined by the points $A_0, A_i, B_{pp} = b^\alpha_p A_\alpha$.

Suppose now that $m_1 \geq 2$, i.e. the submanifold V^m_r has at least two independent forms $\Phi^\alpha_{(2)}$. In the same way as above we can prove that if the focus cone Φ with the vertex $T^{(1)}_x$ of the tangentially degenerate submanifold V^m of rank r does not have multiple components, then the focus variety $F \subset E^l$ of this submanifold decomposes into r planes of dimension $l-1$.

Thus we have established the following result.

Theorem 4.4 *If one of two focal images of a tangentially degenerate submanifold V^m_r of rank r does not have multiple components, then another focal image decomposes in the way described above provided that in one of these two cases the dimension is $l \geq 1$, and in the other case $m_1 \geq 2$.* ■

Consider now the cases excluded in Theorem 4.4. Suppose first that $l < 1$, i.e. $l = 0$. Then $r = m$, and the submanifold V^m is tangentially nondegenerate. Its focal variety F is reduced to the point A_0, and its focus cone Φ is defined by the equation:

$$\det(\xi_\alpha b^\alpha_{ij}) = 0,$$

i.e. this cone is a hypercone whose vertex is the tangent subspace $T_x^{(1)}(V^m)$.

Next suppose that $m_1 < 2$, i.e. $m_1 = 1$. Then the submanifold V_r^m of rank r has only one second fundamental form $\Phi_{(2)}^{m+1}$. If this form is not a perfect square, then by the Segre theorem (see Theorem 2.2 in Section **2.5**), the submanifold V_r^m is a hypersurface in the projective space $P^{m+1} = T_x^{(2)}(V_r^m)$. The focus cone Φ of this hypersurface is the tangent subspace $T_x^{(1)}(V_r^m)$ itself, and its focus variety F is a general algebraic hypersurface in a plane generator E^{n-r-1}.

On the other hand, if the form $\Phi_{(2)}^{m+1}$ is the square of a linear form, then the submanifold V^m is the envelope of a one-parameter family of m-dimensional planes. The submanifold in question is a submanifold V_1^m of rank 1, and its focus cone Φ decomposes into a bundle of hyperplanes whose center is the tangent subspace $T_x^{(1)}(V_1^m)$, and the focus variety F coincides with an $(m-1)$-dimensional plane generator of the submanifold V_1^m.

Suppose now that the focal images of a tangentially degenerate submanifold V_r^m of rank r do not have multiple components. Then each focus variety F of this submanifold decomposes into r different $(l-1)$-planes, and each focus cone Φ decomposes into r bundles of hyperplanes with $(m+1)$-dimensional centers. In this case all the matrices (b_{pq}^α) and (c_{aq}^p) are simultaneously reduced to diagonal form. This implies that equations (4.5) and (4.12) take the following form:

$$\omega_p^\alpha = b_p^\alpha \omega^p, \quad \omega_a^p = c_a^p \omega^p, \tag{4.28}$$

where b_p^α and c_a^p denote the diagonal entries of these matrices, and there is no summation over the index p.

Now the equation of the focus cone Φ has the form (4.27), and the equation of the focus variety F has the form

$$\prod_{p=l+1}^{m} (x^0 + x^a c_a^p) = 0, \tag{4.29}$$

and thus the variety F decomposes into r planes of dimension $l-1$.

For the case in question, the matrices of the diagonal entries of the tensors $\{b_p^\alpha\}$ and $\{c_a^p\}$ have the form:

$$B = \begin{pmatrix} b_{l+1}^{m+1} & b_{l+2}^{m+1} & \dots & b_m^{m+1} \\ \dots & \dots & \dots & \dots \\ b_{l+1}^n & b_{l+2}^n & \dots & b_m^n \end{pmatrix}, \quad C = \begin{pmatrix} 1 & 1 & \dots & 1 \\ c_1^{l+1} & c_1^{l+2} & \dots & c_1^m \\ \dots & \dots & \dots & \dots \\ c_l^{l+1} & c_l^{l+2} & \dots & c_l^m \end{pmatrix}. \tag{4.30}$$

4.4 The Holonomicity of the Focal Net

By (4.28), the equations (4.18) and (4.20) take the following forms:

$$(x^0 + x^a c_a^p)\omega^p = 0, \tag{4.31}$$

$$(\xi_\alpha b_p^\alpha)\omega^p = 0. \tag{4.32}$$

It follows from these equations that to the pth plane component

$$C^p := x^0 + x^a c_a^p = 0, \quad p \text{ is fixed}, \tag{4.33}$$

of the focus variety F there corresponds the following direction on V^m:

$$\omega^q = 0, \quad q \neq p. \tag{4.34}$$

For displacement along this direction, the $(l-1)$-plane (4.33) is the characteristic of a one-parameter family of l-dimensional generators E^l. This family is a torse, i.e. a tangentially degenerate manifold of rank one, whose tangent subspace is the $(l+1)$-plane determined by the plane E^l and the point A_p.

The same direction (4.34) corresponds to the pth bundle of hyperplanes of the focus cone Φ. The $(m+1)$-dimensional center of this bundle is the tangent subspace to the $(m+1)$-dimensional manifold described by the subspace $E^m = T_x^{(1)}(V_r^m)$ for the displacement defined by equation (4.34). This tangent subspace is the linear span of the plane E^m and the point $B_p = b_p^\alpha A_\alpha$. As in Chapter 3, we denote by $\widetilde{A}_\alpha$ the projectivization of the point A_α with the center $T_x^{(1)} = E^m$. Then the columns of the matrix B (see (4.30)) uniquely define the points

$$\widetilde{B}_\alpha = b_p^\alpha \widetilde{A}_\alpha. \tag{4.35}$$

Thus for every generator E^l there are defined r distinct directions called *focal directions* of the tangentially degenerate submanifold V_r^m. It is easy to see that if $r = m$ (i.e. $l = 0$), these focal directions coincide with the conjugate directions introduced in Chapter 3. The difference is that while in Chapter 3 we required the existence of these conjugate directions on V^m, the existence of the focal directions here follows from the fact that the focal images do not have multiple components.

If the index p is fixed, the system (4.34) defines an $(r-1)$-parameter family of torses on V_r^m. For all values of p this system defines a *net of torses* which is called the *focal net* Σ of the submanifold V_r^m.

It is easy to show that a correlative transformation $\mathcal{K}$ maps the focal net of a tangentially degenerate submanifold V^m of rank r into the focal net of the submanifold $\mathcal{K}(V_r^m)$.

If each of the equations

$$\omega^p = 0, \quad p \text{ is fixed}, \quad p = l+1, \ldots, m, \tag{4.36}$$

is completely integrable, then the focal net on the submanifold V_r^m is called *holonomic*. Otherwise, the focal net is called *nonholonomic*.

Note also that the characteristic plane of a generator E^l of a tangentially degenerate submanifold V_r^m of rank r is the intersection of r planes C^p of dimension $l-1$, and thus the dimension of this characteristic plane is equal to $l - l_1$, where $l_1 \leq r$. The osculating subspace $T_x^{(2)}(V_r^m)$ is determined by its tangent subspace $E^m = T_x^{(1)}$ and the points B_p. Thus the dimension of this osculating subspace is equal to $m + m_1$ where $m_1 \leq r$.

Since the focal net Σ is fixed on the tangentially degenerate submanifold V^m, the forms ω_q^p can be expressed in terms of the basis forms ω^q of V^m:

$$\omega_q^p = \sum_{s=l+1}^{m} l_{qs}^p \omega^s. \tag{4.37}$$

Since by (4.28) and (4.31) we have

$$d\omega^p \equiv l_{qs}^p \omega^p \wedge \omega^s \pmod{\omega^p}, \tag{4.38}$$

the condition for the focal net Σ on V^m to be holonomic has the form:

$$l_{[qs]}^p = 0, \quad q \neq s, p, \; s \neq p. \tag{4.39}$$

Let us now establish the relationship between the holonomicity condition for the focal net and the properties of matrix (4.30). Exterior differentiation of equations (4.28) gives the following exterior quadratic equations:

$$\begin{aligned} &\Delta b_p^\alpha \wedge \omega^p - \sum_{q\neq p} (b_q^\alpha \omega_p^q + b_p^\alpha \omega_q^p) \wedge \omega^q = 0, \\ &\Delta c_a^p \wedge \omega^p + \sum_{q \neq p} (c_a^q - c_a^p) \omega_q^p \wedge \omega^q = 0, \end{aligned} \tag{4.40}$$

where

$$\begin{aligned} \Delta b_p^\alpha &= db_p^\alpha + b_p^\beta \omega_\beta^\alpha + b_p^\alpha(\omega_0^0 - 2\omega_p^p) + b_p^\alpha c_a^p \omega^a, \\ \Delta c_a^p &= dc_a^p - c_b^p \omega_a^b + c_a^p \omega_0^0 - \omega_a^0 + c_a^p c_b^p \omega^b. \end{aligned}$$

Substituting the forms ω_q^p by their expressions taken from (4.37), and using the linear independence of the products $\omega^s \wedge \omega^q$, where $s, q \neq p$, we find that

$$b_p^\alpha (l_{qs}^p - l_{sq}^p) + b_q^\alpha l_{ps}^q - b_s^\alpha l_{pq}^s = 0 \tag{4.41}$$

and

$$(c_a^q - c_a^p) l_{qs}^p - (c_a^s - c_a^p) l_{sq}^p = 0. \tag{4.42}$$

Contracting equations (4.41) with the points $\widetilde{A}_\alpha$ and using notation (4.35), we obtain

$$\widetilde{B}_p (l_{qs}^p - l_{sq}^p) + \widetilde{B}_q l_{ps}^q - \widetilde{B}_s l_{pq}^s = 0. \tag{4.43}$$

These equations are similar to equations (3.24) which we obtained for a submanifold carrying a conjugate net. But for a tangentially degenerate submanifold

V_r^m we also have some additional conditions to which the left-hand sides C^p of the equations of $(l-1)$-planes (4.34) satisfy. To get these conditions, we supplement equations (4.32) by the trivial identities

$$(1-1)l_{qs}^p - (1-1)l_{sq}^p = 0. \tag{4.44}$$

and add these identities to equations (4.42) contracted with the coordinates x^a. This gives the following equations:

$$(C^q - C^p)l_{qs}^p - (C^s - C^p)l_{sq}^p = 0, \tag{4.45}$$

where C^p is the $(l-1)$-plane defined by equation (4.33).

Now it is not difficult to prove the following result.

Theorem 4.5 *Let the conditions of Theorem 4.3 hold on a tangentially degenerate submanifold V_r^m of rank r. Suppose that for any three focal directions, one of the following two conditions holds:*

a) *The $(l-1)$-planes C^p, C^q and C^s (p, q and s are distinct) corresponding to these three directions do not belong to a pencil, or*

b) *The points $\widetilde{B}_p, \widetilde{B}_q$ and $\widetilde{B}_s$ (p, q and s are distinct) do not lie on a straight line.*

Then the focal net of the submanifold V_r^m is holonomic, the submanifold V_r^m is foliated by r families of $(m-1)$-dimensional submanifolds $V_{(p)}^{m-1}$, and each of the submanifolds $V_{(p)}^{m-1}$ is tangentially degenerate of rank $r-1$ and inherits the focal net from the submanifold V_r^m.

Proof. First suppose that condition b) of the theorem holds. Then it follows from (4.43) that

$$l_{pq}^s = 0 \ \text{ for } q \neq s, p, \ s \neq p. \tag{4.46}$$

So the holonomicity condition (4.39) holds. Moreover, the additional second fundamental forms arising on the submanifolds $V_{(p)}^{m-1}$ also are sums of squares. This means that the submanifolds $V_{(p)}^{m-1}$ carry the focal nets.

If condition a) of the theorem holds, then conditions (4.45) also imply conditions (4.46), and we again arrive at the conclusion of the theorem. ■

Corollary 4.6 *If $l = 0$ and $n - m > 2$ or $n - m = 1$ and $l > 1$, and the submanifold V_r^m of rank $r = m - l$ carries a conjugate net, then in general this net is holonomic.*

Proof. In fact, under the hypotheses of the theorem we have one of the equations, (4.45) or (4.43), and each of them lead to equations (4.46) which imply the holonomicity of the conjugate net on V^m. ■

Let us prove the existence of a tangentially degenerate submanifold V^m of rank r carrying a holonomic focal net.

Theorem 4.7 *A tangentially degenerate submanifold V^m of rank r carrying a holonomic focal net exists, and depends on $r(r-1)$ arbitrary functions of two variables.*

Proof. The submanifolds V^m in question are defined by a system of Pfaffian equations (2.5), (4.4), (4.28), and the equations

$$\omega_q^p = l_{qp}^p \omega^p + l_{qq}^p \omega^q, \tag{4.47}$$

which follow from equations (4.39) and (4.46). Exterior differentiation of equations (2.5) and (4.4) leads to identities. By (4.40) and (4.47), exterior differentiation of equations (4.28) gives the following equations:

$$\Delta b_p^\alpha \wedge \omega^p = 0, \quad \Delta c_a^p \wedge \omega^p = 0, \tag{4.48}$$

where Δ_p^α and Δ_a^p are the same as in equations (4.40). Finally, the exterior differentials of equations (4.47) have the form:

$$\Delta l_{qp}^p \wedge \omega^p + \Delta l_{qq}^p \wedge \omega^q = 0,$$

where

$$\begin{aligned}
\Delta l_{qp}^p = {} & dl_{qp}^p + l_{qp}^p(\omega_0^0 - \omega_q^q) + c_a^p(c_{qp}^p \omega^a - \omega_q^a) - \omega_q^0 \\
& + \sum_{s \neq p} l_{qp}^p l_{sp}^p \omega^s - \sum_{s \neq p,q} l_{sp}^p (l_{qs}^s + l_{qq}^s)\omega^s, \\
\Delta l_{qq}^p = {} & dl_{qq}^p + l_{qq}^p(\omega_0^0 - 2\omega_q^q + \omega_p^p) + b_q^\alpha \omega_\alpha^p + l_{qq} c_a^q \omega^a \\
& + \sum_{s \neq q} l_{qq}^p l_{sq}^q \omega^s - \sum_{s \neq p,q} l_{ss}^p l_{qq}^s \omega^s.
\end{aligned}$$

Calculating the integers q, s_1, s_2, Q and N for this system in the same way as we did this in Section **3.4**, we find that

$$\begin{aligned}
& q = lr + (n-m)r + r(r-1), \\
& s_1 = lr + (n-m)r, \quad s_2 = r(r-1), \\
& Q = N = lr + (n-m)r + 2r(r-1).
\end{aligned}$$

Thus, the system in question is in involution, and its general integral manifold depends on $r(r-1)$ arbitrary functions of two variables. ■

4.5 Some Other Classes of Tangentially Degenerate Submanifolds

1. Consider a tangentially degenerate submanifold V_r^m, one of the focal images of which has an r-fold component.

Theorem 4.8 *If the focus variety F of a tangentially degenerate submanifold V_r^m of rank r is an r-fold $(l-1)$-plane, and $r \geq 2$, then the submanifold V_r^m is a cone with an $(l-1)$-dimensional vertex and l-dimensional plane generators.*

Proof. We place the vertices A_b, $b = 1, \dots, l$, of our moving frame into the r-fold component of the focus variety F. Then this variety has the equation $x^0 = 0$. Since the equation of the focus variety F has the form (4.19), it follows that the quasitensor c^p_{aq} vanishes:

$$c^p_{aq} = 0. \tag{4.49}$$

This implies that equation (4.12) takes the form

$$\omega^p_a = 0. \tag{4.50}$$

Exterior differentiation of this equation gives

$$\omega^0_a \wedge \omega^p = 0.$$

Since $r \geq 2$, these exterior quadratic equations yield

$$\omega^0_a = 0. \tag{4.51}$$

Using equations (4.50) and (4.51), we find that

$$dA_b = \omega^c_b A_c,$$

i.e. the $(l-1)$-plane $A_1 \wedge \dots \wedge A_l$ is constant. Therefore, the submanifold V^m_r is a cone with an $(l-1)$-dimensional vertex and l-dimensional plane generators. ■

Note that by conditions (4.49), equations (4.10) are identically satisfied and do not impose any condition on the tensor b^α_{pq}. This shows that the dimension of the osculating subspace $T^{(2)}_x(V^m_r)$ can be greater than $m + r$ but does not exceed $m + \frac{1}{2}r(r+1)$.

Theorem 4.9 *If the focus cone Φ of a tangentially degenerate submanifold V^m_r of rank r is an r-fold $(n-m-2)$-bundle of hyperplanes, and $r \geq 2$, then the submanifold V^m_r is a hypersurface belonging to the fixed $(m+1)$-dimensional center of this bundle.*

Proof. Since the focus cone Φ is an r-fold bundle of hyperplanes, equation (4.21) must be the rth power of a linear expression $\xi_\alpha b^\alpha$. This implies that the second fundamental tensors of the submanifold V^m_r under consideration have the form:

$$b^\alpha_{pq} = b^\alpha b_{pq},$$

i.e. there is only one independent form among the second fundamental forms of the submanifold V^m_r. Thus all points of the submanifold V^m_r are axial (see Section **2.5**). Applying the same considerations which we used in the proof of the Segre theorem (Theorem 2.2 of Section **2.5**) and using the condition $r \geq 2$, we find that the submanifold V_r^m belongs to its constant $(m+1)$-dimensional osculating subspace $T_x^{(2)}$ and has the structure indicated in the theorem. ■

2. Consider a tangentially degenerate submanifold V_r^m for which all the points $\widetilde{B}_p$ defined by equations (4.35) belong to a straight line, but are mutually distinct. Suppose also that all the planes C^p defined by equations (4.34) belong to a pencil, and are also mutually distinct.

Then the osculating subspace $T_x^{(2)}(V_r^m)$ is of dimension $m_1 = m+2$, and the dimension of the characteristic plane (the plane of intersection of all the planes C^p) is equal to $l-2$. In this case the conditions of Theorem 4.4 are not satisfied, and the focal net can be nonholonomic.

Moreover, if $r \geq 3$, then applying the method similar to that which we used in the proof of the generalized Segre theorem (Theorem 3.10 in Section **3.3**), we can prove that the submanifold V_r^m belongs to its $(m+2)$-dimensional osculating subspace $T_x^{(2)}$ and is a cone with an $(l-2)$-dimensional vertex. (We recall that $m-l=r$.)

Following the same line of reasoning, we can consider tangentially degenerate submanifolds V_r^m of rank r whose focal images have one or more multiple components where the sum of multiplicities is equal to r. Under some natural additional conditions, these submanifolds are foliated into a family of cones or a family of submanifolds belonging to their second osculating subspaces (see [A 62b] for more details).

3. There are special families of l-dimensional planar subspaces of the space P^n (called congruences and pseudocongruences) which are closely related to tangentially degenerate submanifolds V^m of rank r.

A *congruence C of l-planes* E^l in the space P^n is a family of l-planes that depends on $r=n-l$ parameters and does not lie in a subspace P^m of dimension $m<n$.

With each l-plane E^l of a congruence C we associate the family of projective frames $\{A_u\}$, $u=0,\dots,n$, such that the vertices $A_0,\dots,A_l$ of each frame of this family belong to E^l. The equations of infinitesimal displacement of each of these frames have the following form:

$$\begin{aligned} dA_0 &= \omega_0^0 A_0 + \omega_0^b A_b + \omega_0^p A_p, \\ dA_b &= \omega_b^0 A_0 + \omega_b^c A_c + \omega_b^p A_p, \\ dA_p &= \omega_p^0 A_0 + \omega_p^b A_b + \omega_p^q A_q, \end{aligned} \tag{4.52}$$

where $a,b,c=1,\dots,l$ and $p,q=l+1,\dots,n$. Since the 1-forms ω_0^p and ω_b^p are the basis forms on the Grassmannian $G(l,n)$, on the congruence C these forms must be expressed in terms of r linearly independent forms. It is not difficult to prove that a generator E^l of the congruence C contains points possessing a neighborhood $U_\delta \subset P^n$ such that through any of its points there passes a generator $'E^l \subset C$, sufficiently close to E^l. Such points of E^l are called regular. If we take the point A_0 to be a regular point of a generator E^l, then the forms ω_0^p are linearly independent and can be taken as basis forms. The equations of a congruence C then have the form:

$$\omega_b^p = c_{bq}^p \omega^q, \tag{4.53}$$

where $\omega^q = \omega_0^q$.

As in the case of tangentially degenerate submanifolds V_r^m, each plane E^l of a congruence C possesses a focus variety. The equations of this variety can be found from the condition

$$dx \in E^l, \tag{4.54}$$

where $x = x^0 A_0 + x^b A_b$ is a point in the plane E^l (cf. Section **4.2**). Conditions (4.54) lead to the equations:

$$(x^0 \delta_q^p + x^a c_{aq}^p)\omega^q = 0, \tag{4.55}$$

which give the following equation of the focus variety F:

$$\det(x^0 \delta_q^p + x^a c_{aq}^p) = 0. \tag{4.56}$$

We can see from this equation that the point $A_0 \notin F$, and that this results from the linear independence of the basis forms ω_0^p. As in Section **4.2**, the focus variety F is an algebraic variety of degree r, where now the number r is: $r = n - l$.

In contrast to tangentially degenerate submanifolds V_r^m, for a congruence C the components of the quasitensor c_{aq}^p are not related by equations (4.10) and are completely arbitrary. Thus, in general, the focus variety F does not decompose.

To each point of the focus variety F there corresponds a focal direction defined by the system of equations (4.55). However, in general, a congruence C does not possess a net of focal directions.

The exception is a congruence of straight lines, i.e. the case where $l = 1$. For these, the focus variety F has the equation:

$$\det(x^0 \delta_q^p + x^1 c_{1q}^p) = 0, \tag{4.57}$$

and decomposes into r points called the *foci*. For each of r foci, the system (4.55) defines focal directions. If all the foci are distinct, then the *focal net* arises on a rectilinear congruence, and this net is formed by r families of torses.

A dual image of a congruence is a *pseudocongruence* C^* whose element is an $(r-1)$-plane E^{r-1}. As was the case for a congruence C, a pseudocongruence C^* depends on r parameters. Thus, a pseudocongruence C^* is an arbitrary r-parameter family of $(r-1)$-dimensional planes.

Without trying to reach a complete duality, we associate with a plane $E^{r-1} \subset C^*$ a family of point frames such that the vertices A_p, $p = n - r + 1, \ldots, n$, of each frame of this family belong to E^{r-1}.

The equations of infinitesimal displacement of each of these frames have the following form:

$$dA_p = \omega_p^q A_q + \omega_p^\alpha A_\alpha, \quad dA_\alpha = \omega_\alpha^p A_p + \omega_\alpha^\beta A_\beta, \tag{4.58}$$

where $\alpha, \beta = 0, 1, \ldots, n - r$. The forms ω_p^α are expressed in terms of r basis forms of the pseudocongruence. If we denote the basis forms by θ^p, we have:

$$\omega_p^\alpha = b_{pq}^\alpha \theta^q, \tag{4.59}$$

where the coefficients b_{pq}^α no longer have to be symmetric.

The focal image of a pseudocongruence is a focus cone Φ in the bundle of hyperplanes $\xi_\alpha x^\alpha = 0$ passing through the planes $E^{r-1} \subset C^*$. The equations of the cone Φ can be derived in the same way as equations (4.21); they have the following form:

$$\det(\xi_\alpha b_{pq}^\alpha) = 0. \tag{4.60}$$

Each hyperplane belonging to the cone Φ defines a focal direction, which can be found from the system of equations:

$$\xi_\alpha b_{pq}^\alpha \theta^q = 0, \tag{4.61}$$

which is similar to system (4.20).

The cone Φ is an algebraic cone of degree r. In general, this cone does not decompose. But if $r = n - 1$, the index α in equation (4.60) takes only two values, 0 and 1, and this equation determines exactly r focus hyperplanes into which the cone Φ decomposes. Each of these hyperplanes determines a focal direction on the pseudocongruence C^*. This direction can be found from the system (4.61). This proves that if $r = n - 1$, then the pseudocongruence C^* possesses the *focal net*.

4.6 Manifolds of Hypercones

We now give a geometric interpretation of the determinant submanifolds defined by the symmetric tensors x^{ij} (see Section **1.7**) and prove that these manifolds are tangentially degenerate. However, it is more convenient to use the symmetric tensors a_{uv}, $u, v = 0, 1, \ldots, n$, since each of these tensors defines a hyperquadric Q in the space P^n. The equation of Q is

$$a_{uv} x^u x^v = 0, \quad a_{uv} = a_{vu}. \tag{4.62}$$

To each hyperquadric Q there corresponds a point in the space P^N, where $N = \frac{1}{2}(n+1)(n+2) - 1$. The degenerate hyperquadrics—the hypercones—are defined by the condition:

$$\det(a_{uv}) = 0, \tag{4.63}$$

which determines a hypersurface V^{N-1} in P^N. Since the degree of degeneracy of a hyperquadric can vary, in the space P^N we obtain the sequence of submanifolds defined by the equations

$$\text{rank } (a_{uv}) = r, \tag{4.64}$$

where $2 \leq r \leq n$. Each term of this sequence defines a submanifold of hypercones with $(n-r+1)$-dimensional plane generators and an $(n-r)$-dimensional vertex in P^N.

If $r = 1$, the submanifold (4.64) is a Veronese variety (see Section **1.7**). In this case

$$a_{uv} = a_u a_v \tag{4.65}$$

(cf (1.106)), and the hyperquadric (4.62) defined by the tensor (a_{uv}) becomes a double hyperplane $a_u x^u = 0$.

Let us study the structure of the determinant submanifold (4.63) in the general case. We consider a family of moving frames $\{A^{uv}\}$ in P^N, such that $A^{uv} = \alpha^u \alpha^v$, where α^u is a basis hyperplane of the space P^{n*}. Since the equations of infinitesimal displacement of a tangential moving frame in the space P^{n*} have the form (1.72):

$$d\alpha^u = -\omega_v^u \alpha^v, \tag{4.66}$$

for the moving frames in the space P^N we get:

$$dA^{uv} = -\omega_w^u A^{wv} - \omega_w^v A^{uw}. \tag{4.67}$$

Consider a hypercone Q in the space P^n and associate with Q a family of moving point frames $\{A_u\}$ in such a way that their point A_0 belongs to the vertex of the hypercone Q. Then the equation of the hypercone Q has the form:

$$a_{ij} x^i x^j = 0, \quad i, j = 1, \ldots, n. \tag{4.68}$$

To a hypercone Q there corresponds a point on the hypersurface $V^{N-1} \in P^N$.

The family of hypercones with the common vertex A_0 is a vector space, and therefore a plane generator E^L of the hypersurface V^{N-1} corresponds to this family. The dimension L of this generator is equal to $\frac{1}{2}n(n+1) - 1$, and the set of all these generators depends on n parameters.

In the space P^N the points A^{ij} of our moving frame lie on a generator E^L of the hypersurface V^{N-1}. Applying formulas (4.67), we calculate the differentials of these points A^{ij}:

$$dA^{ij} = -\omega_k^i A^{kj} - \omega_k^j A^{ki} - \omega_0^i A^{0j} - \omega_0^j A^{0i}. \tag{4.69}$$

Let $x = x_{ij} A^{ij}$ be a point of the generator E^L. Then

$$dx = (dx_{ij} - x_{ik}\omega_j^k - x_{kj}\omega_i^k) A^{ij} - 2x_{ij}\omega_0^i A^{0j}. \tag{4.70}$$

This equation shows that at all points of the generator E^L for which

$$\det(x_{ij}) \neq 0, \tag{4.71}$$

the tangent subspace $T_x^{(1)}$ to the hypersurface V^{N-1} is determined by the points A^{ij} and A^{0j}. Hence this subspace is of dimension $N-1$ and is constant for all points $x \in E^L$ for which inequality (4.71) holds. Therefore, the hypersurface V^{N-1} is a tangentially degenerate submanifold V_n^{N-1} of rank n.

At the points of the generator E^L for which

$$\det(x_{ij}) = 0, \tag{4.72}$$

the dimension of the tangent subspace $T_x^{(1)}(V_n^{N-1})$ is reduced. Thus these points are foci of the generator, and singular points of the hypersurface V_n^{N-1}.

Similarly, in the space P^N we can consider a submanifold defined by equation (4.64), where $r = n-2$. This submanifold is also a tangentially degenerate submanifold whose generators are of dimension $\frac{1}{2}n(n-1)-1$ and depend on $2(n-1)$ parameters. Each plane generator of such a submanifold has a focus variety defined by the equation $\det(x_{ab}) = 0,\ a, b = 2, \ldots, n$.

Finally, if $r = 1$, the submanifold defined by equation (4.64) is a Veronese variety, which does not carry plane generators, and which is obviously is not a tangentially degenerate submanifold. As we already noted, a Veronese variety does not have singular points.

4.7 Parabolic Submanifolds without Singularities in Euclidean and Non-Euclidean Spaces

1. With tangentially degenerate submanifolds of a projective space P^n there are associated the so-called *parabolic submanifolds* in Riemannian spaces of constant curvature.

Let V^n be a Riemannian manifold of dimension n, V^m be a smooth submanifold of dimension m, $m < n$, in V^n, x be a point of V^m, and $T_x(V^m)$ be the tangent subspace to V^m at the point x. Denote by $T_x^{\perp}(V^m)$ the orthogonal complement of the subspace $T_x(V^m)$ in the tangent space $T_x(V^n)$ to the Riemannian manifold V^n. Then for each vector $\xi \in T_x^{\perp}(V^m)$ the second fundamental form

$$\Phi_{(2)}(\xi) = (\xi, d^2x) \tag{4.73}$$

is defined, where the parentheses (,) denote the scalar product of vectors in the space $T_x(V^n)$.

Denote by $\Phi_{(2)}(\xi; X, Y)$, where $X, Y \in T_x(V^m)$, the bilinear form corresponding to the second fundamental form (4.73), and consider the subspace:

$$\widetilde{T}_x(V^m) = \begin{array}{l}\{X \in T_x(V^m) | \Phi_{(2)}(\xi; X, Y) = 0 \\ \text{for all } \xi \in T_x^{\perp}(V^m) \text{ and } Y \in T_x(V^m)\}.\end{array} \tag{4.74}$$

The dimension of the space $\widetilde{T}_x(V^m)$ is called the *index of relative nullity* of the submanifold V^m at the point x and is denoted by $\nu(x)$. This notion was introduced by Chern and Kuiper in their joint paper [CK 52] (see also the book [KN 63], vol. 2, p. 348). If $\nu(x) > 0$, then the point x is called a *parabolic* point of the submanifold V^m. If all points of a submanifold V^m are parabolic, then the submanifold V^m is called *parabolic.*

However, the second fundamental forms (4.73) are connected not so much to the metric structure of the submanifold V^m as to its projective structure, since these forms are preserved under projective transformations of the Riemannian submanifold V^m. If a Riemannian manifold V^n admits a mapping f into a domain in n-dimensional projective space P^n, and if this mapping transforms the geodesics of V^n into straight lines of P^n, then the second fundamental forms (4.73) coincide with the second fundamental forms of the submanifold $\overline{V}^m = f(V^m)$ of the projective space which were defined in Chapter 2 by formula (2.32). We will call the mapping $f : V^n \to P^n$ introduced above a *geodesic mapping.*

If we apply the notations of Chapter 2, then condition (4.74) defining the subspace $\widetilde{T}_x(V^m)$ can be written in the form:

$$b^{\alpha}_{ij} x^i = 0. \tag{4.75}$$

If we place the points $A_a, a = 1, \ldots, \nu(x)$, of our moving frames, associated with a point $x = A_0$ of the submanifold $\overline{V}^m = f(V^m)$, in the subspace $f^*\widetilde{T}_x(V^m)$, then from condition (4.75) we find that

$$b^{\alpha}_{aj} = 0, \tag{4.76}$$

and thus the matrices (b^{α}_{ij}) of the second fundamental forms of the submanifold $\overline{V}^m$ take the form (4.3). This proves that the submanifold $\overline{V}^m$ is tangentially degenerate, and its rank r is connected to the index $\nu(x)$ of relative nullity of the submanifold V^m by the relation

$$r = m - \nu(x).$$

It follows that the index $\nu(x)$ of relative nullity of the submanifold V^m coincides with the dimension l of the plane generators of the submanifold $\overline{V}^m = f(V^m)$. Thus we have proved the following result.

Theorem 4.10 *If a Riemannian manifold V^n admits a geodesic mapping f into the space P^n, and if a submanifold $V^m \subset V^n$ has the constant index $\nu(x)$ of relative nullity, then the image $\overline{V}^m = f(V^m)$ of the submanifold V^m is a tangentially degenerate submanifold V^m_r of rank $r = m - \nu(x)$ whose plane generators are of dimension $l = \nu(x)$.* ■

Submanifolds V^m of a Riemannian space V^n of a constant index $\nu(x) = l$ of relative nullity are called *l-parabolic submanifolds* (cf. papers [Bori 82] and [Bori 85] by Borisenko).

2. We will now study complete l-parabolic submanifolds in real simply connected Riemannian spaces V_c^n of constant curvature c. If $c = 0$, then V_c^n is the Euclidean space E^n. If $c > 0$, then V_c^n is the elliptic space S^n. If $c < 0$, then V_c^n is the hyperbolic space H^n. Each of these spaces admits a geodesic mapping into the space P^n, which is usually called the *projective realization* of the corresponding space V_c^n.

The Euclidean space E^n is realized in the projective space P^n from which a hyperplane E_∞ has been removed (this hyperplane is called *improper* or *the hyperplane at infinity*), and the proper domain of the space E^n can be identified with the open simply connected manifold $P^n \backslash E_\infty$. The elliptic space S^n is realized in the entire projective space P^n, since the absolute of S^n is an imaginary hyperquadric and its proper domain coincides with the entire space P^n. Finally, the hyperbolic space H^n is realized in the part of the projective space P^n lying within the convex hyperquadric that is the absolute of this space. This open simply connected domain is the proper domain of the hyperbolic space H^n. We denote by G the proper domain of the simply connected space V_c^n in all these cases.

Let V^m be a complete parabolic submanifold of a space V_c^n of constant curvature. Suppose that V^m has a constant index $\nu(x) = l$ of relative nullity. Let $\overline{V}^m$ be the image of V^m in the domain G of the space P^n in which the space V_c^n is realized, and let $\widetilde{V}^m$ be the natural extension of this image in the space P^n, so that $\overline{V}^m = \widetilde{V}^m \bigcap G$. In this extension, l-dimensional plane generators of the submanifold $\overline{V}^m$ are complemented by improper elements from the complement $P^n \backslash G$. By Theorem 4.9, the submanifold $\widetilde{V}^m$ is a tangentially degenerate submanifold of rank $r = n - l$. The focus variety F belonging to a plane generator E^l of $\widetilde{V}^m$ is the set of all singularities of this generator.

One of the important problems of multidimensional differential geometry is the finding of complete l-parabolic submanifolds V^m without singularities in spaces V_c^n of constant curvature. Theorems 4.3 and 4.9 imply the following result.

Theorem 4.11 *Let V^m be a complete l-parabolic submanifold of a simply connected space V_c^n of constant curvature. Let $\overline{V}^m = f(V^m)$ be the image of V^m in the proper domain G of the space P^n in which the space V_c^n is realized, and let $\widetilde{V}^m$ be the natural extension of this image in the space P^n. The submanifold V^m is regular if and only if the real parts* Re *F of the focus varieties F belonging to generators E^l of the submanifold $\widetilde{V}^m$ lie outside of the proper domain $G \subset P^n$.* ■

3. Let us examine the content of Theorem 4.11 for the different kinds of spaces V_c^n of constant curvature.

If $c = 0$, then V_c^n is the Euclidean space E^n, and $P^n \backslash E_\infty$ is the proper domain of its projective realization. Thus a complete l-parabolic submanifold V^m of the space E^n is regular if and only if the real part Re F of the focus variety F of each plane generator E^l of the corresponding submanifold $\widetilde{V}^m \subset P^n$ coincides with the intersection $E^l \bigcap E_\infty$ and constitutes a ρ-fold $(l-1)$-plane where $0 < \rho \leq r$.

If $c > 0$, then V_c^n is the elliptic space S^n, and its proper domain coincides with the entire space P^n. Thus a complete l-parabolic submanifold V^m of the space S^n is regular if and only if the focus variety F of each plane generator E^l of the corresponding submanifold $\widetilde{V}^m$ is pure imaginary.

If $c < 0$, then V_c^n is the hyperbolic space H^n, and the proper domain of its realization lies inside of the absolute of this space. Thus a complete l-parabolic submanifold V^m of the space H^n is regular if and only if the real part Re F of the focus variety F of each plane generator E^l of the corresponding submanifold $\widetilde{V}^m$ lies outside of or on the absolute of this space.

Parabolic surfaces of a three-dimensional space V_c^3 of constant curvature allow an especially simple description. In P^3, to each parabolic surface V^2 there corresponds a torse, each rectilinear generator of which possesses a focus point. The locus of these focus points constitutes an edge of regression of the surface V^2. If $c = 0$, then this edge of regression must belong to the improper plane E_∞, i.e. the edge of regression is a plane curve. But this is possible if and only if the edge of regression degenerates into a point. Therefore, a projective realization of a hyperbolic surface V^2 of a three-dimensional Euclidean space E^3 is a cone with its vertex in the improper plane E_∞. Thus the surface V^2 itself is a cylinder. Hence in the space E^3 there are no other regular parabolic surfaces except the cylinders.

If $c > 0$, i.e. if we have the elliptic space S^3, then there are no regular parabolic surfaces, since the edge of regression of the torse V^2 is always real. Finally, if $c < 0$, i.e. if we have the hyperbolic space H^3, then there are regular parabolic surfaces, since the real edge of regression of the torse V^2 can be located outside of the absolute.

Thus we have proved the following result.

Theorem 4.12 *In the Euclidean space E^3, only cylinders are regular parabolic surfaces. In the space S^3, there are no regular parabolic surfaces at all, and in the space H^3, regular parabolic surfaces exist and depend on of two arbitrary functions of one variable.* ■

The last statement follows from the fact that a torse in P^3 is completely defined by its edge of regression, i.e. by an arbitrary space curve, but these curves are defined by two arbitrary functions of one variable, as indicated in the theorem.

4. We will now consider some examples of regular parabolic submanifolds with a constant index of relative nullity in Riemannian spaces of constant curvature and dimension $n > 3$.

One such example was pointed out by Sacksteder in his paper [Sac 60]. It is the hypersurface $V^3 \subset E^4$, defined by the equation

$$x^4 = x^1 \cos(x^3) + x^2 \sin(x^3). \tag{4.77}$$

This hypersurface is everywhere regular in E^4, and its index of relative nullity $\nu(x) = 1$. As a result, its rank $r = 2$. We will investigate this example from the standpoint of subsection **4.7.3**.

We introduce in the space E^4 homogeneous coordinates $(\overline{x}^0, \overline{x}^1, \overline{x}^2, \overline{x}^3, \overline{x}^4)$ such that $x^\alpha = \frac{\overline{x}^\alpha}{\overline{x}^0}$, $\alpha = 1, 2, 3, 4$, and extend this space to a projective space P^4 by means of the improper hyperplane $\overline{x}^0 = 0$. Denote by $\widetilde{V}^3$ the natural extension of the hypersurface V^3 in the space P^4. The equations for the hypersurface $\widetilde{V}^3$ can be represented in the parametric form:

$$\left\{\begin{array}{l} \overline{x}^0 = s, \\ \overline{x}^1 = -sv \sin u + t \cos u, \\ \overline{x}^2 = sv \cos u + t \sin u, \\ \overline{x}^3 = su, \\ \overline{x}^4 = t. \end{array}\right. \tag{4.78}$$

These equations can be written in the form:

$$x = sA_0 + tA_1,$$

where

$$A_0 = \{1, -v \sin u, v \cos u, u, 0\}, \quad A_1 = \{0, \cos u, \sin u, 0, 1\}$$

are points of the space P^4. The straight lines $A_0 \wedge A_1$ are the generators of the hypersurface $\widetilde{V}^3$. Differentiating the points A_0 and A_1, we obtain

$$dA_0 = A_2 du + A_3 dv, \quad dA_1 = A_3 du,$$

where

$$A_2 = \{0, -v \cos u, -v \sin u, 1, 0\}, \quad A_3 = \{0, -\sin u, \cos u, 0, 0\}.$$

It can be easily verified that the points A_0, A_1, A_2 and A_3 are linearly independent. As a consequence, the tangent hyperplane $T_x = A_0 \wedge A_1 \wedge A_2 \wedge A_3$ remains constant along the straight line $A_0 \wedge A_1$. This hyperplane, like a rectilinear generator of the hypersurface $\widetilde{V}^3$, depends solely on the parameters u and v.

We find the singularities (foci) of a generator $A_0 \wedge A_1$ of the hypersurface $\widetilde{V}^3 \subset P^4$ in the same manner as for the general case in Section **4.2**. A point $x = \overline{x}^0 A_0 + \overline{x}^1 A_1$ is the focus of this generator if $dx \in A_0 \wedge A_1$, whence it follows that, for the focus,

$$\overline{x}^0 (A_2 du + A_3 dv) + \overline{x}^1 A_3 du = 0.$$

Since the points A_2 and A_3 are linearly independent, it follows that

$$\begin{aligned} \overline{x}^0 du &= 0, \\ \overline{x}^1 du + \overline{x}^0 dv &= 0. \end{aligned} \tag{4.79}$$

This system should have a nontrivial solution relative to du and dv, which defines a focal direction on $\widetilde{V}^3$. Consequently,

$$\det \begin{pmatrix} \overline{x}^0 & 0 \\ \overline{x}^1 & \overline{x}^0 \end{pmatrix} = 0$$

and $(\overline{x}^0)^2 = 0$. This means that the point A_1 (the point at infinity of a rectilinear generator of the hypersurface V^3) is the double focus of the line $A_0 \wedge A_1$. It follows from (4.79) that the focal direction defined by the equation $du = 0$ corresponds to this focus. The torses on the hypersurface V^3 are therefore defined by the equation $u = \text{const}$. Each of these torses is a pencil of straight lines which is located in the 2-plane

$$\overline{x}^3 = u\overline{x}^0, \quad \overline{x}^4 = \overline{x}^1 \cos u + \overline{x}^2 \sin u, \tag{4.80}$$

and whose center is the point A_1. The point A_1 describes the curve in E_∞ defined by the equations

$$\overline{x}^0 = 0, \quad \overline{x}^3 = 0, \quad (\overline{x}^1)^2 + (\overline{x}^2)^2 = (\overline{x}^4)^2. \tag{4.81}$$

Besides the point A_1, the plane (4.80) contains the point $A_3 = \frac{\partial A_1}{\partial u}$ and is therefore tangent to curve (4.81).

Thus, the hypersurface V^3 defined by equation (4.77) in the space E^4, has no singularities in the proper domain of this space, since they have "retreated" to the improper hyperplane E_∞ of this space.

The example discussed above can easily be generalized. Let γ be an arbitrary complete smooth curve in the improper plane E_∞ of an Euclidean space E^n. Suppose that this curve is described by the point $A_1 = A_1(u)$. We set $A_3 = \frac{\partial A_1}{\partial u}$, and let $\pi = \pi(u)$ be the smooth family of proper tangent 2-planes of the curve γ. These 2-planes form a complete regular submanifold V_2^3 of rank $r = 2$ with a constant index of relative nullity $\nu(x) = 1$. The proof of this assertion differs little from our above investigation of the structure of hypersurface (4.77) in E^4.

5. In order to construct other examples, in P^n, $n \geq 4$, we will consider a three-dimensional submanifold V_2^3 of rank $r = 2$ with imaginary focus variety F. Equations (4.4), (4.12) and (4.5) defining this submanifold in P^n take the form

$$\omega_0^\alpha = \omega_1^\alpha = 0, \quad \alpha = 4, \ldots, n, \tag{4.82}$$

$$\omega_1^p = c_q^p \omega_0^q, \quad \omega_p^\alpha = b_{pq}^\alpha \omega_0^q, \quad p, q = 2, 3, \tag{4.83}$$

while equation (4.19) defining the foci on the generator $A_0 \wedge A_1$ of this submanifold, is written as

$$\det \begin{pmatrix} x^0 + x^1 c_2^2 & x^1 c_3^2 \\ x^1 c_2^3 & x^0 + x^1 c_3^3 \end{pmatrix} = 0.$$

Setting $\frac{x^0}{x^1} = -\lambda$, we reduce this equation to the form

$$\lambda^2 - (c_2^2 + c_3^3)\lambda + (c_2^2 c_3^3 - c_3^2 c_2^3) = 0.$$

Since the focus surface F is assumed to be imaginary, this equation has complex-conjugate roots $\lambda = c_2 \pm i c_3$, where $c_3 \neq 0$. As a result, a real transformation converts the matrix $A = (c_q^p)$ to the form

$$C = \begin{pmatrix} c_2 & c_3 \\ -c_3 & c_2 \end{pmatrix}.$$

Substituting these values for the components of the matrix A into equations (4.10), and taking into account that $c_3 \neq 0$, we find that

$$b_{22}^\alpha + b_{33}^\alpha = 0.$$

In view of this, the symmetric matrices B^α can be written in the form

$$B^\alpha = \begin{pmatrix} b_2^\alpha & b_3^\alpha \\ b_3^\alpha & -b_2^\alpha \end{pmatrix}.$$

Equations (4.83) then assume the form

$$\begin{cases} \omega_1^2 = c_2 \omega_0^2 + c_3 \omega_0^3, \\ \omega_1^3 = -c_3 \omega_0^2 + c_2 \omega_0^3, \end{cases} \tag{4.84}$$

$$\begin{cases} \omega_2^\alpha = b_2^\alpha \omega_0^2 + b_3^\alpha \omega_0^3, \\ \omega_3^\alpha = b_3^\alpha \omega_0^2 - b_2^\alpha \omega_0^3. \end{cases} \tag{4.85}$$

We will now find the osculating subspace $T_x^{(2)}$ of our submanifold $V^3 \subset P^n$. Its tangent subspace $T_x^{(1)}$ is spanned by the points A_0, A_1, A_2 and A_3. Since by (4.85),

$$dA_2 \equiv (b_2^\alpha \omega_0^2 + b_3^\alpha \omega_0^3) A_\alpha \pmod{T_x^{(1)}},$$
$$dA_3 \equiv (b_3^\alpha \omega_0^2 - b_2^\alpha \omega_0^3) A_\alpha \pmod{T_x^{(1)}},$$

the subspace $T_x^{(2)}$ comprises the linear hull of the subspace $T_x^{(1)}$ and the points $B_2 = b_2^\alpha A_\alpha$ and $B_3 = b_3^\alpha A_\alpha$.

Two cases are possible:

(a) The points B_2 and B_3 are linearly independent. Then $\dim T_x^{(1)} = 5$, and the dimension of the space $n \geq 5$.

(b) The points B_2 and B_3 are linearly dependent. Then $\dim T_x^{(1)} = 4$, and $n \geq 4$.

We will examine these two cases in turn. In case (a), we specialize the moving frames in P^n in such fashion that $A_4 = B_2$ and $A_5 = B_3$. Then equations (4.85) take the form

$$\begin{aligned} \omega_2^4 &= \omega_0^2, & \omega_2^5 &= \omega_0^3, & \omega_2^\lambda &= 0, \\ \omega_3^4 &= -\omega_0^3, & \omega_3^5 &= \omega_0^2, & \omega_3^\lambda &= 0, \end{aligned} \tag{4.86}$$

where $\lambda = 6, \ldots, n$. Therefore the submanifold V^3 in case (a) is determined by the system of Pfaffian equations (4.82), (4.84) and (4.86). We will investigate the consistency of this system by means of the Cartan test (see Section **1.2**). For this purpose we adjoin to the above Pfaffian equations the exterior quadratic equations obtained as the result of exterior differentiation of these Pfaffian equations. Exterior differentiation of the equations (4.82) leads to identities, by virtue of (4.84) and (4.86). Exterior differentiation of the equations (4.84) yields

$$\begin{aligned} &(\Delta c_2 - c_3(\omega_2^3 + \omega_3^2)) \wedge \omega_0^2 + (\Delta c_3 + c_3(\omega_2^2 - \omega_3^3)) \wedge \omega_0^3 &= 0, \\ - &(\Delta c_3 - c_3(\omega_2^2 - \omega_3^3)) \wedge \omega_0^2 + (\Delta c_2 + c_2(\omega_2^3 - \omega_3^2)) \wedge \omega_0^3 &= 0, \end{aligned} \tag{4.87}$$

where

$$\begin{aligned} \Delta c_2 &= dc_2 + c_2(\omega_0^0 - \omega_1^1) - \omega_1^0 + ((c_2)^2 - (c_3)^2)\omega_0^1, \\ \Delta c_3 &= dc_3 + c_3(\omega_0^0 - \omega_1^1) + 2c_2c_3\omega_0^1. \end{aligned}$$

Exterior differentiation of the equations (4.86) gives

$$\begin{aligned} &(\omega_0^0 + \omega_4^4 + c_2\omega_0^1 - 2\omega_2^2) \wedge \omega_0^2 + (\omega_2^3 - \omega_3^2 + \omega_5^4 + c_3\omega_0^1) \wedge \omega_0^3 = 0, \\ &(\omega_2^3 - \omega_3^2 + \omega_5^4 + c_3\omega_0^1) \wedge \omega_0^2 - (\omega_0^0 + \omega_4^4 + c_2\omega_0^1 - 2\omega_2^2) \wedge \omega_0^3 = 0, \\ &(\omega_4^5 - c_3\omega_0^1 - 2\omega_2^3) \wedge \omega_0^2 + (\omega_0^0 - \omega_2^2 - \omega_3^3 + \omega_5^5 + c_2\omega_0^1) \wedge \omega_0^3 = 0, \\ &(\omega_0^0 - \omega_2^2 - \omega_3^3 + \omega_5^5 + c_2\omega_0^1) \wedge \omega_0^2 - (\omega_4^5 - c_3\omega_0^1 - 2\omega_3^2) \wedge \omega_0^3 = 0, \\ &\omega_4^\lambda \wedge \omega_0^2 + \omega_5^\lambda \wedge \omega_0^3 = 0, \\ &\omega_5^\lambda \wedge \omega_0^2 - \omega_4^\lambda \wedge \omega_0^3 = 0, \end{aligned} \tag{4.88}$$

where $\lambda = 6, \ldots, n$. The system (4.87)–(4.88) contains $s_1 = 2n - 4$ independent equations that include the following independent characteristic forms:

$$\begin{aligned} &\Delta c_2, \Delta c_3, \omega_2^3 + \omega_3^2, \omega_2^2 - \omega_3^3, \\ &\omega_0^0 + \omega_4^4 + c_2\omega_0^1 - 2\omega_2^2, \omega_2^3 - \omega_3^2 + \omega_5^4 + c_3\omega_0^1, \\ &\omega_4^5 - c_3\omega_0^1 - 2\omega_2^3, \omega_0^0 - \omega_2^2 - \omega_3^3 + \omega_5^5 + c_2\omega_0^1, \\ &\omega_4^\lambda, \omega_5^\lambda, \quad \lambda = 6, \ldots, n. \end{aligned}$$

Their number is $q = 2n - 2$. The second character of the system is therefore $s_2 = q - s_1 = 2$, and the Cartan number $Q = s_1 + 2s_2 = 2n$. The number N of parameters on which the most general integral element depends is computed from the formula $N = 2q - s_1 = 2n$. Since $Q = N$, by the Cartan test, the system of Pfaffian equations (4.82), (4.84) and (4.86) is in involution, and its general integral manifold depends on two arbitrary functions of two variables.

In case (b), we have $b_2^\alpha = b_2 b^\alpha$ and $b_3^\alpha = b_3 b^\alpha$. Equations (4.85) can therefore be written in the form

$$\begin{aligned} \omega_2^\alpha &= b^\alpha(b_2\omega_0^2 + b_3\omega_0^3), \\ \omega_3^\alpha &= b^\alpha(b_3\omega_0^2 - b_2\omega_0^3). \end{aligned}$$

Consequently,

$$\begin{aligned} dA_2 &\equiv (b_2\omega_0^2 + b_3\omega_0^3)B \pmod{T_x^{(1)}}, \\ dA_3 &\equiv (b_3\omega_0^2 - b_2\omega_0^3)B \pmod{T_x^{(1)}}, \end{aligned}$$

where $B = b^\alpha A_\alpha$. We specialize our moving frame assuming $A_4 = B$. Then equations (4.85) take the form

$$\omega_2^4 = b_2\omega_0^2 + b_3\omega_0^3, \quad \omega_3^4 = b_3\omega_0^2 - b_2\omega_0^3, \tag{4.89}$$

$$\omega_2^\lambda = 0, \quad \omega_3^\lambda = 0, \tag{4.90}$$

where $\lambda = 5, \ldots, n$. Exterior differentiation of the last two equations gives the following quadratic equations:

$$\omega_2^4 \wedge \omega_4^\lambda = 0, \quad \omega_3^4 \wedge \omega_4^\lambda = 0.$$

Since by (4.89), 1-forms ω_2^4 and ω_3^4 are linearly independent, it follows from the last equations that

$$\omega_4^\lambda = 0, \quad \lambda = 5, \ldots, n.$$

This means that the submanifold V^3 belongs to the four-dimensional space P^4 spanned by the points A_0, A_1, A_2, A_3 and A_4. In case (b), the submanifold V^3 is thus a hypersurface in the space P^4, being defined in this space by the system of equations (4.82) (with $\alpha = 4$), (4.84) and (4.89).

We will investigate the consistency of the last system. For this purpose we apply exterior differentiation to equations (4.89). As a result, we obtain the following quadratic equations:

$$\begin{aligned} &(\Delta b_2 - 2(b_2\omega_2^2 + b_3\omega_2^3)) \wedge \omega_0^2 + \Delta b_3 \wedge \omega_0^3 = 0, \\ &\Delta b_3 \wedge \omega_0^2 - (\Delta b_2 - 2(b_2\omega_3^3 - b_3\omega_3^2)) \wedge \omega_0^3 = 0, \end{aligned} \tag{4.91}$$

where

$$\begin{aligned} \Delta b_2 &= db_2 + b_2(\omega_0^0 + \omega_4^4) + (c_2 b_2 - c_3 b_3)\omega_0^1, \\ \Delta b_3 &= db_3 + b_3(\omega_0^0 - \omega_2^2 - \omega_3^3 + \omega_4^4) + b_2(\omega_2^3 - \omega_3^2) + (c_2 b_3 + c_3 b_2)\omega_0^1. \end{aligned}$$

The system of exterior equations (4.87) and (4.91) consists of $s_1 = 4$ independent equations. They contain $q = 6$ characteristic forms. As a result, the character $s_2 = q - s_1 = 2$, and the Cartan number $Q = s_1 + 2s_2 = 8$. The number N of parameters, on which the most general integral element depends, is also equal to 8. Since $Q = N$, by the Cartan test, the system of Pfaffian equations (4.82), (4.84) and (4.89) is in involution, and its general integral manifold depends on two arbitrary functions of two variables.

Thus, three-dimensional parabolic submanifolds V_2^3 of rank 2 in P^n that have no real singularities exist, in both cases (a) and (b), and a general integral manifold, defining such submanifolds, depends on two arbitrary functions of two variables.

Now let a simply connected space V_c^n of constant curvature be realized in a projective space P^n and let G be its proper domain. If V_2^3 is a three-dimensional parabolic submanifold of rank 2 in P^n that has no real singularities, then the intersection $V^3 \bigcap G$ is a submanifold having the same properties in V_c^n. Such submanifolds consequently also exist in V_c^n, and a general integral manifold of the system, defining such submanifolds, depends on two arbitrary functions of two variables.

NOTES

4.1–4.4. Tangentially degenerate submanifolds V^m of rank $r < n$ were considered by É. Cartan in [Ca 16] in connection with his study of metric deformation of hypersurfaces, and in [Ca 19] in connection with his study of manifolds of constant curvature; by Yanenko in [Ya 53] in connection with his study of metric deformation of submanifolds of arbitrary class; by Akivis in [A 57] and [A 62b]; by Savelyev in [Sav 57] and [Sav 60]; and by Ryzhkov in [Ry 60] (see also the survey paper [AR 64]). Brauner in [Br 38] studied such submanifolds in Euclidean n-space.

Most of the results presented in these sections are due to Akivis (see [A 57]).

4.5. The results presented in this section are due to Akivis (see [A 62b]).

4.6. The results presented in this section are due to Safaryan (see [Sa 70]).

4.7. The notion of the index of relative nullity was introduced by Chern and Kuiper in [CK 52] (see also [KN 63], vol. 2, p. 348). Complete parabolic submanifolds in an Euclidean n-dimensional space were studied by Borisenko in [Bori 82] and [Bori 85]. In [Bori 92], he used the notion of parabolicity to formulate and prove a theorem on the unique determination of $V^m \subset E^n$ from its Grassmann image.

Akivis recognized that the problem of finding singularities on complete parabolic submanifolds in a Riemannian space V_c^n of constant curvature and of distinguishing those submanifolds that have no singularities, is related not so much to the metric as to the projective structure of the spaces V_c^n. In this section we follow Akivis' paper [A 87b] written on this subject.

The hypersurface defined by equations (4.78) was studied by Sacksteder in [Sac 60].

Chapter 5

Submanifolds with Asymptotic and Conjugate Distributions

5.1 Distributions on Submanifolds of a Projective Space

Consider a submanifold V^m in a projective space P^n. If a subspace Δ_x^p of dimension p is given in each tangent subspace $T_x^{(1)}(V^m)$, we say that a *distribution* Δ^p is defined on V^m (cf. Section **3.2** where distributions were already discussed). As always, we assume that the subspaces Δ_x^p depend differentiably on the point x.

With this distribution we associate a bundle of first order moving frames, which is a subbundle of the bundle of first order moving frames associated with the submanifold V^m: the points A_b, $b = 1, \ldots, p$, of moving frames of this subbundle are in the subspace Δ_x^p passing through the point $x = A_0$.

If at a point x a vector $\xi \in \Delta_x^p$ is given, then relative to the above moving frame associated with the point x we have the following equations:

$$\omega^u(\xi) = 0, \quad u = p+1, \ldots, m.$$

Thus, the equations $\omega^u = 0$ are the equations of the subspace Δ_x^p at the point x.

Since the subspace Δ_x^p is constant if the point x is fixed, and since the point x is fixed if the equations $\omega^i = 0$, $i = 1, \ldots, m$, hold, then under this condition the equations of infinitesimal displacements of the points A_0 and A_b of the subspace Δ_x^p have the form:

$$\delta A_0 = \pi_0^0 A_0, \quad \delta A_b = \pi_b^0 A_0 + \pi_b^c A_c. \tag{5.1}$$

Hence the forms ω_a^u, $u = p+1, \ldots, m$, vanish if $\omega^i = 0$, $i = 1, \ldots, m$. Therefore the forms ω_a^u are linear combinations of the forms ω^i:

$$\omega_a^u = l_{ai}^u \omega^i. \tag{5.2}$$

Equations (5.2) are the equations of the distribution Δ^p in the chosen subbundle of frames of first order.

A distribution Δ^p is said to be *integrable* or *holonomic* (cf. subsection **1.2.4**) if there exists a family of submanifolds V^p of dimension p such that

a) At any point $x \in V^p$ the tangent subspace $T_x^1(V^p) = \Delta_x^p$, and

b) Through any point $x \in V^m$ there passes a unique submanifold V^p.

The submanifolds V^p are called *integral submanifolds* of the distribution Δ^p. An integrable distribution Δ^p defines a foliation of the submanifold V^m into an $(m-p)$-parameter family of submanifolds V^p.

Let us find the conditions of integrability of a distribution Δ^p. By the Frobenius theorem, this condition is that the exterior differentials $d\omega^u$ of the forms ω^u must vanish modulo these forms, i.e. the following relation must hold:

$$d\omega^u \equiv 0 \pmod{\omega^u}. \tag{5.3}$$

Since

$$d\omega^u = \omega_0^0 \wedge \omega^u + \omega^a \wedge \omega_a^u + \omega^v \wedge \omega_v^u,$$

then, by (5.2), we have:

$$d\omega^u = \omega_0^0 \wedge \omega^u + \omega^v \wedge \omega_v^u + l_{ab}^u \omega^a \wedge \omega^b + l_{av}^u \omega^a \wedge \omega^v. \tag{5.4}$$

Thus, by (5.3), for a distribution Δ^p to be integrable, the term in equations (5.4) which does not contain the forms ω^v must vanish, i.e. we must have:

$$l_{[ab]}^u = 0. \tag{5.5}$$

We have arrived at the following result.

Theorem 5.1 *A distribution Δ^p is integrable if and only if condition* (5.5) *holds.* ■

The *integral curves* of a distribution Δ^p are the curves that are tangent to the tangent subspaces $\Delta_x^p \subset \Delta^p$ at each of points $x \in V^m$. All such curves satisfy the equation

$$\omega^u = 0.$$

If a distribution Δ^p is integrable, then its integral curves belong to its integral submanifolds V^p. Note that integral curves of a distribution exist independently on the fact whether the distribution Δ^p is integrable or not.

5.2 Asymptotic Distributions on Submanifolds

1. In Chapter 2 the asymptotic directions on a submanifold V^m were defined as directions annihilating the second fundamental form $\Phi_{(2)}$:

$$\Phi_{(2)}(\xi) = 0.$$

Since $\Phi_{(2)} = b^\alpha_{ij}\omega^i\omega^j \widetilde{A}_\alpha$, the asymptotic direction ξ satisfies the following system of equations:

$$b^\alpha_{ij}\xi^i\xi^j = 0. \tag{5.6}$$

Suppose that the tangent subspace $T^{(1)}_x(V^m)$ has a subspace $\Delta^p_x \subset T^{(1)}_x(V^m)$ with each of its directions asymptotic. We will call such subspace Δ^p_x *asymptotic*. If there are asymptotic subspaces Δ^p_x at any point $x \in V^m$, then we say that the submanifold V^m carries an *asymptotic distribution* Δ^p of p dimensions.

Integral curves of an asymptotic distribution are asymptotic lines of the submanifold V^m. The osculating 2-planes of these curves belong to the tangent subspace $T^{(1)}_x(V^m)$.

We associate a bundle of frames to a submanifold V^m carrying an asymptotic distribution in the same way as in Section **5.1**. Then for any direction $\xi \subset \Delta^p_x$ we have

$$\Phi_{(2)}(\xi) = b^\alpha_{ab}\xi^a\xi^b = 0. \tag{5.7}$$

These equations give the equations:

$$b^\alpha_{ab} = 0. \tag{5.8}$$

This implies that $\widetilde{B}_{au}$ and $\widetilde{B}_{uv}$ are the only nonvanishing points among the points $\widetilde{B}_{ij} = b^\alpha_{ij}\widetilde{A}_\alpha$ defining the normal subspace $N^{(1)}_x(V^m)$. The number of these nonvanishing points is

$$p(m-p) + \frac{1}{2}(m-p)(m-p+1) = \frac{1}{2}(m-p)(m+p+1).$$

It follows that for submanifolds V^m carrying an asymptotic distribution Δ^p, the dimension $m + m_1$ of the osculating subspace $T^{(2)}_x$ satisfies the inequality:

$$m + m_1 \leq m + \frac{1}{2}(m-p)(m+p+1).$$

By (5.8), equations (2.21) take the form:

$$\omega^\alpha_a = b^\alpha_{au}\omega^u, \quad \omega^\alpha_u = b^\alpha_{ua}\omega^a + b^\alpha_{uv}\omega^v. \tag{5.9}$$

Taking exterior derivatives of the first group of these equations and using equations (5.9) and (5.2), we obtain the following exterior quadratic equations:

$$\begin{aligned}&\{\nabla b^\alpha_{au} + (-l^w_{ab} b^\alpha_{wu} + l^w_{au} b^\alpha_{wb} + l^w_{bu} b^\alpha_{aw})\omega^b - l^w_{av} b^\alpha_{wu}\omega^v\} \wedge \omega^u \\ &+(-l^u_{ab} b^\alpha_{uc} + l^u_{bc} b^\alpha_{au})\omega^b \wedge \omega^c = 0,\end{aligned} \tag{5.10}$$

where

$$\nabla b^\alpha_{au} = db^\alpha_{au} + b^\alpha_{au}\omega^0_0 - b^\alpha_{av}\omega^v_u - b^\alpha_{bu}\omega^b_a + b^\beta_{au}\omega^\alpha_\beta.$$

It follows that the last term in (5.10) must vanish. Thus,

$$b^\alpha_{au}(l^u_{bc} - l^u_{cb}) - b^\alpha_{cu} l^u_{ab} + b^\alpha_{bu} l^u_{ac} = 0. \tag{5.11}$$

Equations (5.11) will play an important role in our further considerations. Contracting these equations with the points $\widetilde{A}_\alpha \in N^{(1)}_x$, we find that

$$\widetilde{B}_{au}(l^u_{bc} - l^u_{cb}) - \widetilde{B}_{cu} l^u_{ab} + \widetilde{B}_{bu} l^u_{ac} = 0. \tag{5.12}$$

Equations (5.11) and (5.12) are similar to equations (3.24) and (4.43).

Following Lumiste (see [Lu 59]), we will prove the following result.

Theorem 5.2 *Let a submanifold V^m of a projective space P^n carry an asymptotic distribution Δ^p, $p > 1$, and the bundle of moving frames be chosen as above. If the points $\widetilde{B}_{au}$ and $\widetilde{B}_{bu}$ are linearly independent for any fixed a and b, $a \neq b$, then the asymptotic distribution Δ^p on V^m is integrable, and its integral manifolds are p-planes.*

Proof. Setting $c = a$ in relations (5.12), we obtain

$$\widetilde{B}_{au}(l^u_{ba} - 2l^u_{ab}) + \widetilde{B}_{bu} l^u_{aa} = 0. \tag{5.13}$$

By the linear independence of the points $\widetilde{B}_{au}$ and $\widetilde{B}_{bu}$, equations (5.13) imply that

$$l^u_{ba} - 2l^u_{ab} = 0, \;\; l^u_{aa} = 0. \tag{5.14}$$

Interchanging a and b in the first of these equations, we find that

$$l^u_{ab} - 2l^u_{ba} = 0. \tag{5.15}$$

Thus, it follows from equations (5.14) and (5.15) that

$$l^u_{ab} = 0. \tag{5.16}$$

By Theorem 5.1, it follows from equation (5.16) that the distribution Δ^p is integrable.

Further, by (5.16), we have

$$\omega^u_a = l^u_{av}\omega^v.$$

This and equations (5.9) give the following equations:

$$dA_0 \equiv \omega_0^0 A_0 + \omega^b A_b \pmod{\omega^u}, \quad dA_b \equiv \omega_b^0 A_0 + \omega_b^c A_c \pmod{\omega^u}. \tag{5.17}$$

Equations (5.17) mean that if $\omega^u = 0$, the point A_0 describes a p-dimensional plane. Thus, the submanifold V^m is foliated by an $(m-p)$-parameter family of p-dimensional planes. ■

2. In the second osculating subspace $T_x^{(2)}(V^m)$, we now consider the subspace E_x defined by the points A_0, A_i and B_{au}, where $B_{au} = b_{au}^\alpha A_\alpha$. Let us prove that this subspace is invariantly connected with the asymptotic subspace Δ_x^p. In fact, it follows from equations (5.10) that if the point x is fixed (i.e. $\omega^i = 0$), we have

$$\nabla_\delta b_{au}^\alpha = \delta b_{au}^\alpha + b_{au}^\alpha \pi_0^0 - b_{av}^\alpha \pi_u^v - b_{bu}^\alpha \pi_a^b + b_{au}^\beta \pi_\beta^\alpha = 0, \tag{5.18}$$

where, as above, the symbol δ denotes differentiation with respect to secondary parameters, i. e. δ is the restriction of the differential d to a fiber of the frame bundle associated with the submanifold V^m. We set $\pi_v^u = \omega_v^u(\delta)$.

For the points $\widetilde{B}_{au} = b_{au}^\alpha \widetilde{A}_\alpha$, we get from (5.18) and the obvious equation $\nabla_\delta \widetilde{A}_\alpha = 0$, we have:

$$\nabla_\delta \widetilde{B}_{au} = \nabla_\delta b_{au}^\alpha \widetilde{A}_\alpha + b_{au}^\alpha \nabla_\delta \widetilde{A}_\alpha = 0. \tag{5.19}$$

Conditions (5.19) mean that the points $\widetilde{B}_{au}$ determine an invariant subspace $\widetilde{E}_x$ in the normal subspace $N_x^{(1)}$. Together with the tangent subspace $T_x^{(1)}$, the subspace $\widetilde{E}_x$ determines an invariant subspace E_x in the second osculating subspace $T_x^{(2)}(V^m)$. It is easy to see that the dimension of the subspace E_x does not exceed $m + p(n-p)$.

The hypotheses of Theorem 5.2 are equivalent to the condition that *the $(m-p-1)$-dimensional subspaces $\widetilde{E}_a \subset \widetilde{E}_x$ determined by the points $\widetilde{B}_{au}$, where a is fixed, are mutually in general position in $\widetilde{E}_x$*. Since $\dim \widetilde{E}_a = m+p-1$, for this conditions to be satisfied, it is necessary that the dimension of the subspace $\widetilde{E}_x$ be not less than $2(m-p-1)+1$. In particular, if the dimension of the subspace $\widetilde{E}_x$ is equal to $p(m-p)-1$, the hypotheses of Theorem 5.2 will be automatically satisfied.

If $p = m-1$, the subspaces $\widetilde{E}_a$ are reduced to the points $\widetilde{B}_{am}$. If these points lie on a straight line but mutually are linearly independent, then the hypotheses of Theorem 5.2 will hold. Thus, *if a submanifold V^m carries an $(m-1)$-dimensional distribution Δ^{m-1} of asymptotic directions, the subspaces E_x are of dimension greater than or equal to $m+2$, and the subspaces E_a are mutually distinct, then the submanifold V^m is foliated by a one-parameter family of $(m-1)$-dimensional planes.*

Let us establish a geometric meaning for the subspace E_x for a submanifold V^m which is foliated by an $(m-p)$-parameter family of p-planes Δ_x^p.

Take an arbitrary point $y \in \Delta_x^p$:

$$y = y^0 A_0 + y^a A_a. \tag{5.20}$$

If we differentiate the point y and projectivize the result by Δ_x^p, then by (5.2), (5.9) and (5.16), we find that

$$dy/\Delta_x^p = \big((y^0 \delta_v^u + y^a l_{av}^u)A_u + y^a B_{av}\big)\omega^v. \tag{5.21}$$

Thus, the tangent subspace to the submanifold V^m at a point $y \in E^p$ is determined by the subspace Δ_x^p and the points

$$C_v(y) = (y^0 \delta_v^u + y^a l_{av}^u)A_u + y^a B_{av}. \tag{5.22}$$

We can see from this that the linear span of all such tangent subspaces is defined by the points A_0, A_a, A_u and B_{au}, and thus coincides with the subspace E_x. Moreover, in fact the subspace E_x depends not on the point x but on the plane generator Δ_x^p of the submanifold V^m. Therefore, the subspace E_x depends on the same $m-p$ parameters on which the generator Δ_x^p depends.

5.3 Submanifolds with a Complete System of Asymptotic Distributions

1. Suppose that the submanifold V^m carries a system of asymptotic distributions $\Delta^{p_1}, \ldots, \Delta^{p_k}$. This system is called *complete* if any pair of subspaces $\Delta_x^{p_\lambda}$ and $\Delta_x^{p_\mu}$, $\lambda, \mu = 1, \ldots, k, \lambda \neq \mu$, are in general position, i.e. $\Delta_x^{p_\lambda} \cap \Delta_x^{p_\mu} = x$, and $p_1 + \ldots + p_k = m$.

For simplicity we consider the case $k = 2$. We will now prove the following result.

Theorem 5.3 *If a submanifold V^m carries a complete two-component system of asymptotic distributions Δ^p and Δ^{m-p}, then this submanifold is contained in its second osculating subspace $T_x^{(2)}$ of dimension $m+m_1$ where $m_1 \leq p(m-p)$.*

Proof. For the submanifold V^m in question, the second fundamental tensor (b_{ij}^α) has the form:

$$\begin{pmatrix} 0 & b_{au}^\alpha \\ b_{ua}^\alpha & 0 \end{pmatrix}.$$

Thus the second osculating subspace $T_x^{(2)}(V^m)$ is determined by the points A_0, A_i and $B_{ua} = b_{ua}^\alpha A_\alpha$. Therefore, the dimension $m + m_1$ of this subspace is not greater than $m + p(m-p)$.

We choose a moving frame in the same way as in Section **2.3**, by taking the points $A_{m+1}, \ldots, A_{m+m_1}$ in the subspace $T_x^{(2)}$. Then

$$b_{ua}^{\alpha_1} = 0, \quad \alpha_1 = m + m_1 + 1, \ldots, n.$$

As in Section **2.3**, we can prove that, as a result of the choice of moving frames indicated above, the forms $\omega_{i_1}^{\alpha_1}$ are expressed in terms of the basis forms (2.41):

$$\omega_{i_1}^{\alpha_1} = c_{i_1 k}^{\alpha_1} \omega^k, \tag{5.23}$$

where the coefficients $c_{i_1 k}^{\alpha_1}$ are connected with the second fundamental tensor by equation (2.43). Since we now have

$$b_{ab}^{i_1} = 0, \quad b_{uv}^{i_1} = 0,$$

equations (2.43) take the form:

$$b_{au}^{i_1} c_{i_1 b}^{\alpha_1} = 0, \quad b_{au}^{i_1} c_{i_1 v}^{\alpha_1} = 0. \tag{5.24}$$

Since the rank of the $m_1 \times p(m-p)$ matrix $(b_{au}^{i_1})$ is equal to m_1, equations (5.24) imply

$$c_{i_1 b}^{\alpha_1} = 0, \quad c_{i_1 v}^{\alpha_1} = 0.$$

By (5.23), these equations give

$$\omega_{i_1}^{\alpha_1} = 0. \tag{5.25}$$

Using equations (5.25), we obtain

$$dA_{i_1} = \omega_{i_1}^0 A_0 + \omega_{i_1}^j A_j + \omega_{i_1}^{j_1} A_{j_1}.$$

It follows that the second osculating subspace defined by the points A_0, A_i and A_{i_1} is constant, and the submanifold V^m lies completely in this subspace. ■

Theorem 5.1 also implies the following result.

Theorem 5.4 *Let a submanifold V^m carry a complete two-component system of asymptotic distributions Δ^p and Δ^{m-p}, where $p > 1$ and $m - p > 1$. If in addition the following inequality holds:*

$$m_1 \geq 2 \max\{p, m-p\},$$

and each of the systems of points, $\{\widetilde{B}_{au}, \widetilde{B}_{bu}\}$ where a and b are fixed and $a \neq b$ and $\{\widetilde{B}_{au}, \widetilde{B}_{av}\}$ where u and v are fixed and $u \neq v$, are linearly independent, then the submanifold V^m is doubly foliated by families of plane generators of dimensions p and $m - p$. ■

The hypotheses of Theorem 5.4 are automatically satisfied if $m_1 = p(m-p)$ and $p > 1, m - p > 1$. Thus a submanifold V^m which carries a complete two-component system of asymptotic distributions and whose second osculating subspace $T_x^{(2)}$ is of maximum dimension, is foliated by two families of plane generators of dimensions p and $m - p$.

An example of a submanifold V^m, which has a double foliation by families of plane generators of dimensions p and $m-p$ and for which $m_1 = p(m-p)$, is the Segre variety $S(p, m-p)$. This variety is embedded into a projective space P^n of dimension $n = m+p(m-p)$, which coincides with the second osculating subspace $T_x^{(2)}$ of $S(p, m-p)$ (see Section **2.2**, cf. [Bert 24]). If $m_1 < p(m-p)$, the submanifolds in question can be obtained by projecting a Segre variety $S(p, m-p)$ into a projective space P^n of dimension $n < m+p(m-p)$ (cf. [M 58]).

We will now prove that these examples exhaust all those submanifolds V^m, which are doubly foliated by families of plane generators. In fact, let us write the equations of such a submanifold in the parametric form:

$$x = x(\lambda^a, \lambda^u), \quad a = 1, \ldots, p, \quad u = p+1, \ldots, m,$$

where λ^a, λ^u are coordinates on V^m chosen in such a way that if $\lambda^u = \text{const}$, the point x describes a p-dimensional plane generator Δ_x^p, and if $\lambda^a = \text{const}$, this point describes a q-dimensional plane generator Δ_x^q, where $q = m - p$. If we denote by A_0 and A_a the basis points of the generator Δ_x^p, and by B_0 and B_u the basis points of the generator Δ_x^q of the submanifold V^m, then the equation of this submanifold can be written in the form

$$x = A_0 + \lambda^u A_a,$$

where the points A_0 and A_a depend on λ^u, or in the form

$$x = B_0 + \lambda^u B_u,$$

where the points B_0 and B_u depend on λ^a. But the last two equations imply that the equation of the submanifold V^m has the form

$$x = C + C_a\lambda^a + C_u\lambda^u + C_{au}\lambda^a\lambda^u, \tag{5.26}$$

where C, C_a, C_u and C_{au} are fixed points of the space P^n. Introduce homogeneous coordinates on the generators Δ_x^p and Δ_x^q by setting $\lambda^a = \frac{x^a}{x^0}$ and $\lambda^u = \frac{x^u}{x^0}$, respectively. Then equation (5.26) takes the form

$$x = C_{\hat{a}\hat{u}} x^{\hat{a}} y^{\hat{u}},$$

where the indices $\hat{a}$ and $\hat{u}$ take the values from 0 to p and from 0 to q, respectively. If the points $C_{\hat{a}\hat{u}}$ are linearly independent, then the submanifold V^m is the Segre variety $S(p, q) = P^p \times P^q$ imbedded into a projective space P^n of dimension $n = p + q + pq = m + p(m-p)$. On the other hand, if the points $C_{\hat{a}\hat{u}}$ are linearly dependent and the maximum number of linearly independent among them is equal to $n+1$, where $m < n < m + p(m-p)$, then the submanifold V^m is a projection of the Segre variety $S(p, q)$ onto a projective space P^n.

Multicomponent systems of asymptotic distributions can be investigated in a manner similar to our considerations for complete two-component systems of asymptotic distributions, (see [Lu 59]). In particular, Theorem 5.3 can be generalized as follows.

Theorem 5.5 *If a submanifold V^m carries a complete k-component system of asymptotic distributions $\Delta^{p_1}, \ldots, \Delta^{p_k}$, then this submanifold is contained in its osculating subspace $T_x^{(k)}$ of order k, whose dimension satisfies the following inequality:*

$$\dim T_x^{(k)} \leq \sum_{\kappa=1}^{k} p_{\lambda_1} p_{\lambda_2} \cdots p_{\lambda_\kappa},$$

where summation is taken over the combinations $(\lambda_1, \ldots, \lambda_\kappa)$ of numbers $1, 2, \ldots, \kappa$, i.e.

$$\dim T_x^{(k)} \leq p_1 + \ldots + p_k + p_1 p_2 + \ldots + p_{k-1} p_k + \ldots + p_1 p_2 \cdots p_k. \blacksquare$$

5.4 Three-Dimensional Submanifolds Carrying a Net of Asymptotic Lines

1. Consider a three-dimensional submanifold V^3 in a projective space P^n which carries a complete system of one-dimensional asymptotic distributions. Since such distributions are always integrable, their integral curves form a net of asymptotic lines on V^3. Such submanifolds were briefly considered by Cartan in [Ca 18]. In this section we will study this kind of submanifold in more detail.

We place the vertices A_i of a moving frame associated with a point $x \in V^3$ onto the tangents to the asymptotic lines. Then the second fundamental forms $\Phi^\alpha_{(2)}$ of V^3 take the form:

$$\Phi^\alpha_{(2)} = b^\alpha_{12}\omega^1\omega^2 + b^\alpha_{23}\omega^2\omega^3 + b^\alpha_{31}\omega^3\omega^1. \tag{5.27}$$

Equations (5.27) imply that the number m_1 of linearly independent forms $\Phi^\alpha_{(2)}$ does not exceed three. We will consider the case $m_1 = 3$, since other cases were already considered in Section **2.5**.

So, let $m_1 = 3$. We can choose the points A_α, $\alpha = 4, \ldots, n$, so that the forms $\Phi^\alpha_{(2)}$ take the form:

$$\Phi^4_{(2)} = 2\omega^2\omega^3, \;\; \Phi^5_{(2)} = 2\omega^3\omega^1, \;\; \Phi^6_{(2)} = 2\omega^1\omega^2, \;\; \Phi^{\alpha_1}_{(2)} = 0, \tag{5.28}$$

where $\alpha_1 = 7, \ldots, n$. The forms $\Phi^4_{(2)}, \Phi^5_{(2)}$ and $\Phi^6_{(2)}$ constitute a basis of the bundle of second fundamental forms associated with the point $x \in V^3$. Relations (5.28) imply that equations (2.41) and (2.42) are reduced to the form:

$$\begin{array}{lll} \omega^4_1 = 0, & \omega^4_2 = \omega^3, & \omega^4_3 = \omega^2, \\ \omega^5_1 = \omega^3, & \omega^5_2 = 0, & \omega^5_3 = \omega^1, \\ \omega^6_1 = \omega^2, & \omega^6_2 = \omega^1, & \omega^6_3 = 0, \end{array} \tag{5.29}$$

$$\omega^{\alpha_1}_i = 0, \;\; \alpha_1 = 7, \ldots, n. \tag{5.30}$$

Exterior differentiation of these equations, and application of Cartan's lemma to the obtained exterior quadratic equations, lead to equation (2.50):

$$\omega_{i_1}^{\alpha_1} = c_{i_1 k}^{\alpha_1} \omega^k, \tag{5.31}$$

where the quantities $c_{i_1 k}^{\alpha_1}$ are connected with the second fundamental tensor $b_{ij}^{i_1}$ by equations (2.52), which (by the structure of the tensor $b_{jk}^{i_1}$ following from (5.29)) lead here to the following equations:

$$c_{41}^{\alpha_1} = c_{52}^{\alpha_1} = c_{63}^{\alpha_1} \stackrel{\text{def}}{=} c^{\alpha_1}, \tag{5.32}$$

$$c_{42}^{\alpha_1} = c_{43}^{\alpha_1} = c_{51}^{\alpha_1} = c_{53}^{\alpha_1} = c_{61}^{\alpha_1} = c_{62}^{\alpha_1} = 0. \tag{5.33}$$

If in equations (5.31) we replace the quantities $c_{i_1 k}^{\alpha_1}$ by their values (5.32) and (5.33),then equations (5.31) take the form:

$$\omega_4^{\alpha_1} = c^{\alpha_1} \omega^1, \;\; \omega_5^{\alpha_1} = c^{\alpha_1} \omega^2, \;\; \omega_6^{\alpha_1} = c^{\alpha_1} \omega^3, \tag{5.34}$$

and, by (5.28) and (5.34), the third fundamental forms $\Phi_{(3)}^{\alpha_1}$ defined by equations (2.55) take the form:

$$\Phi_{(3)}^{\alpha_1} = 6 c^{\alpha_1} \omega^1 \omega^2 \omega^3. \tag{5.35}$$

Two cases are possible: all the coefficients c^{α_1} are equal to zero or at least one of these coefficients is different from zero.

2. In the first case, when $c^{\alpha_1} = 0$, equations (5.34) give:

$$\omega_{i_1}^{\alpha_1} = 0. \tag{5.36}$$

By (5.36), we have

$$dA_{i_1} = \omega_{i_1}^0 A_0 + \omega_{i_1}^k A_k + \omega_{i_1}^{j_1} A_{j_1}.$$

This implies that *the six-dimensional second osculating subspace $T_x^{(2)}(V^3)$ is constant, and the submanifold V^3 lies in this subspace.*

The submanifolds V^3 of this type are defined by the system of Pfaffian equations (2.5) and (5.29). Let us investigate this system. Exterior differentiation of equations (2.5) leads to identities, and exterior differentiation of equations (5.29) gives the following exterior quadratic equations:

$$\begin{array}{l}
(\omega_1^3 - \omega_6^4) \wedge \omega^2 + (\omega_1^2 - \omega_5^4) \wedge \omega^3 = 0, \\
(\omega_2^1 - \omega_4^5) \wedge \omega^3 + (\omega_2^3 - \omega_6^5) \wedge \omega^1 = 0, \\
(\omega_3^2 - \omega_5^6) \wedge \omega^1 + (\omega_3^1 - \omega_4^6) \wedge \omega^2 = 0,
\end{array} \tag{5.37}$$

$$\begin{aligned}
&(\omega_1^3 - \omega_6^4) \wedge \omega^1 + 2\omega_2^3 \wedge \omega^2 + (\omega_2^2 + \omega_3^3 - \omega_0^0 - \omega_4^4) \wedge \omega^3 = 0,\\
&(\omega_1^1 + \omega_3^3 - \omega_0^0 - \omega_5^5) \wedge \omega^1 + (\omega_2^1 - \omega_4^5) \wedge \omega^2 + 2\omega_3^1 \wedge \omega^3 = 0,\\
&2\omega_1^2 \wedge \omega^1 + (\omega_1^1 + \omega_2^2 - \omega_0^0 - \omega_6^6) \wedge \omega^2 + (\omega_3^2 - \omega_5^6) \wedge \omega^3 = 0,\\
&(\omega_1^2 - \omega_5^4) \wedge \omega^1 + (\omega_2^2 + \omega_3^3 - \omega_0^0 - \omega_4^4) \wedge \omega^2 + 2\omega_3^2 \wedge \omega^3 = 0,\\
&2\omega_1^3 \wedge \omega^1 + (\omega_2^3 - \omega_6^5) \wedge \omega^2 + (\omega_1^1 + \omega_3^3 - \omega_0^0 - \omega_5^5) \wedge \omega^3 = 0,\\
&(\omega_1^1 + \omega_2^2 - \omega_0^0 - \omega_6^6) \wedge \omega^1 + 2\omega_2^1 \wedge \omega^2 + (\omega_3^1 - \omega_4^6) \wedge \omega^3 = 0.
\end{aligned} \tag{5.38}$$

In addition to the basis forms ω^i, quadratic equations (5.37) and (5.38) contain the following independent 1-forms: ω_j^i, $\omega_j^i - \omega_{3+i}^{3+j}$, $i \neq j$, $\omega_2^2 + \omega_3^3 - \omega_0^0 - \omega_4^4$, $\omega_3^3 + \omega_1^1 - \omega_0^0 - \omega_5^5$, and $\omega_1^1 + \omega_2^2 - \omega_0^0 - \omega_6^6$. Their number is $q = 15$. The system consists of $s_1 = 9$ independent exterior quadratic equations. Thus the successive characters of this system are $s_2 = 6$ and $s_3 = 0$. The Cartan number $Q = s_1 + 2s_2 + 3s_3 = 21$.

Applying Cartan's lemma to equations (5.37), we find that

$$\begin{aligned}
&\omega_1^3 - \omega_6^4 = 2l_{22}^1\omega^2 + 2l_{23}^1\omega^3,\\
&\omega_1^2 - \omega_5^4 = 2l_{23}^1\omega^2 + 2l_{33}^1\omega^3,\\
&\omega_2^1 - \omega_4^5 = 2l_{33}^2\omega^3 + 2l_{31}^2\omega^1,\\
&\omega_2^3 - \omega_6^5 = 2l_{31}^2\omega^3 + 2l_{11}^2\omega^1,\\
&\omega_3^2 - \omega_5^6 = 2l_{11}^3\omega^1 + 2l_{12}^3\omega^2,\\
&\omega_3^1 - \omega_4^6 = 2l_{12}^3\omega^1 + 2l_{22}^3\omega^2.
\end{aligned} \tag{5.39}$$

Replacing the forms $\omega_j^i - \omega_{3+j}^{3+i}$ in quadratic equations (5.38) by their values taken from equations (5.39), and applying Cartan's lemma to the quadratic equations obtained after this replacement, we arrive at the following equations:

$$\begin{aligned}
&\omega_2^3 = l_{22}^1\omega^1 + a\omega^2 + a_2\omega^3,\\
&\omega_2^2 + \omega_3^3 - \omega_0^0 - \omega_4^4 = 2l_{23}^1\omega^1 + 2a_2\omega^2 + 2a_3\omega^3,\\
&\omega_3^2 = l_{33}^1\omega^1 + a_3\omega^2 + A\omega^3,\\
&\omega_3^1 = l_{33}^2\omega^2 + b\omega^3 + b_2\omega^1,\\
&\omega_1^1 + \omega_3^3 - \omega_0^0 - \omega_5^5 = 2l_{31}^2\omega^2 + 2b_2\omega^3 + 2b_3\omega^1,\\
&\omega_1^3 = l_{11}^2\omega^2 + b_3\omega^3 + B\omega^1,\\
&\omega_1^2 = l_{11}^3\omega^3 + c\omega^1 + c_2\omega^2,\\
&\omega_1^1 + \omega_2^2 - \omega_0^0 - \omega_6^6 = 2l_{12}^3\omega^3 + 2c_2\omega^1 + 2c_3\omega^2,\\
&\omega_2^1 = l_{22}^3\omega^3 + c_3\omega^1 + C\omega^2.
\end{aligned} \tag{5.40}$$

Equations (5.39) and (5.40) prove that the number N of parameters on which the most general three-dimensional integral element depends, is equal to 21: $N = 21$. Since $N = Q$, *the system is in involution, and its general integral manifold* $V^3 \subset P^6$ *depends on six arbitrary functions of two variables.*

3. Suppose now that at least one of the coefficients c^{α_1} is different from zero. Then there is only one independent form among the third fundamental forms $\Phi_{(3)}^{\alpha_1}(V^3)$, and $\dim T_x^{(3)}(V^3) = 7$. If we take the point A_7 of our moving frame in this subspace $T_x^{(3)}(V^3)$, we obtain:

$$\Phi^7_{(3)} = 6\omega^1\omega^2\omega^3, \quad \Phi^{\alpha_2}_{(3)} = 0, \quad \alpha_2 = 8, \dots, n. \tag{5.41}$$

In the case under consideration, equations (5.34) take the form:

$$\omega^7_4 = \omega^1, \quad \omega^7_5 = \omega^2, \quad \omega^7_6 = \omega^3, \tag{5.42}$$

$$\omega^{\alpha_2}_{i_1} = 0, \quad i_1 = 4, 5, 6. \tag{5.43}$$

Let us prove that in this case the subspace $T^{(3)}_x(V^3)$ is constant, and thus *the submanifold V^3 lies in this 7-dimensional subspace.*

In fact, exterior differentiation of equations (5.43) gives the following exterior quadratic equations:

$$\omega^i \wedge \omega^7_{\alpha_2} = 0,$$

which yield

$$\omega^7_{\alpha_2} = 0. \tag{5.44}$$

Therefore, *the subspace $T^{(3)}_x(V^3)$ defined by the points A_0, A_i, A_{3+i} and A_7 is constant, and the submanifold V^3 lies in this subspace.*

This result matches Theorem 5.4 stated above.

4. The submanifold V^3 in question is defined in the space P^7 by Pfaffian equations (2.5), (5.29), (5.30), where $\alpha_1 = 7$, and (5.42). By (5.42), exterior differentiation of equations (5.30) leads to identities, and exterior differentiation of equations (5.29) give exterior quadratic equations (5.37) and (5.38). If we find exterior derivatives of equations (5.42) and apply (5.39) and (5.40), we arrive at the following three exterior quadratic equations:

$$\begin{aligned}
&\{\theta + 2(l^2_{31} - a_2 - c_3)\omega^2 + 2(l^3_{12} - b_2 - a_3)\omega^3\} \wedge \omega^1 + 4(l^2_{33} - l^3_{22})\omega^2 \wedge \omega^3 = 0,\\
&\{\theta + 2(l^3_{12} - b_2 - a_3)\omega^3 + 2(l^1_{23} - c_2 - b_3)\omega^1\} \wedge \omega^2 + 4(l^3_{11} - l^1_{33})\omega^3 \wedge \omega^1 = 0,\\
&\{\theta + 2(l^1_{23} - c_2 - b_3)\omega^1 + 2(l^2_{31} - a_2 - c_3)\omega^2\} \wedge \omega^3 + 4(l^1_{22} - l^2_{11})\omega^1 \wedge \omega^2 = 0,
\end{aligned} \tag{5.45}$$

where

$$\theta = \omega^1_1 + \omega^2_2 + \omega^3_3 - 2\omega^0_0 - \omega^7_7. \tag{5.46}$$

Equations (5.45) yield the following equations:

$$l^2_{33} = l^3_{22}, \quad l^3_{11} = l^1_{33}, \quad l^1_{22} = l^2_{11} \tag{5.47}$$

and

$$\theta = 2(c_2 + b_3 - l^1_{23})\omega^1 + 2(a_2 + c_3 - l^2_{31})\omega^2 + 2(b_2 + a_3 - l^3_{12})\omega^3. \tag{5.48}$$

We can now establish a geometric meaning for conditions (5.47). By (5.40), we have the following relations:

$$\begin{aligned} d\omega^1 &\equiv (l^3_{22} - l^2_{33})\omega^2 \wedge \omega^3 \pmod{\omega^1}, \\ d\omega^2 &\equiv (l^1_{33} - l^3_{11})\omega^3 \wedge \omega^1 \pmod{\omega^2}, \\ d\omega^3 &\equiv (l^2_{11} - l^1_{22})\omega^1 \wedge \omega^2 \pmod{\omega^3}. \end{aligned}$$

Thus, conditions (5.47) mean that the two-dimensional distributions defined by three asymptotic directions on a submanifold V^3 are integrable. Therefore, *an asymptotic net on a submanifold V^3 is holonomic.*

5. Let us find under what conditions the asymptotic lines of a submanifold V^3 in question are straight lines. First of all, note that equations (5.37) and (5.38), and therefore equations (5.39) and (5.40), are valid even in case the submanifold V^3 is not contained in a space P^6. Next, we find the differentials of the points A_1, A_2 and A_3 located on the tangents to the asymptotic lines of the submanifold V^3 passing through a point $x = A_0$. Applying equations (5.39), (5.40) and (5.47), we find that

$$dA_1 = \omega^0_1 A_0 + (\omega^1_1 - c_3\omega^2 - b_2\omega^3)A_1 + \omega^1(cA_2 + BA_3) + \omega^2\widetilde{A}_6 + \omega^3\widetilde{A}_5, \quad (5.49)$$

$$dA_2 = \omega^0_2 A_0 + (\omega^2_2 - a_3\omega^3 - c_2\omega^1)A_2 + \omega^2(aA_3 + CA_1) + \omega^3\widetilde{A}_4 + \omega^1\widetilde{A}_6, \quad (5.50)$$

and

$$dA_3 = \omega^0_3 A_0 + (\omega^3_3 - b_3\omega^1 - a_2\omega^2)A_3 + \omega^3(bA_1 + AA_2) + \omega^1\widetilde{A}_5 + \omega^2\widetilde{A}_4, \quad (5.51)$$

where

$$\begin{aligned} \widetilde{A}_4 &= A_4 + l^3_{22}A_1 + a_3A_2 + a_2A_3, \\ \widetilde{A}_5 &= A_5 + l^1_{33}A_2 + b_3A_3 + b_2A_1, \\ \widetilde{A}_6 &= A_6 + l^2_{11}A_3 + c_3A_1 + c_2A_2. \end{aligned} \quad (5.52)$$

Consider these equations together with the equation

$$dA_0 = \omega^0_0 A_0 + \omega^1 A_1 + \omega^2 A_2 + \omega^3 A_3. \quad (5.53)$$

Then from equations (5.49) and (5.53) we find that

$$d(A_0 \wedge A_1) \equiv (\omega^0_0 + \omega^1_1)A_0 \wedge A_1 + A_0 \wedge (cA_2 + BA_3)\omega^1 \pmod{\omega^2, \omega^3}.$$

It follows that the first family of asymptotic lines of a submanifold V^3, defined by the equations $\omega^2 = \omega^3 = 0$, are straight lines if and only if the conditions

$$c = B = 0 \quad (5.54)$$

are satisfied.

Similarly, we can prove that the second and third families of asymptotic lines of the submanifold V^3 are straight lines if and only if the conditions

$$a = C = 0 \tag{5.55}$$

and

$$b = A = 0 \tag{5.56}$$

are satisfied, respectively.

We now return to the general case. Consider the nondevelopable ruled surface λ_{12} generated by the straight lines $A_0 \wedge A_1$ and defined by the equations $\omega^1 = \omega^3 = 0$. Since by (5.49) and (5.53) we have

$$\begin{aligned} dA_0 &\equiv \omega_0^0 A_0 + \omega^1 A_1 \pmod{\omega^1, \omega^3}, \\ dA_1 &\equiv \omega_1^0 A_0 + (\omega_1^1 - c_3\omega^2) A_2 + \omega^2 \widetilde{A}_6 \pmod{\omega^1, \omega^3}, \end{aligned}$$

the point $\widetilde{A}_6$ belongs to the 3-plane $A_0 \wedge A_1 \wedge A_2 \wedge \widetilde{A}_6 = E_{12}$ which contains the tangent 3-planes to the surface λ_{12} at the points of the straight line $A_0 \wedge A_1$. In a similar manner, the point $\widetilde{A}_5$ belongs to the 3-plane $A_0 \wedge A_1 \wedge A_2 \wedge \widetilde{A}_5 = E_{13}$ which contains the tangent 3-planes to the ruled surface λ_{13} defined by the equations $\omega^1 = \omega^2 = 0$ at the points of the straight line $A_0 \wedge A_1$.

Similarly equations (5.50) and (5.53) imply that if $\omega^2 = \omega^1 = 0$, the straight line $A_0 \wedge A_2$ describes the ruled surface λ_{23} whose tangent planes at the points of this straight line belong to the 3-plane $A_0 \wedge A_2 \wedge A_3 \wedge \widetilde{A}_4 = E_{23}$, and if $\omega^2 = \omega^3 = 0$, this straight line $A_0 \wedge A_2$ describes the ruled surface λ_{21} whose tangent planes at the points of this straight line belong to the 3-plane $A_0 \wedge A_2 \wedge A_1 \wedge \widetilde{A}_6 = E_{21}$. Finally, by (5.51) and (5.53), the tangent planes to the ruled surfaces λ_{31} and λ_{32}, described by the straight line $A_0 \wedge A_3$ when $\omega^3 = \omega^2 = 0$ and $\omega^3 = \omega^1 = 0$ respectively, belong to the 3-planes $E_{31} = A_0 \wedge A_3 \wedge A_1 \wedge \widetilde{A}_5$ and $E_{32} = A_0 \wedge A_3 \wedge A_2 \wedge \widetilde{A}_4$ respectively.

Moreover, it is clear that $E_{12} = E_{21}$, i.e. the tangent 3-planes to the surfaces λ_{12} and λ_{21} coincide. Similarly, we have $E_{23} = E_{32}$ and $E_{31} = E_{13}$.

From equation (5.52) it follows that the points $A_0, A_1, A_2, A_3, \widetilde{A}_4, \widetilde{A}_5$ and $\widetilde{A}_6$ are linearly independent. Thus we can specialize our moving frames by taking $A_4 = \widetilde{A}_4, A_5 = \widetilde{A}_5$ and $A_6 = \widetilde{A}_6$. Then the point A_4 belongs to the plane E_{23}, which contains first order differential neighborhoods of the straight lines $A_0 \wedge A_2$ and $A_0 \wedge A_3$ of the surfaces λ_{23} and λ_{32}; the point A_5 belongs to the plane E_{31}, which contains first order differential neighborhoods of the straight lines $A_0 \wedge A_3$ and $A_0 \wedge A_1$ of the surfaces λ_{31} and λ_{13}; and the point A_6 belongs to the plane E_{12}, which contains first order differential neighborhoods of the straight lines $A_0 \wedge A_1$ and $A_0 \wedge A_2$ of the surfaces λ_{12} and λ_{21}.

Having made the choice of moving frames indicated above, we obtain

$$l_{11}^2 = l_{22}^3 = l_{33}^1 = 0, \quad a_2 = a_3 = b_2 = b_3 = c_2 = c_3 = 0. \tag{5.57}$$

By (5.57), equations (5.39) and (5.40) take the form:

$$\begin{array}{lll} \omega_2^3 = a\omega^2, & \omega_3^1 = b\omega^3, & \omega_1^2 = c\omega^1, \\ \omega_3^2 = A\omega^3, & \omega_1^3 = B\omega^1, & \omega_2^1 = C\omega^2, \end{array} \tag{5.58}$$

$$\begin{array}{ll} \omega_1^3 - \omega_6^4 = 2l_{23}^1\omega^3, & \omega_1^2 - \omega_5^4 = 2l_{23}^1\omega^2, \\ \omega_2^1 - \omega_4^5 = 2l_{31}^2\omega^1, & \omega_2^3 - \omega_6^5 = 2l_{31}^2\omega^3, \\ \omega_3^2 - \omega_5^6 = 2l_{12}^3\omega^2, & \omega_3^1 - \omega_4^6 = 2l_{12}^3\omega^1, \end{array} \tag{5.59}$$

$$\begin{array}{l} \omega_2^2 + \omega_3^3 - \omega_0^0 - \omega_4^4 = 2l_{23}^1\omega^1, \\ \omega_3^3 + \omega_1^1 - \omega_0^0 - \omega_5^5 = 2l_{31}^2\omega^2, \\ \omega_1^1 + \omega_2^2 - \omega_0^0 - \omega_6^6 = 2l_{12}^3\omega^3. \end{array} \tag{5.60}$$

Let us further prove that the coefficients l_{23}^1, l_{31}^2 and l_{12}^3 in the right-hand sides of equations (5.59) and (5.60) can be reduced to 0. By formulas (5.58)–(5.60), the differentials of the points A_4, A_5 and A_6 satisfy the following relations:

$$\begin{array}{ll} dA_4 & \equiv C\omega^2 A_5 + b\omega^3 A_6 + (\omega_4^4 + 2l_{23}^1\omega^1)A_4 \\ & +\omega^1(A_7 - 2l_{23}^1 A_4 - 2l_{31}^2 A_5 - 2l_{12}^3 A_6) \pmod{T_x(V^3)}, \\ dA_5 & \equiv A\omega^3 A_6 + c\omega^1 A_4 + (\omega_5^5 + 2l_{31}^2\omega^2)A_5 \\ & +\omega^2(A_7 - 2l_{23}^1 A_4 - 2l_{31}^2 A_5 - 2l_{12}^3 A_6) \pmod{T_x(V^3)}, \\ dA_6 & \equiv B\omega^1 A_4 + a\omega^2 A_5 + (\omega_6^6 + 2l_{12}^3\omega^3)A_6 \\ & +\omega^3(A_7 - 2l_{23}^1 A_4 - 2l_{31}^2 A_5 - 2l_{12}^3 A_6) \pmod{T_x(V^3)}. \end{array}$$

It follows that if we replace the point A_7 of our moving frame by the point

$$\widetilde{A}_7 = A_7 - 2l_{23}^1 A_4 - 2l_{31}^2 A_5 - 2l_{12}^3 A_6,$$

then the coefficients l_{23}^1, l_{31}^2 and l_{12}^3 will be reduced to 0, i.e. in addition to relations (5.57), we obtain the following relations:

$$l_{23}^1 = l_{31}^2 = l_{12}^3 = 0. \tag{5.61}$$

After this specialization of moving frames, equations (5.59) and (5.60) take the form

$$\omega_{3+i}^{3+j} = \omega_j^i, \quad i \neq j, \tag{5.62}$$

and

$$\begin{array}{l} \omega_4^4 = \omega_2^2 + \omega_3^3 - \omega_0^0, \\ \omega_5^5 = \omega_3^3 + \omega_1^1 - \omega_0^0, \\ \omega_6^6 = \omega_1^1 + \omega_2^2 - \omega_0^0, \end{array} \tag{5.63}$$

respectively. In addition, by (5.61), equations (5.46) and (5.48) imply

$$\omega_7^7 = \omega_1^1 + \omega_2^2 + \omega_3^3 - 2\omega_0^0. \tag{5.64}$$

6. The remaining coefficients in equations (5.58) cannot be reduced to 0 since, as we will show in subsection **5.4.7**, they are relative invariants of the submanifold V^3 in question. As we proved above, the vanishing of these coefficients is related to the rectilinearity of asymptotic lines of the submanifold V^3.

Consider the case where all these coefficients are equal to 0, i.e. where the submanifold V^3 carries three families of rectilinear asymptotic lines. In this case the ruled surfaces λ_{12} and λ_{21}, which we considered in subsection **5.4.5**, are two families of rectilinear generators of the same quadric located in the 3-plane E_{12}, the surfaces λ_{23} and λ_{32} are two families of rectilinear generators of the same quadric located in the 3-plane E_{23}, and the surfaces λ_{31} and λ_{13} are two families of rectilinear generators of the same quadric located in the 3-plane E_{31}. Thus, the submanifold V^3 in question is foliated by three families of quadrics, and its parametric equation can be written in the form:

$$x = A_0+u^1A_1+u^2A_2+u^3A_3+u^2u^3A_4+u^3u^1A_5+u^1u^2A_6+u^1u^2u^3A_7. \quad (5.65)$$

Introduce homogeneous coordinates $(x^0, x^1), (y^0, y^1)$ and (z^0, z^1) so that $u^1 = \frac{x^1}{x^0}, u^2 = \frac{y^1}{y^0}$ and $u^3 = \frac{z^1}{z^0}$, and homogeneous coordinates $w^{ijk}, i, j, k = 0, 1$, in the space P^7. Then equation (5.65) can be written in the form:

$$w^{ijk} = x^i y^j z^k. \quad (5.66)$$

This equation proves that the submanifold (5.65) is the Segre variety $S(1,1,1)$ which is an imbedding of the direct product of three projective straight lines into the space P^7.

We will now return to the general case and establish another geometric meaning of the relative invariants a, b, c, A, B and C.

Consider one of the two-dimensional surfaces by which the submanifold V^3 is foliated. Let us take, for example, the surface V_3^2, defined by the equation $\omega^3 = 0$, and find the second fundamental forms of this surface. By equation (5.28), one of the nonvanishing second fundamental forms of this surface is the form

$$\Phi^6_{(2)} = 2\omega^1\omega^2. \quad (5.67)$$

Besides this form, another nonvanishing second fundamental form of this surface is the form

$$\Phi^3_{(2)} = \omega^1\omega^3_1 + \omega^2\omega^3_2,$$

which by (5.58) takes the form:

$$\Phi^3_{(2)} = B(\omega^1)^2 + a(\omega^2)^2. \quad (5.68)$$

Therefore, the relative invariants B and a are the coefficients of the second nonvanishing second fundamental form of the surface V_3^2. Since each of the surfaces V_3^2 has only two nonvanishing second fundamental forms, the second osculating subspaces of these surfaces are four-dimensional, and each of these surfaces carries a conjugate net. It is easy to prove that this conjugate net is real.

Similarly, the relative invariants C and b are the coefficients of the second nonvanishing second fundamental form of the surface V_1^2, defined by the equation $\omega^1 = 0$, and the relative invariants A and c are the coefficients of the second nonvanishing second fundamental form of the surface V_2^2, defined by the equation $\omega^2 = 0$.

Our previous considerations can be easily complemented by the following result.

Theorem 5.6 *If the asymptotic lines of a submanifold V^3 are asymptotic lines on all two-dimensional surfaces by which the submanifold V^3 is foliated, then these asymptotic lines are rectilinear, and the two-dimensional surfaces are quadrics.*

Proof. Consider, for example, the surface V_3^2 defined by the equation $\omega^3 = 0$. This surface carries asymptotic lines if and only if its second fundamental form (5.68) is proportional to the form (5.67), i.e. if $B = a = 0$. Similarly, surfaces V_1^2 and V_2^2 carry asymptotic lines if and only if the conditions $C = b = 0$ and $A = c = 0$ hold, respectively. But as we proved earlier, the simultaneous vanishing of all these coefficients leads to the rectilinearity of all asymptotic lines of a submanifold V^3, and consequently, V^3 is foliated by three families of quadrics. ■

7. Let us again return to the general case. Exterior differentiation of equations (5.58) and (5.62)–(5.64) gives the following exterior quadratic equations:

$$
\begin{array}{ll}
\Omega_6^3 \wedge \omega^1 + \Delta a \wedge \omega^2 + \Omega_4^3 \wedge \omega^3 = 0, & 2\Omega_6^3 \wedge \omega^2 + \Omega_7^4 \wedge \omega^3 = 0, \\
\Omega_4^1 \wedge \omega^2 + \Delta b \wedge \omega^3 + \Omega_5^1 \wedge \omega^1 = 0, & 2\Omega_4^1 \wedge \omega^3 + \Omega_7^5 \wedge \omega^1 = 0, \\
\Omega_5^2 \wedge \omega^3 + \Delta c \wedge \omega^1 + \Omega_6^2 \wedge \omega^2 = 0, & 2\Omega_5^2 \wedge \omega^1 + \Omega_7^6 \wedge \omega^2 = 0, \\
\Omega_5^2 \wedge \omega^1 + \Omega_4^2 \wedge \omega^2 + \Delta A \wedge \omega^3 = 0, & 2\Omega_5^2 \wedge \omega^3 + \Omega_7^4 \wedge \omega^2 = 0, \\
\Omega_6^3 \wedge \omega^2 + \Omega_5^3 \wedge \omega^3 + \Delta B \wedge \omega^1 = 0, & 2\Omega_6^3 \wedge \omega^1 + \Omega_7^5 \wedge \omega^3 = 0, \\
\Omega_4^1 \wedge \omega^3 + \Omega_6^1 \wedge \omega^1 + \Delta C \wedge \omega^2 = 0, & 2\Omega_4^1 \wedge \omega^2 + \Omega_7^6 \wedge \omega^1 = 0, \\
\Omega_7^4 \wedge \omega^1 + 2\Omega_4^3 \wedge \omega^2 + 2\Omega_4^2 \wedge \omega^3 = 0, & \\
\Omega_7^5 \wedge \omega^2 + 2\Omega_5^1 \wedge \omega^3 + 2\Omega_5^3 \wedge \omega^1 = 0, & \\
\Omega_7^6 \wedge \omega^3 + 2\Omega_6^2 \wedge \omega^1 + 2\Omega_6^1 \wedge \omega^2 = 0, &
\end{array}
\tag{5.69}
$$

where

$$\begin{array}{ll}
\Delta a = da + a(\omega_0^0 - 2\omega_2^2 + \omega_3^3), & \Omega_4^3 = \omega_4^3 - \omega_2^0, \\
\Delta b = db + b(\omega_0^0 - 2\omega_3^3 + \omega_1^1), & \Omega_5^1 = \omega_5^1 - \omega_3^0, \\
\Delta c = dc + c(\omega_0^0 - 2\omega_1^1 + \omega_2^2), & \Omega_6^2 = \omega_6^2 - \omega_1^0, \\
\Delta A = dA + A(\omega_0^0 - 2\omega_3^3 + \omega_2^2), & \Omega_4^2 = \omega_4^2 - \omega_3^0, \\
\Delta B = dB + B(\omega_0^0 - 2\omega_1^1 + \omega_3^3), & \Omega_5^3 = \omega_5^3 - \omega_1^0, \\
\Delta C = dC + C(\omega_0^0 - 2\omega_2^2 + \omega_1^1), & \Omega_6^1 = \omega_6^1 - \omega_2^0, \\
\Omega_4^1 = \omega_4^1 - ba\omega^2 - CA\omega^3, & \Omega_7^4 = \omega_5^3 + \omega_6^2 - \omega_1^0 - \omega_7^4, \\
\Omega_5^2 = \omega_5^2 - cb\omega^3 - AB\omega^1, & \Omega_7^5 = \omega_6^1 + \omega_4^3 - \omega_2^0 - \omega_7^5, \\
\Omega_6^3 = \omega_6^3 - ac\omega^1 - BC\omega^2, & \Omega_7^6 = \omega_4^2 + \omega_5^1 - \omega_3^0 - \omega_7^6.
\end{array}$$

Note that by (5.69), exterior differentiation of equation (5.64) gives an identity.

Applying Cartan's lemma to equations (5.69), we obtain the following Pfaffian equations:

$$\begin{array}{rll}
\Delta a & = & \alpha_1\omega^2 + x_1\omega^3, \\
\omega_4^3 - \omega_2^0 & = r\omega^1 + & x_1\omega^2 + y_1\omega^3, \\
\omega_4^2 - \omega_3^0 & = q\omega^1 + & y_1\omega^2 + z_1\omega^3, \\
\Delta A & = & z_1\omega^2 + \beta_1\omega^3, \\
\Delta b & = & \alpha_2\omega^3 + x_2\omega^1, \\
\omega_5^1 - \omega_3^0 & = p\omega^2 + & x_2\omega^3 + y_2\omega^1, \\
\omega_5^3 - \omega_1^0 & = r\omega^2 + & y_2\omega^3 + z_2\omega^1, \\
\Delta B & = & z_2\omega^3 + \beta_2\omega^1, \\
\Delta c & = & \alpha_3\omega^1 + x_3\omega^2, \\
\omega_6^2 - \omega_1^0 & = q\omega^3 + & x_3\omega^1 + y_3\omega^2, \\
\omega_6^1 - \omega_2^0 & = p\omega^3 + & y_3\omega^1 + z_3\omega^2, \\
\Delta C & = & z_3\omega^1 + \beta_3\omega^2,
\end{array} \tag{5.70}$$

$$\begin{array}{l}
\omega_4^1 = p\omega^1 + ab\omega^2 + AC\omega^3, \\
\omega_5^2 = q\omega^2 + bc\omega^3 + BA\omega^1, \\
\omega_6^3 = r\omega^3 + ca\omega^1 + CB\omega^2,
\end{array} \tag{5.71}$$

$$\begin{array}{llll}
\Omega_7^4 = & & 2r\omega^2 & +2q\omega^3, \\
\Omega_7^5 = & 2r\omega^1 + & & 2p\omega^3, \\
\Omega_7^6 = & 2q\omega^1 + & 2p\omega^2. &
\end{array} \tag{5.72}$$

It follows from equations (5.69) and (5.70)–(5.72) that we have:

$$q = 18, s_1 = 15, s_2 = 3, s_2 = 0, Q = s_1 + 2s_2 + 3s_3 = 21, N = 18.$$

Since $Q \neq N$, the system is not in involution, and we need to find its prolongation.

Note that equations (5.70) imply that the quantities a, b, c, A, B and C are relative invariants.

8. We will now prove that some of the coefficients in expansions (5.70)–(5.72) can be reduced to 0. To prove this, we apply (5.58), (5.62), and (5.64) to find the differentials of the points A_4, A_5 and A_6. As a result, we find that

$$\begin{aligned} dA_4 = {} & \omega_4^0 A_0 + (ab\omega^2 + AC\omega^3)A_1 + (\omega_4^2 - q\omega^1)A_2 + (\omega_4^3 - r\omega^1)A_3 \\ & + \omega_4^4 A_4 + C\omega^2 A_5 + b\omega^3 A_6 + \omega^1(A_7 + pA_1 + qA_2 + rA_3), \\ dA_5 = {} & \omega_5^0 A_0 + (bc\omega^3 + BA\omega^1)A_2 + (\omega_5^3 - r\omega^2)A_3 + (\omega_5^1 - p\omega^2)A_1 \\ & + \omega_5^5 A_5 + A\omega^3 A_6 + c\omega^1 A_4 + \omega^2(A_7 + pA_1 + qA_2 + rA_3), \\ dA_6 = {} & \omega_6^0 A_0 + (ca\omega^1 + CB\omega^2)A_3 + (\omega_6^1 - p\omega^3)A_1 + (\omega_6^2 - q\omega^3)A_2 \\ & + \omega_6^6 A_6 + B\omega^1 A_4 + a\omega^2 A_5 + \omega^3(A_7 + pA_1 + qA_2 + rA_3). \end{aligned}$$

It follows that if we replace the vertex A_7 of our moving frame by the point

$$\widetilde{A}_7 = A_7 + pA_1 + qA_2 + rA_3,$$

we reduce the coefficients p, q and r to 0, i.e. we obtain:

$$p = q = r = 0. \tag{5.73}$$

Note that the above choice of the vertex A_7 of moving frames is consistent with our previous choice made in subsection **5.4.5**, since there we used a linear combination of this point with the points A_{3+i}, while here we use its linear combination with the points A_i.

Let us write the expression of dA_4 using equations (5.73) and replacing the forms ω_4^2 and ω_4^3 by their values taken from equations (5.70):

$$\begin{aligned} dA_4 = {} & \omega_4^0 A_0 + (ab\omega^2 + AC\omega^3)A_1 + (\omega_3^0 + y_1\omega^2 + z_1\omega^3)A_2 \\ & + (\omega_2^0 + x_1\omega^2 + y_1\omega^3)A_3 + \omega_4^4 A_4 + C\omega^2 A_5 + b\omega^3 A_6 + \omega^1 A_7. \end{aligned}$$

Consider the point $\widetilde{A}_4 = A_4 - y_1 A_0$. Its differential has the form:

$$\begin{aligned} d\widetilde{A}_4 = {} & (\omega_4^0 + y_1(\omega_4^4 - \omega_0^0) - dy_1)A_0 + (ab\omega^2 + AC\omega^3 - y_1\omega^1)A_1 \\ & + (\omega_3^0 + z_1\omega^3)A_2 + (\omega_2^0 + x_1\omega^2)A_3 + \omega_4^4 \widetilde{A}_4 + C\omega^2 A_5 + b\omega^3 A_6 + \omega^1 A_7. \end{aligned}$$

If we change our moving frames by replacing the vertex A_4 by the point $\widetilde{A}_4$, then we reduce the coefficient y_1 to 0. In a similar manner, by replacing the vertices A_5 and A_6 by the points $\widetilde{A}_5 = A_5 - y_2 A_0$ and $\widetilde{A}_6 = A_6 - y_3 A_0$ respectively, we reduce o the coefficients y_2 and y_3. to 0. As a result of this choice of moving frame, we obtain:

$$y_1 = y_2 = y_3 = 0. \tag{5.74}$$

9. As we showed earlier, the system of Pfaffian equations (5.58), (5.59), (5.62) and (5.64) is not in involution. However, this does not mean that this system has no solution. It follows from the general theory (see Section **1.5**) that the further investigation of this system is connected with its successive differential prolongations. As a straightforward calculation shows, in our case even after double prolongation of the system (5.58), (5.59), (5.62) and (5.64) the prolonged system is still not in involution. Therefore, we consider some particular cases of the system (5.58), (5.59), (5.62) and (5.64) which are characterized by the vanishing of some of the coefficients a, b, c, A, B and C. In subsection **5.4.7** we already considered the case when all these coefficients vanish. Consider now the following case:

$$B = c = 0, \quad C = a = 0. \tag{5.75}$$

First of all note that since $B = a = 0$, the considerations in subsection **5.4.6** prove that *the submanifold V^3 in question is foliated by a family of quadrics V_3^2.*

Next, by (5.70), equations (5.75) imply

$$\alpha_1 = \alpha_3 = \beta_2 = \beta_3 = 0, \quad x_1 = x_3 = z_2 = z_3 = 0. \tag{5.76}$$

By (5.75) and (5.76), four equations of system (5.70) are identically satisfied. By (5.75), (5.76), (5.73), and (5.74), the remaining eight equations of system (5.70), and equations (5.71) and (5.72) take the following form:

$$\begin{array}{ll} \omega_4^3 - \omega_2^0 = 0, & \omega_4^2 - \omega_3^0 = z_1\omega^3, \\ \omega_5^3 - \omega_1^0 = 0, & \Delta A = z_1\omega^2 + \beta_1\omega^3, \\ \omega_6^2 - \omega_1^0 = 0, & \Delta b = \alpha_2\omega^3 + x_2\omega^1, \\ \omega_6^1 - \omega_2^0 = 0, & \omega_5^1 - \omega_3^0 = x_2\omega^3, \end{array} \tag{5.77}$$

$$\omega_4^1 = 0, \quad \omega_5^2 = 0, \quad \omega_6^3 = 0, \tag{5.78}$$

$$\omega_7^4 - \omega_1^0 = 0, \quad \omega_7^5 - \omega_2^0 = 0, \quad \omega_7^6 - \omega_3^0 = (z_1 + x_2)\omega^3. \tag{5.79}$$

Exterior differentiation of equations (5.77)–(5.79) gives the following quadratic equations:

$$\begin{array}{l} \Omega_7^3 \wedge \omega^1 - 2\omega_4^0 \wedge \omega^3 = 0, \\ \Omega_7^2 \wedge \omega^1 - 2\omega_4^0 \wedge \omega^2 + \Delta z_1 \wedge \omega^3 = 0, \\ \Delta z_1 \wedge \omega^2 + \Delta\beta_1 \wedge \omega^3 = 0, \\ \Delta\alpha_2 \wedge \omega^3 + \Delta x_2 \wedge \omega^1 = 0, \\ \Omega_7^1 \wedge \omega^2 + \Delta x_2 \wedge \omega^3 - 2\omega_5^0 \wedge \omega^1 = 0, \\ \Omega_7^3 \wedge \omega^2 - 2\omega_5^0 \wedge \omega^3 = 0, \\ \Omega_7^2 \wedge \omega^3 - 2\omega_6^0 \wedge \omega^2 = 0, \\ \Omega_7^1 \wedge \omega^3 - 2\omega_6^0 \wedge \omega^1 = 0, \end{array} \tag{5.80}$$

$$\Omega_7^1 \wedge \omega^1 = 0, \quad \Omega_7^2 \wedge \omega^2 = 0, \quad \Omega_7^3 \wedge \omega^3 = 0, \tag{5.81}$$

and

$$\begin{array}{l} \Omega_7^3 \wedge \omega^2 + \Omega_7^2 \wedge \omega^3 = 0, \\ \Omega_7^1 \wedge \omega^3 + \Omega_7^3 \wedge \omega^1 = 0, \\ \Omega_7^2 \wedge \omega^1 + \Omega_7^1 \wedge \omega^2 = 0, \end{array} \tag{5.82}$$

where

$$\begin{array}{ll}
\Delta z_1 = dz_1 + z_1(2\omega_0^0 - 2\omega_3^3) - 2A\omega_2^0, & \Omega_7^1 = \omega_7^1 - \omega_4^0, \\
\Delta x_2 = dx_2 + x_2(2\omega_0^0 - 2\omega_3^3) - 2b\omega_1^0, & \Omega_7^2 = \omega_7^2 - \omega_5^0, \\
\Delta \alpha_2 = d\alpha_2 + \alpha_2(2\omega_0^0 - 3\omega_3^3 + \omega_2^2) + 4A\omega_3^0, & \Omega_7^3 = \omega_7^3 - \omega_6^0. \\
\Delta \beta_1 = d\beta_1 + \beta_1(2\omega_0^0 - 3\omega_3^3 + \omega_1^1) + 4b\omega_3^0. &
\end{array}$$

Application of Cartan's lemma to equations (5.80)–(5.82) gives

$$\begin{array}{ll}
2\omega_4^0 = -\rho\omega^1 - s_1\omega^3, & \Delta\alpha_2 = p_2\omega^3 + q_2\omega^1, \\
\Delta z_1 = s_1\omega^2 + t_1\omega^3, & \Delta x_2 = q_2\omega^3 + r_2\omega^1, \\
\Delta\beta_1 = t_1\omega^2 + u_1\omega^3, & 2\omega_5^0 = -\rho\omega^2 - r_2\omega^3, \\
2\omega_6^0 = -\rho\omega^3 &
\end{array} \tag{5.83}$$

and

$$\begin{aligned}
\omega_7^1 - \omega_4^0 &= \rho\omega^1, \\
\omega_7^2 - \omega_5^0 &= \rho\omega^2, \\
\omega_7^3 - \omega_6^0 &= \rho\omega^3.
\end{aligned} \tag{5.84}$$

Let us prove that by an appropriate change of the vertex A_7 of our moving frames, we can reduce the quantity ρ to 0.

We write the expression of dA_7 using relations (5.77), (5.79), (5.83), and (5.84):

$$\begin{aligned}
dA_7 = \ & \omega_7^0 A_0 + (-\tfrac{1}{2}s_1\omega^3 + \rho\omega^1)A_1 + (-\tfrac{1}{2}r_2\omega^3 + \rho\omega^2)A_2 + \rho\omega^3 A_3 \\
& + \omega_1^0 A_4 + \omega_2^0 A_5 + (\omega_3^0 + z_1\omega^3)A_6 + \omega_7^7 A_7.
\end{aligned}$$

Consider the point $\widetilde{A}_7 = A_7 - \rho A_0$. Its differential has the form:

$$\begin{aligned}
d\widetilde{A}_7 = \ & (\omega_7^0 - d\rho + \rho(\omega_7^7 - \omega_0^0))A_0 - \tfrac{1}{2}s_1\omega^3 A_1 - \tfrac{1}{2}r_2\omega^3 A_2 \\
& + \omega_1^0 A_4 + \omega_2^0 A_5 + (\omega_3^0 + z_1\omega^3)A_6 + \omega_7^7 \widetilde{A}_7.
\end{aligned}$$

If we change our moving frames by replacing the vertex A_7 by the point $\widetilde{A}_7$, then we reduce the quantity ρ to 0:

$$\rho = 0. \tag{5.85}$$

By (5.85), equations (5.83) and (5.84) take the form:

$$\begin{array}{ll}
2\omega_4^0 = -s_1\omega^3, & \Delta\alpha_2 = p_2\omega^3 + q_2\omega^1, \\
\Delta z_1 = s_1\omega^2 + t_1\omega^3, & \Delta x_2 = q_2\omega^3 + r_2\omega^1, \\
\Delta\beta_1 = t_1\omega^2 + u_1\omega^3, & 2\omega_5^0 = -r_2\omega^3, \\
2\omega_6^0 = 0. &
\end{array} \tag{5.86}$$

$$\omega_7^1 - \omega_4^0 = 0, \quad \omega_7^2 - \omega_5^0 = 0, \quad \omega_7^3 - \omega_6^0 = 0. \tag{5.87}$$

Exterior differentiation of equations (5.87) gives the following three exterior quadratic equations:

$$\omega_7^0 \wedge \omega^1 = 0, \quad \omega_7^0 \wedge \omega^2 = 0, \quad \omega_7^0 \wedge \omega^3 = 0. \tag{5.88}$$

It follows from equations (5.88) that

$$\omega_7^0 = 0. \tag{5.89}$$

Exterior differentiation of equation (5.89) leads to an identity.

Exterior differentiation of equations (5.86) gives the following quadratic equations:

$$\begin{array}{ll} \Delta s_1 \wedge \omega^3 = 0, & \Delta p_2 \wedge \omega^3 + \Delta q_2 \wedge \omega^1 = 0, \\ \Delta s_1 \wedge \omega^2 + \Delta t_1 \wedge \omega^3 = 0, & \Delta q_2 \wedge \omega^3 + \Delta r_2 \wedge \omega^1 = 0, \\ \Delta t_1 \wedge \omega^2 + \Delta u_1 \wedge \omega^3 = 0, & \Delta r_2 \wedge \omega^3 = 0, \end{array} \tag{5.90}$$

where

$$\begin{array}{ll} \Delta s_1 & = ds_1 + s_1(3\omega_0^0 - \omega_2^2 - 2\omega_3^3) - 2z_1\omega_2^0, \\ \Delta t_1 & = dt_1 + t_1(3\omega_0^0 - 3\omega_3^3) + 4z_1\omega_3^0 - 2\beta_1\omega_2^0, \\ \Delta u_1 & = du_1 + u_1(3\omega_0^0 - 4\omega_3^3 + \omega_2^2) \\ & \quad +10\beta_1\omega_3^0 + 4A(b\omega_1^0 + A\omega_2^0) + 2A(s_1\omega^2 + r_2\omega^1), \\ \Delta p_2 & = dp_2 + p_2(3\omega_0^0 - 4\omega_3^3 + \omega_1^1) \\ & \quad +10\alpha_2\omega_3^0 + 4b(b\omega_1^0 + A\omega_2^0) + 2b(s_1\omega^2 + r_2\omega^1), \\ \Delta q_2 & = dq_2 + q_2(3\omega_0^0 - 3\omega_3^3) + 4x_2\omega_3^0 - 2\alpha_2\omega_1^0, \\ \Delta r_2 & = dr_2 + r_2(3\omega_0^0 - \omega_1^1 - 2\omega_3^3) - x_2\omega_1^0. \end{array}$$

Application of Cartan's lemma to equations (5.90) gives:

$$\begin{array}{ll} \Delta s_1 = S_1\omega^3, & \Delta p_2 = P_2\omega^3 + Q_2\omega^1, \\ \Delta t_1 = S_1\omega^2 + T_1\omega^3, & \Delta q_2 = Q_2\omega^3 + R_2\omega^1, \\ \Delta u_1 = T_1\omega^2 + U_1\omega^3, & \Delta r_2 = R_2\omega^3. \end{array} \tag{5.91}$$

Exterior equations (5.90) are independent and they contain six unknown 1-forms $\Delta s_1, \Delta t_1, \Delta u_1, \Delta p_2, \Delta q_2$ and Δr_2.

Thus, we have:

$$q = 6, \; s_1 = 6, \; s_2 = 0, \; Q = s_1 + 2s_2 = 6.$$

Equations (5.91) prove that the number N of parameters, on which the most general three-dimensional integral element depends, is equal to 6: $N = 6$. Since $N = Q$, *the system is in involution, and its general integral manifold* $V^3 \subset P^7$, *carrying three one-dimensional asymptotic distributions with two of them being distributions of straight lines, depends on six arbitrary functions of one variable.*

This result implies that the general submanifolds V^3 carrying a complete system of one-dimensional asymptotic distributions, are not necessarily reduced to submanifolds with all asymptotic distributions consisting of straight lines.

5.5 Submanifolds with a Complete System of Conjugate Distributions

1. In Chapter 3 we considered a submanifold V^m having m independent one-dimensional conjugate directions at each of its points. In this section we consider submanifolds carrying systems of conjugate directions of arbitrary dimension.

Consider a complete system of distributions $\Delta^{p_1}, \ldots, \Delta^{p_k}$ on V^m. This system is called a *conjugate system of distributions* if any two directions $\xi \in \Delta_x^{p_\lambda}$ and $\eta \in \Delta_x^{p_\mu}$, $\lambda \neq \mu$, at any point $x \in V^m$ are conjugate with respect to the system of the second fundamental forms $\Phi^\alpha_{(2)}$ of V^m, i. e.

$$\Phi^\alpha_{(2)}(\xi, \eta) = 0. \tag{5.92}$$

We will associate a subbundle of moving frames with the submanifold V^m in such a way that the vertex A_0 of each frame of the subbundle coincides with the point $x \in V^m$: $x = A_0$, and the vertices $A_{a_\lambda} \in \Delta_x^{p_\lambda}, \lambda = 1, \ldots, k$. Then it follows from (5.92) that the components of the second fundamental tensors $\{b^\alpha_{ij}\}$ of V^m satisfy the condition:

$$b^\alpha_{a_\lambda b_\mu} = 0, \ \ \lambda \neq \mu. \tag{5.93}$$

Equations (5.93) imply that the second fundamental forms of a submanifold V^m are reduced to the form:

$$\Phi^\alpha_{(2)} = \sum_\lambda b^\alpha_{a_\lambda b_\lambda} \omega^{a_\lambda} \omega^{b_\lambda}. \tag{5.94}$$

It follows that the maximal number of linearly independent forms among the forms $\Phi^\alpha_{(2)}$ satisfies the inequality:

$$m_1 \leq \frac{1}{2} \sum_\lambda p_\lambda (p_\lambda + 1). \tag{5.95}$$

Thus the dimension of the second osculating subspace $T_x^{(2)}(V^m)$ does not exceed the number $m + \frac{1}{2}\sum_\lambda p_\lambda(p_\lambda + 1)$.

We consider a two-component conjugate system in more detail. Suppose that $A_a \in \Delta_x^p$ and $A_u \in \Delta_x^{m-p}$ where $a = 1, \ldots, p$ and $u = p+1, \ldots, m$. Then conjugacy condition (5.93) takes the form:

$$b^\alpha_{au} = 0, \tag{5.96}$$

and the second fundamental forms of a submanifold V^m take the form:

$$\Phi^\alpha_{(2)} = b^\alpha_{ab} \omega^a \omega^b + b^\alpha_{uv} \omega^u \omega^v. \tag{5.97}$$

By (5.96) and (2.21), the expressions for the forms ω_i^α defining a displacement of the tangent subspace $T_x^{(1)}(V^m)$ take the form:

$$\omega_a^\alpha = b_{ab}^\alpha \omega^b, \quad \omega_u^\alpha = b_{uv}^\alpha \omega^v. \tag{5.98}$$

By (5.2), the 1-forms ω_a^u and ω_u^a are expressed as linear combinations of the basis forms ω^a and ω^u:

$$\omega_a^u = l_{ab}^u \omega^b + l_{av}^u \omega^v, \quad \omega_u^a = l_{ub}^a \omega^b + l_{uv}^a \omega^v. \tag{5.99}$$

Let us clarify the geometric meaning of the coefficients in equations (5.99). First of all, from Theorem 5.1 it follows that the conditions

$$l_{[ab]}^u = 0 \tag{5.100}$$

and

$$l_{[uv]}^a = 0 \tag{5.101}$$

are necessary and sufficient for the integrability of the distributions Δ^p and Δ^{m-p} respectively.

Suppose that the distribution Δ^p is integrable on the submanifold V^m. Consider further a family of subspaces Δ_x^{m-p} along one of the integral surfaces V^p of the distribution Δ^p, $x \in V^p$. This family of subspaces forms an m-parameter submanifold W^m in the space P^n with plane generators of dimension $m-p$. Let us prove that *this submanifold W^m is tangentially degenerate of rank p*. We take an arbitrary point $Y \in \Delta_x^{m-p}$: $Y = y^0 A_0 + y^u A_u$ and find the differential of this point modulo ω^u. By (5.98) and (5.99), we find that

$$dY \equiv (dy^0 + y^0\omega_0^0 + y^u\omega_u^0)A_0 + (dy^u + y^v\omega_v^u)A_u + (y^0\delta_a^b + y^u l_{ub}^a)A_a \pmod{\omega^u}. \tag{5.102}$$

This implies that the tangent subspace to the submanifold W^m at the point Y is determined by the points A_0, A_u and A_a and does not depend on the location of the point Y in the plane Δ_x^{m-p}. By the definition of a tangentially degenerate submanifold (see Section **4.1**), this means that the submanifold W^m is tangentially degenerate of rank p.

By (4.19), the focus variety of the generator Δ_x^{m-p} of the submanifold W^m has the equation:

$$\det(y^0\delta_a^b + y^u l_{ub}^a) = 0. \tag{5.103}$$

Thus the geometric meaning of the coefficients l_{ub}^a appearing in equations (5.99) is that these coefficients determine the focus variety of the generator Δ_x^{m-p}. The coefficients l_{au}^v have a similar geometric meaning.

Note that if the distribution Δ^p on a submanifold V^m is not integrable, we still can consider the family of integral curves of this distribution which pass through a point x and consider equation (5.102) along these integral curves. This brings us again to the focus variety (5.103) of the plane Δ_x^{m-p}.

2. To continue our consideration of two-component conjugate systems, we need the prolongation of equations (5.98). Taking their exterior derivatives and applying equations (5.99), we obtain:

$$\begin{array}{l}\{\nabla b^\alpha_{ab} + (b^\alpha_{ac} l^c_{ub} + b^\alpha_{vu} l^v_{ab})\omega^u\} \wedge \omega^b + (b^\alpha_{ab} l^b_{uv} + b^\alpha_{wu} l^w_{av})\omega^u \wedge \omega^v = 0, \\ \{\nabla b^\alpha_{uv} + (b^\alpha_{uw} l^w_{av} + b^\alpha_{ba} l^b_{uv})\omega^a\} \wedge \omega^v + (b^\alpha_{uv} l^v_{ab} + b^\alpha_{ca} l^c_{ub})\omega^a \wedge \omega^b = 0, \end{array} \quad (5.104)$$

where

$$\begin{array}{lll}\nabla b^\alpha_{ab} & = & db^\alpha_{ab} + b^\alpha_{ab}\omega^0_0 - b^\alpha_{ac}\omega^c_b - b^\alpha_{cb}\omega^c_a + b^\beta_{ab}\omega^\alpha_\beta, \\ \nabla b^\alpha_{uv} & = & db^\alpha_{uv} + b^\alpha_{uv}\omega^0_0 - b^\alpha_{uw}\omega^w_v - b^\alpha_{wv}\omega^w_u + b^\beta_{uv}\omega^\alpha_\beta, \end{array}$$

It is easy to see that the 1-forms in the curly braces in equations (5.104) can be expressed in terms of the basis forms ω^b and ω^v. If we substitute these expressions into equations (5.104), we must obtain identities. In this process the second terms on the left-hand sides of equations (5.104) will not have similar terms, and therefore they must vanish. This implies that the alternated coefficients in the products $\omega^u \wedge \omega^v$ and $\omega^a \wedge \omega^b$ in equations (5.104) are equal to 0:

$$\begin{array}{l} b^\alpha_{ab}(l^b_{uv} - l^b_{vu}) + b^\alpha_{uw} l^w_{av} - b^\alpha_{vw} l^w_{au} = 0, \ u \neq v, \\ b^\alpha_{uv}(l^v_{ab} - l^v_{ba}) + b^\alpha_{ac} l^c_{ub} - b^\alpha_{bc} l^c_{ua} = 0, \ a \neq b. \end{array} \quad (5.105)$$

Let us contract these relations with the basis points $\widetilde{A}_\alpha$ of the normal subspace $N^{(1)}_x(V^m)$. As above, we use the notations:

$$\widetilde{B}_{ab} = b^\alpha_{ab}\widetilde{A}_\alpha, \quad \widetilde{B}_{uv} = b^\alpha_{uv}\widetilde{A}_\alpha. \quad (5.106)$$

As a result, we find that

$$\widetilde{B}_{ab}(l^b_{uv} - l^b_{vu}) + \widetilde{B}_{uw} l^w_{av} - \widetilde{B}_{vw} l^w_{au} = 0, \ u \neq v, \quad (5.107)$$

$$\widetilde{B}_{uv}(l^v_{ab} - l^v_{ba}) + \widetilde{B}_{ac} l^c_{ub} - \widetilde{B}_{bc} l^c_{ua} = 0, \ a \neq b. \quad (5.108)$$

The points $\widetilde{B}_{ab}$ and $\widetilde{B}_{uv}$ define the reduced normal subspaces $\widetilde{N}^{(1)}_x(V^m)$ in the normal subspace $N^{(1)}_x(V^m)$, and thus these points define the second osculating subspace $T^{(2)}_x(V^m)$.

Theorem 5.7 *Let V^m be a submanifold carrying a complete two-component conjugate system $\{\Delta^p, \Delta^{m-p}\}$ and suppose that the dimension of the second osculating subspace $T^{(2)}_x(V^m)$ at any point x reaches its maximum, which is equal to*

$$m + \frac{1}{2}p(p+1) + \frac{1}{2}(m-p)(m-p+1).$$

Then:

1) *The submanifold V^m is foliated by an $(m-p)$-parameter family of submanifolds V^p and a p-parameter family of submanifolds V^{m-p}.*

2) *The tangentially degenerate submanifolds $'W^m$ and $''W^m$, formed by the planes Δ_x^{m-p} along each of the submanifolds V^p and by the planes Δ_x^p along each of the submanifolds V^{m-p} respectively, are conic submanifolds, with $(m-p-1)$- and $(p-1)$-dimensional vertices, respectively.*

3) *The second fundamental forms of the submanifolds V^p are the same at all points of their intersection with a fixed submanifold V^{m-p} and vice versa.*

Proof. By the conditions of the theorem, relations (5.107) and (5.108) imply conditions (5.100) and (5.101) of integrability of the distributions Δ^p and Δ^{m-p}. This proves the first conclusion of the theorem.

By condition (5.100), condition (5.108) takes the form:

$$\widetilde{B}_{ac} l^c_{ub} - \widetilde{B}_{bc} l^c_{ua} = 0, \ a \neq b. \tag{5.109}$$

Collecting similar terms in this equation, we can write it in the form:

$$\sum_{c \neq b} \widetilde{B}_{ac} l^c_{ub} - \sum_{c \neq a} \widetilde{B}_{bc} l^c_{ua} + \widetilde{B}_{ab}(l^b_{ub} - l^a_{ua}) = 0. \tag{5.110}$$

Since the points $\widetilde{B}_{ab}$ are linearly independent, equations (5.110) give

$$l^a_{ub} = 0, \ a \neq b, \ l^a_{ua} = l^b_{ub}. \tag{5.111}$$

Equations (5.111) can be written as follows:

$$l^a_{ub} = \delta^a_b l_u. \tag{5.112}$$

Substituting these values of l^a_{ub} into equations (5.103) of the focus variety, we arrive at the following equation:

$$(y^o + y^u l_u)^p = 0. \tag{5.113}$$

This means that the focus variety of the generator Δ_x^p of a tangentially degenerate submanifold $''W^m$ is a p-fold $(m-p-1)$-plane. Thus, by Theorem 4.6, the submanifold $''W^m$ is a cone with a $(m-p-1)$-dimensional vertex. The conclusion 2) of the theorem for the submanifold $'W^m$ can be proved in a similar manner.

Next we specialize the moving frames associated with a point $x \in V^m$ by choosing the points A_u to lie in the $(m-p-1)$-dimensional vertex of the cone $''W^m$ and the points A_a to lie in the $(p-1)$-dimensional vertex of the cone $'W^m$. Then we obtain $l^u = l^a = 0$, and equation (5.106) and the equation analogous to (5.106) take the form:

$$l^a_{ub} = 0, \ l^v_{au} = 0. \tag{5.114}$$

By (5.114), formulas (5.99) become

$$\omega_a^u = l_{ab}^u \omega^b, \quad \omega_u^a = l_{uv}^a \omega^v, \tag{5.115}$$

where the coefficients l_{ab}^u and l_{uv}^a are symmetric.

We will now prove the third conclusion of the theorem. Consider the second fundamental forms of the submanifold V^p. They have the form:

$$\Phi_{(2)}^\alpha = b_{ab}^\alpha \omega^a \omega^b, \quad \Phi_{(2)}^u = l_{ab}^u \omega^a \omega^b. \tag{5.116}$$

Let us find the differential equations which the coefficients of these forms satisfy. For the coefficients b_{ab}^α we find from equation (5.104) that

$$(\nabla b_{ab}^\alpha + l_{ab}^v \omega_v^\alpha) \wedge \omega^b = 0. \tag{5.117}$$

Exterior differentiation of the first group of equations (5.115) leads to the following quadratic equations:

$$(\nabla l_{ab}^u + b_{ab}^\alpha \omega_\alpha^u) \wedge \omega^b = 0, \tag{5.118}$$

where

$$\nabla l_{ab}^u = dl_{ab}^u + l_{ab}^u \omega_0^0 - l_{ac}^u \omega_b^c + l_{ab}^v \omega_v^u.$$

Equations (5.117) and (5.118) imply that the system of second fundamental forms with the basis (5.116) remains unchanged if $\omega^b = 0$, i.e. along a submanifold V^{m-p}. In a similar way we can prove that the system of the second fundamental forms of a submanifold V^{m-p} remains unchanged along a submanifold V^p. ■

Since the submanifolds $'W^m$ and $''W^m$, formed by the planes Δ_x^{m-p} along each of the submanifolds V^p, and by the planes Δ_x^p along each of the submanifolds V^{m-p} respectively, are conic submanifolds, the two-component conjugate system described in Theorem 5.6 is called *conic*.

The reader can find more details on two-component conjugate systems in the paper [A 66] by Akivis.

3. A two-component integrable conjugate system on a tangentially nondegenerate hypersurface V^m of a projective space P^{m+1} can be constructed in the following way. Let E^q be a fixed q-dimensional subspace in $P^{m+1}, 1 \leq q \leq m-1$, which is in general position with the hypersurface V^m. This means that the subspace E^q does not contain points of the hypersurface V^m and does not belong to any of its tangent hyperplanes. We construct a subspace E^{q+1} of dimension $q+1$ passing through a point $x \in V^m$ and the subspace E^q. The subspace E^{q+1} intersects the hypersurface V^m along a q-dimensional submanifold V^q, $V^q = V^m \bigcap E^{q+1}$. The submanifolds V^q form the first family of the two-component conjugate system which we are constructing. This family depends on $p = m - q$ parameters.

In order to construct the second family of submanifolds of the two-component conjugate system, at every point of the hypersurface V^m we consider the

p-dimensional direction Δ_x^p which passes through the point x and is conjugate to the q-dimensional direction Δ_x^q tangent to the submanifold V^q. We will prove that the distribution Δ^p is involutive on V^m. To prove this, we associate with a point x a family of moving frames in the manner indicated above in this section. On the hypersurface V^m, equations (5.98) take the form:

$$\omega_a^{m+1} = b_{ab}\omega^b, \quad \omega_u^{m+1} = b_{uv}\omega^v. \tag{5.119}$$

We will also assume that the points A_u and A_n of our moving frames belong to the fixed subspace E^q. Then we find that

$$\omega_u^0 = \omega_u^a = \omega_n^0 = \omega_n^a = 0. \tag{5.120}$$

The distribution Δ^p is defined on the hypersurface V^m by the system of equations $\omega_0^u = 0$, and the condition of complete integrability of this system is (5.100). It follows from the equation $\omega_u^a = 0$ of the system (5.120) that $l_{ub}^a = 0$. This implies that the second subsystem of the system (5.105) takes the form:

$$b_{uv}(l_{ab}^v - l_{ba}^v) = 0.$$

Since the hypersurface V^m is tangentially nondegenerate, we have $\det(b_{uv}) \neq 0$. It follows that condition (5.100) holds, and the distribution Δ^p is involutive. Denote the integral submanifolds of this distribution by V^p.

Consider now the manifold described by the plane Δ_x^q (which is tangent to the submanifold V^q) when the point x moves along the submanifold V^p. The plane Δ_x^q is spanned by the points A_0 and A_u. Since equations (5.119) imply that $\omega_u^{m+1} \equiv 0 \pmod{\omega^v}$, by (5.120), we find that

$$dA_u \equiv \omega_u^v A_v \pmod{\omega^v}.$$

This means that when the point x moves along the submanifold V^p, the plane $A_{p+1} \wedge \ldots \wedge A_m = E^{q-1}$ remains constant, and the plane Δ_x^q describes a cone with the vertex E^{q-1}. Thus the submanifold V^p is a "shadow submanifold" from a light source which is uniformly distributed along the plane E^{q-1}.

Therefore, we have proved the following result.

Theorem 5.8 *Suppose that V^m is a tangentially nondegenerate hypersurface of a projective space P^{m+1}, and E^q is a subspace of P^{m+1} of dimension q, $1 \leq q \leq m-1$, which is in general position with V^m. Then E^q induces a two-component conjugate system. One family of submanifolds of this system is formed on V^m by $(q+1)$-planes passing through the subspace E^q, and the second family is formed by the shadow submanifolds from light sources which are uniformly distributed along the planes of dimension $q-1$ belonging to E^q.*

■

Note that the two-component conjugate system described in this theorem is a multidimensional generalization of the Koenigs net on a two-dimensional surface of a three-dimensional projective space (see, for example, the book [Vy 49]). In this case we have $m = 2, q = 1$ and the subspace E^q becomes the axis of the Koenigs net.

NOTES

5.2. Asymptotic distributions on multidimensional submanifolds were studied by Lumiste in [Lu 59]. In particular, Theorem 5.2 is due to him.

5.3. Multicomponent systems of asymptotic distributions were considered by Lumiste (see [Lu 59]). In particular, he proved Theorem 5.3.

Submanifolds $V^m \subset P^n$ carrying a family of rectilinear generators were considered by Waksman in [Wak 65], Svoboda in [Svo 67] and [Svo 69], and Kruglyakov and Rybchenko in [KR 73].

Submanifolds V^3 with several families of rectilinear generators were considered by Muracchini in [M 54] and by Ruscior in [Ru 59]. Ruscior in [Ru 63] and Ivanova-Karatopraklieva in [IK 68] considered ruled hypersurfaces $V^3 \subset P^4$. Barner and Walter in [BW 69] studied ruled submanifolds $V^2 \subset P^4$. Baimuratov in [Ba 75a] investigated the structure of $V^3 \subset P^5$ with four families of rectilinear generators. Kruglyakov studied ruled submanifolds $V^3 \subset P^5$ (see [Kr 67a], [Kr 67b], [Kr 68] and [Kr 73]). Gluzdov in [Gl 70] considered hypersurfaces $V^4 \subset P^5$ carrying a family of two-dimensional generators. Juza in [Ju 63a] and [Ju 63b] considered monosystems $V^3 \subset P^6$ (one-parameter families of 2-planes) and introduced asymptotic and quasiasymptotic lines of such systems.

Kruglyakov in [Kr 69], [Kr 71a] and [Kr 71b] and in his joint paper [ShK 73] with Shcherbakov investigated submanifolds $V^m \subset P^n$ which carry an $(m-p)$-parameter family of p-dimensional generators and whose rank is greater than $m - p$.

5.4. The example which we consider in this section was briefly described by É. Cartan in his note [Ca 18]. However, Cartan did not give an investigation of the system of differential equations defining such submanifolds $V^3 \subset P^7$. He only indicated that a detailed study of this system is very long and that "it seems that its general solution depends only on arbitrary constants". As a solution, he indicated the submanifolds V^3 defined by the equations

$$x^4 = x^2 x^3, \quad x^5 = x^3 x^1, \quad x^6 = x^1 x^2, \quad x^7 = x^1 x^2 x^3.$$

For such submanifolds, the asymptotic curves of all three families are straight lines.

Our considerations show that besides this solution, there also exist submanifolds $V^3 \subset P^7$ carrying two families of straight lines and a family of curvilinear asymptotic lines. The problem of existence of submanifolds $V^3 \subset P^7$ carrying two or three families of curvilinear asymptotic lines remains open.

5.5. Two-component conjugate systems were investigated by Akivis in [A 66]. Multi-component conjugate systems were considered by Ryzhkov in [Ry 58] and by Akivis in [A 70a].

Degen in [De 67] and Walter in [Wal 68] considered two-component conjugate systems on $V^{n-1} \subset P^n$ with involutive distributions Δ_1 and Δ_2. In particular, in [De 67], Degen introduced the Darboux tensors of Δ_1 and Δ_2, proved that if Δ_1 defines the shadow submanifolds (along which V^{n-1} and the hypercone whose generators define Δ_2 are tangent), then one of the Darboux tensors vanishes; he also considered the case when the shadow submanifolds belong to some $P^s \subset P^n$. Walter in [Wal

68] studied systems on V^{n-1} for which the integral manifolds of the distribution Δ_1 are projectively mapped onto one another by means of the integral manifolds of the distribution Δ_2. In [De 68] Degen kinematically defined a projective analog of a canal hypersurface and established its relation with the hypersurface $V^{n-1} \subset P^n$ possessing a family of generating $(n-2)$-quadrics. (This hypersurface was considered in [Wal 68].)

Two-component conjugate systems generalizing the Koenigs nets were introduced by Akivis in [A 81b]. A particular case of this generalization was considered by Norden and Chebysheva in [NC 74] in their study of the geometry of a hypersurface in a space with degenerate metric.

Chapter 6

Normalized Submanifolds in a Projective Space

6.1 The Problem of Normalization of a Submanifold in a Projective Space

1. For a two-dimensional surface V^2 in a three-dimensional Euclidean space E^3, in addition to the tangent plane $T_x^{(1)}$ at every point x, one can naturally define the normal line n_x which is perpendicular to the tangent plane $T_x^{(1)}$. Moreover, the tangent plane $T_x^{(1)}$ contains the line l_x at infinity not passing through the point x, and the normal line n_x contains the point m_x at infinity. These objects, n_x, l_x and m_x, represent natural normalizing elements of a surface V^2 which are invariant with respect to the group of motions of the Euclidean space E^3. All metric properties of the surface V^2 are connected with this normalization.

The normalization of a submanifold V^m of a Euclidean space E^n is defined in a similar manner: in this case the normal space is a $(n-m)$-dimensional subspace n_x perpendicular to the tangent space $T_x^{(1)}$, the tangent space $T_x^{(1)}$ contains the $(m-1)$-plane l_x at infinity, and the normal space n_x contains the $(n-m-1)$-plane m_x at infinity.

The situation is more complicated in the case of an affine space A^n. For a submanifold $V^m \subset A^n$, we can still define the $(m-1)$-plane $l_x \subset T_x^{(1)}$ at infinity but there is no unique subspace n_x of dimension $n-m$ in a first order neighborhood of a point $x \in V^m$. However, if the subspace n_x is somehow defined, then it is naturally to choose its $(n-m-1)$-plane at infinity as the $(n-m-1)$-plane m_x.

The problem of finding a normalization, i.e. subspaces n_x, l_x and m_x, for a submanifold V^m in a projective space P^n is even more complicated since none of these subspaces is uniquely defined in a first order neighborhood of a point $x \in V^m$.

There are two ways for determining subspaces n_x, l_x and m_x associated with a point x of a submanifold V^m in a projective space P^n:

1. The subspaces n_x, m_x and l_x can be arbitrarily assigned at any point $x \in V^m$.

2. The subspaces n_x, m_x and l_x are intrinsically determined by means of elements connected with neighborhoods of orders higher than one of a point $x \in V^m$. Such a normalization is called *invariant.*

We will discuss both of these ways. First, we will consider an arbitrary normalization of a submanifold V^m by means of families of subspaces n_x, m_x and l_x, and next, for some classes of submanifolds V^m, we will indicate the ways to construct invariant normalizations.

2. Let us give the precise definitions. Let V^m be an m-dimensional submanifold in a projective space P^n and $T_x^{(1)}$ be its tangent subspace at the point $x \in V^m$. A subspace n_x of dimension $n - m$ of the space P^n having with the subspace $T_x^{(1)}$ only one common point x is said to be the *normal of the first kind* of the submanifold V^m at the point x. From here on, for brevity, we will call the normal of the first kind simply "the first normal". A subspace $l_x \subset T_x^{(1)}$ of dimension $m - 1$ not passing through the point x is called the *normal of the second kind* of the submanifold V^m at the point x. From here on, for brevity, we will call the normal of the second kind simply "the second normal". A subspace m_x of dimension $n - m - 1$ belonging to the first normal n_x but not passing through the point x is called the *subnormal of the first kind normal* (the subnormal of the first normal). If at any point $x \in V^m$, there are assigned unique first and second normals n_x and l_x and the subnormal m_x, then the submanifold V^m is called the *equipped* or *normalized* submanifold (see [N 76]). As always, we assume that the normals n_x, l_x and m_x are differentiable functions of a point x.

Since for a normalized submanifold V^m a family of first normals depends on m parameters and each of the first normals is of dimension $n - m$, this family is a *congruence* of $(n - m)$-planes in the space P^n. A family of $(m - 1)$-dimensional second normals of a normalized submanifold V^m also depends on m parameters, and thus this family is a *pseudocongruence* of $(m - 1)$-planes in the space P^n (cf. the corresponding definitions in Section **4.6**).

Let us now study a normalized submanifold V^m in P^n in more detail. With V^m we associate a subbundle of projective frames defined by the conditions that the point A_0 coincides with a variable point x, the points A_i, $i = 1, \ldots, m$, lie in the second normal l_x, and the points A_α, $\alpha = m + 1, \ldots, n$, lie in the subnormal m_x. If a point $x = A_0$ is fixed, then the admissible infinitesimal displacements of projective frames, associated with the point x as indicated above, have the following form:

$$\delta A_0 = \pi_0^0 A_0, \quad \delta A_i = \pi_i^j A_j, \quad \delta A_\alpha = \pi_\alpha^\beta A_\beta. \tag{6.1}$$

In these equations, the forms π_u^v are invariant forms of the stationary subgroup of the element $\{A_0, l_x, n_x, m_x\}$ of the normalized submanifold V^m. Since $\pi_0^0 + \pi_i^i + \pi_\alpha^\alpha = 0$, this subgroup is the direct product of the groups $\mathbf{GL}(m)$ and $\mathbf{GL}(n-m)$.

This implies that in addition to the forms ω_i^α which are expressed in terms of the basis forms ω^i by formulas (2.21):

$$\omega_i^\alpha = b_{ij}^\alpha \omega^j, \quad b_{ij}^\alpha = b_{ji}^\alpha, \tag{6.2}$$

the forms ω_α^i, ω_i^0 and ω_α^0 can also be expressed in terms of the basis forms ω^i:

$$\omega_\alpha^i = c_{\alpha k}^i \omega^k, \tag{6.3}$$

$$\omega_i^0 = l_{ij} \omega^j, \tag{6.4}$$

$$\omega_\alpha^0 = c_{\alpha i} \omega^i. \tag{6.5}$$

Let us find focal elements of the congruence of first normals n_x and the pseudocongruence of second normals l_x of a normalized submanifold V^m. We will apply considerations similar to those in Section **4.6**. Let $Y = y^0 A_0 + y^\alpha A_\alpha$ be an arbitrary point of the normal n_x. Then from the condition $dY \in n_x$, we find the equation of the focus variety F in the form:

$$\det(y^0 \delta_j^i + y^\alpha c_{\alpha j}^i) = 0 \tag{6.6}$$

(cf. (4.56)). This proves that the point A_0 of V^m does not belong to the focus variety F. An arbitrary hyperplane through the second normal l_x of V^m has the equation:

$$\xi_0 y^0 + \xi_\alpha y^\alpha = 0.$$

This implies that equation (4.60) of the focus cone Φ with the $(m-1)$-dimensional vertex l_x has the form:

$$\det(\xi_0 l_{ij} + \xi_\alpha b_{ij}^\alpha) = 0. \tag{6.7}$$

The focus variety $F \subset n_x$ allows us to construct a special normalization for which the subnormal m_x is the polar of the point A_0 with respect to the focus variety F.

If we denote the left-hand side of equation (6.6) by $F = F(y^0, y^\alpha)$, then we can write the equation of the polar of the point A_0 with respect to the focus variety (6.6) in the form:

$$\left.\frac{\partial F}{\partial y^0}\right|_{A_0} y^0 + \left.\frac{\partial F}{\partial y^\alpha}\right|_{A_0} y^\alpha = 0.$$

But it easy to see from (6.6) that

$$\left.\frac{\partial F}{\partial y^0}\right|_{A_0} = m, \quad \left.\frac{\partial F}{\partial y^\alpha}\right|_{A_0} = c^i_{\alpha i},$$

and the equation of the polar takes the form:

$$my^0 + y^\alpha c^i_{\alpha i} = 0.$$

If we place the points A_α into this polar, then its equation has the form $y^0 = 0$, and therefore, under this normalization, we have:

$$c^i_{\alpha i} = 0. \tag{6.8}$$

We will call a normalization of this kind *minimal*.

Exterior differentiation of equations (6.3), (6.4) and (6.5) gives the following exterior quadratic equations:

$$(\nabla c^i_{\alpha j} - \delta^i_j c_{\alpha k}\omega^k) \wedge \omega^j = 0, \tag{6.9}$$

$$(\nabla l_{ij} + b^\alpha_{ij} c_{\alpha k}\omega^k) \wedge \omega^j = 0, \tag{6.10}$$

$$(\nabla c_{\alpha j} - l_{ij} c^i_{\alpha k}\omega^k) \wedge \omega^j = 0, \tag{6.11}$$

where

$$\begin{aligned} \nabla c^i_{\alpha j} &= dc^i_{\alpha j} - c^i_{\alpha k}\theta^k_j + c^k_{\alpha j}\theta^i_k - c^i_{\beta j}\theta^\beta_\alpha, \\ \nabla l_{ij} &= dl_{ij} - l_{ik}\theta^k_j - l_{kj}\theta^i_k, \\ \nabla c_{\alpha j} &= dc_{\alpha j} - c_{\alpha k}\theta^k_j - c_{\beta j}\theta^\beta_\alpha, \end{aligned} \tag{6.12}$$

and

$$\theta^i_j = \omega^i_j - \delta^i_j\omega^0_0, \quad \theta^\alpha_\beta = \omega^\alpha_\beta - \delta^\alpha_\beta\omega^0_0.$$

These equations imply that the quantities $c^i_{\alpha j}, l_{ij}$ and $c_{\alpha j}$ are tensors. Note that this can be foreseen since, as was indicated earlier, the group of admissible transformations of moving frames is the group $\mathbf{GL}(m) \times \mathbf{GL}(n-m)$.

We will say that a normalization of a submanifold V^m is *harmonic* if the tensor l_{ij} is symmetric for this normalization. It is easy to see that a normalization of V^m is harmonic if and only if $d\omega^0_0 = 0$, i.e. the form ω^0_0 is a total differential.

3. We will now indicate other special normalizations of a submanifold V^m. If the first normals belong to a bundle of $(n-m)$-planes with a $(n-m-1)$-dimensional center Z, then by identifying the subnormal m_x with this center, we obtain

$$\omega^i_\alpha = 0, \quad \omega^0_\alpha = 0. \tag{6.13}$$

These equations and equations (6.3) and (6.5) imply

$$c^i_{\alpha j} = 0, \quad c_{\alpha j} = 0. \tag{6.14}$$

Equation (6.6) proves that in this case, the focus variety F coincides with the m-fold center Z. A normalization of this kind is called *central.*

The second special normalization of a submanifold V^m arises if all the second normals l_x and the subnormals m_x belong to a fixed hyperplane ξ. If the hyperplane ξ coincides with the coordinate hyperplane α^0 of the tangential frame (see Section **1.3**), the conditions for this hyperplane to be fixed are

$$\omega^0_i = 0, \quad \omega^0_\alpha = 0, \tag{6.15}$$

and these conditions lead to the equations:

$$l_{ij} = 0, \quad c_{\alpha i} = 0. \tag{6.16}$$

Such a normalization is called *affine*. It is obvious that an affine normalization is harmonic. However, it is not necessarily minimal since, in general, equations (6.8) do not hold. If we have an affine normalization of V^m, then the geometry of an affinely normalized submanifold V^m in a projective space P^n is reduced to the geometry of a normalized submanifold V^m in an affine space A^n whose hyperplane at infinity is the hyperplane ξ.

We will now consider the so-called *quadratic normalization.* This kind normalization is given on a submanifold V^m by means of a fixed hyperquadric Q in the space P^n. Since this hyperquadric induces the geometry of a non-Euclidean space in P^n, the geometry of a submanifold with a quadratic normalization is reduced to the geometry of a submanifold of a non-Euclidean space.

We write the equations of a hyperquadric Q in the form $(Y, Y) = 0$. We assume that the point $A_0 \in V^m$ does not belong to the hyperquadric Q. This allows us to normalize this point by the condition $(A_0, A_0) = 1$. In the tangent space $T^{(1)}_x$, we choose the second normal l_x at the point x of V^m in such a way that this normal is polar-conjugate to the point $A_0 = x$ with respect to the hyperquadric Q. We also choose the first normal n_x in such a way that it is polar-conjugate to the normal l_x with respect to the hyperquadric Q. In addition, we assume that the subnormal m_x is polar-conjugate to the point A_0 with respect to the hyperquadric Q. All these conditions can be expressed in the following form:

$$(A_0, A_0) = 1, \quad (A_0, A_i) = 0, \quad (A_0, A_\alpha) = 0, \quad (A_i, A_\alpha) = 0. \tag{6.17}$$

Define also:

$$(A_i, A_j) = g_{ij}, \quad (A_\alpha, A_\beta) = g_{\alpha\beta}. \tag{6.18}$$

The quantities g_{ij} and $g_{\alpha\beta}$ are nondegenerate symmetric tensors, and the equation of the hyperquadric Q has the form:

$$(y^0)^2 + g_{ij} y^i y^j + g_{\alpha\beta} y^\alpha y^\beta = 0.$$

By differentiating equations (6.17) and (6.18) we obtain:

$$\begin{aligned} &\omega_0^0 = 0, \ \ \omega_i^0 + g_{ij}\omega^j = 0, \ \ \omega_\alpha^0 + g_{\alpha\beta}\omega_0^\beta = 0, \ \ g_{ij}\omega_\alpha^j + g_{\alpha\beta}\omega_i^\beta = 0, \\ &dg_{ij} = g_{ik}\omega_j^k + g_{kj}\omega_i^k, \ \ dg_{\alpha\beta} = g_{\gamma\beta}\omega_\alpha^\gamma + g_{\alpha\gamma}\omega_\beta^\gamma. \end{aligned} \tag{6.19}$$

Since $\omega_0^0 = 0$, the latter equations can be written in the form:

$$\nabla g_{ij} = 0, \ \ \nabla g_{\alpha\beta} = 0. \tag{6.20}$$

Equations (6.19) also imply that

i) $l_{ij} = -g_{ij}$ (see (6.4)). Since the tensor g_{ij} is symmetric, a quadratic normalization is harmonic.

ii) Since $\omega_0^\beta = 0$, we have $\omega_\alpha^0 = 0$, i.e. $c_{\alpha i} = 0$.

iii) $\omega_\alpha^i = -g^{ij} g_{\alpha\beta}\omega_j^\beta$, where g^{ij} is the inverse tensor of the tensor g_{ij}. From this and equations (6.2) and (6.3) it follows that

$$c_{\alpha j}^i = -g^{ik} g_{\alpha\beta} b_{kj}^\beta,$$

i.e. the tensor $c_{\alpha j}^i$ can be expressed in terms of the second fundamental tensor of the submanifold V^m.

Note also that if for a quadratic normalization condition (6.8) holds, then the above written equation implies that

$$g^{ij} b_{ij}^\alpha = 0.$$

But this condition expresses the fact that the manifold V^m is minimal in that non-Euclidean geometry which is induced by the hyperquadric Q in the space P^n. This was the reason for calling "minimal" a normalization satisfying condition (6.8).

6.2 The Affine Connection on a Normalized Submanifold

1. If a submanifold V^m in P^n is normalized by means of subspaces n_x, l_x and m_x, then on V^m two connections are induced: the connection in the tangent bundle with V^m as its base and $T_x^{(1)}$ with a chosen second normal as its typical fiber, and the connection in the normal bundle with the same base V^m and with $\{n_x, m_x\}$ as its typical fiber.

For studying these connections, we first write the exterior differentials of the basis forms ω^i of the submanifold V^m:

$$d\omega^i = \omega_0^0 \wedge \omega^i + \omega^j \wedge \omega_j^i = \omega^j \wedge \theta_j^i, \tag{6.21}$$

where $\theta_j^i = \omega_j^i - \delta_j^i \omega_0^0$. As we indicated in Chapter 2, the forms θ_j^i are related to the differential structure on V^m. If a point x is fixed, the forms θ_j^i are structure forms of the general linear group $\mathbf{GL}(m)$ of admissible transformations of the coframe $\{\omega^i\}$ on the submanifold V^m (see [KN 63] or [Lap 66]). The exterior differentials of these forms are determined by equations:

$$d\theta_j^i = \theta_j^k \wedge \theta_k^i + (\delta_k^i \omega_j^0 + \delta_j^i \omega_k^0) \wedge \omega^k + b_{jk}^\alpha \omega^k \wedge \omega_\alpha^i. \tag{6.22}$$

However, if V^m is normalized, the forms ω_α^i and ω_i^0 have the form (6.3) and (6.4). As a result, equations (6.22) become:

$$d\theta_j^i = \theta_j^k \wedge \theta_k^i + (-\delta_k^i l_{jm} - \delta_j^i l_{km} + b_{jk}^\alpha c_{\alpha m}^i)\omega^k \wedge \omega^m. \tag{6.23}$$

Equations (6.21) and (6.23) imply that the forms ω^i and θ_j^i define an affine connection γ_t (see the end of Section **1.2**) in the tangent bundle of a normalized submanifold V^m. Geometrically, this connection is defined as a projection of the tangent subspace $T_{x+dx}^{(1)}$ to V^m at a point $x + dx$, infinitesimally close to the point x, onto the space $T_x^{(1)}$ from the center m_x. This is the reason that the forms θ_j^i are called the *connection forms* of the connection γ_t. These forms allow us to define a covariant differentiation in the connection γ_t. For example, for $(1,2)$-tensor t_{jk}^i, its covariant differential has the form:

$$\nabla t_{jk}^i = dt_{jk}^i - t_{lk}^i \theta_j^l - t_{jl}^i \theta_k^l + t_{ij}^l \theta_l^k = t_{jkl}^i \omega^l, \tag{6.24}$$

where t_{jkl}^i are covariant derivatives of the tensor t_{jk}^i which themselves form $(1,3)$-tensor. Equations (6.21) and (6.23) prove that the torsion tensor of the connection γ_t is equal to zero, and its curvature tensor is expressed by the following formula:

$$R_{jkh}^i = -\delta_{[k}^i l_{|j|h]} - \delta_j^i l_{[kh]} + b_{j[k}^\alpha c_{|\alpha|h]}^i. \tag{6.25}$$

It is easy to verify that the curvature tensor satisfies the *Ricci identity*: $R_{[jkh]}^i = 0$.

2. We will now write the expression for the curvature tensor in some of the special normalizations of V^m which were introduced earlier.

1. If a normalization of V^m is central, then, by relations (6.14), we have:

$$R_{jkh}^i = -\delta_{[k}^i l_{|j|h]} - \delta_j^i l_{[kh]}. \tag{6.26}$$

2. If a normalization of V^m is affine, then, by relations (6.16), we have:

$$R^i_{jkh} = b^\alpha_{j[k} c^i_{|\alpha|h]}. \tag{6.27}$$

3. If a normalization of V^m is quadratic, then, by relations (6.19), we have:

$$R^i_{jkh} = \delta^i_{[k} g_{h]j} - g^{il} g_{\alpha\beta} b^\alpha_{j[k} b^\beta_{h]l}. \tag{6.28}$$

The affine connection γ_t induced by a harmonic normalization is called *harmonic*.

3. Let us recall the definition of geodesics of a normalized submanifold V^m. A curve $x(t)$ on a normalized submanifold V^m is called a *geodesic* if, at any of its points, its osculating two-plane intersects the first normal n_x along a straight line. A curve $x(t)$ is defined on V^m by the equation:

$$\omega^i = \xi^i dt.$$

This implies that

$$dA_0 = \omega^0_0 A_0 + (\xi^i A_i)dt, \tag{6.29}$$

and the tangent line to the curve $x(t)$ is defined by the points A_0 and $\xi^i A_i$. The osculating two-plane of the curve $x(t)$ is defined by the same two points and the point:

$$d(\xi^i A_i) = (d\xi^i + \xi^j \omega^i_j)A_i + b^\alpha_{ij}\xi^i\xi^j A_\alpha dt. \tag{6.30}$$

This osculating two-plane and the first normal n_x have a common straight line if and only if the first term of the sum on the right-hand side of (6.30) belongs to the tangent $A_0 \wedge \xi^i A_i]$ to the curve $X(t)$. This proves that the equations of a geodesic have the form:

$$d\xi^i + \xi^j \omega^i_j = \theta\xi^i, \tag{6.31}$$

where θ is an 1-form.

4. Note that this definition does not depend on the choice of the second normal l_x. This proves that a change of the second normal implies a transformation of the connection γ_t preserving geodesics. Such a transformation of the connection γ_t is called *geodesic*.

Let us calculate the *tensor of projective curvature* of the connection γ_t. This tensor is invariant relative to geodesic transformations of γ_t. The tensor of projective curvature of γ_t is expressed by the formula (see [S 54], p. 289):

$$P^i_{jkh} = R^i_{jkh} - 2P_{[kh]}\delta^i_j + 2\delta^i_{[k} P_{h]j}. \tag{6.32}$$

This tensor satisfies the condition:

$$P^i_{jki} = 0 \tag{6.33}$$

which enables one to find the tensor P_{jk}. In our notation, the latter tensor has the form:

$$P_{jk} = \frac{1}{m^2 - 1}(R_{jk} + mR_{kj}), \tag{6.34}$$

where

$$R_{jk} = R^i_{jki} \tag{6.35}$$

is the Ricci tensor of the connection γ_t.

Using expression (6.25) for the curvature tensor, we find that

$$2R_{jk} = ml_{jk} - l_{kj} + b^\alpha_{jk} c^l_{\alpha l} - b^\alpha_{jl} c^l_{\alpha k}. \tag{6.36}$$

From formulas (6.34) and (6.36) we obtain:

$$2P_{jk} = l_{kj} + \frac{1}{m-1} b^\alpha_{jk} c^l_{\alpha l} - \frac{1}{m^2-1}(b^\alpha_{jl} c^l_{\alpha k} + mb^\alpha_{kl} c^l_{\alpha j}). \tag{6.37}$$

Substituting this expression into formula (6.32), after rather tedious calculations we obtain:

$$\begin{aligned} P^i_{jkh} = \; & \tfrac{1}{2}(b^\alpha_{jk} c^i_{\alpha h} - b^\alpha_{jh} c^i_{\alpha k}) \\ & + \frac{1}{2(m-1)}(\delta^i_k b^\alpha_{hj} - \delta^i_h b^\alpha_{kj}) c^l_{\alpha l} + \frac{1}{2(m+1)} \delta^i_j (b^\alpha_{hl} c^l_{\alpha k} - b^\alpha_{kl} c^l_{\alpha h}) \\ & + \frac{1}{2(m^2-1)}[c^l_{\alpha j}(\delta^i_h b^\alpha_{kl} - \delta^i_k b^\alpha_{hl}) + mb^\alpha_{jl}(\delta^i_h c^l_{\alpha k} - \delta^i_k c^l_{\alpha h})]. \end{aligned} \tag{6.38}$$

Using formula (6.38), one can again verify condition (6.33).

Formula (6.38) proves that the tensor of projective curvature P^i_{jkh} is expressed in terms of the tensors b^α_{jk} and $c^j_{\alpha k}$ alone and does not depend on the quantities l_{ij}. This proves that *the tensor P^i_{jkh} is completely determined by a congruence of first normals of a submanifold V^m*.

If the tensor $c^j_{\alpha k}$ has the form:

$$c^i_{\alpha j} = c_\alpha \delta^i_j, \tag{6.39}$$

then the direct calculation proves that the tensor P^i_{jkh} of projective curvature vanishes, i.e. the connection γ_t becomes projectively Euclidean (see [S 54]). Geometrically, condition (6.39) means that the congruence of first normals n_x is a bundle of $(n-m)$-planes with a $(n-m-1)$-dimensional center Z defined by the equation $c_\alpha x^\alpha = 0$. Placing the points A_α into the center Z of this bundle, we arrive at the condition $c_\alpha = 0$ which implies relation (6.13).

If we project a submanifold V^m from the $(n-m-1)$-dimensional center Z onto an m-plane having only one common point with normals n_x, then geodesics of the submanifold V^m will be projected onto straight lines (see Figure 6.1). This projection gives a realization of a geodesic mapping of V^m on an m-dimensional plane.

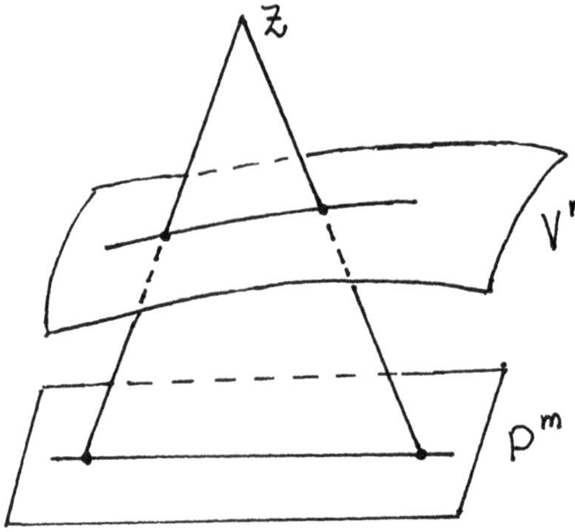

Figure 6.1

6.3 The Connection in the Normal Bundle

1. Suppose that a submanifold V^m is framed by families of first and second normals. However, we will not now assume that the subspaces m_x are fixed in the normals n_x. As earlier, we assume that the points A_i span the second normal l_x, and the points A_α belong to the first normal n_x. Then the equations of infinitesimal displacement of the frames associated with the point A_0 have the form:

$$\delta A_0 = \pi_0^0 A_0, \quad \delta A_i = \pi_i^j A_j, \quad \delta A_\alpha = \pi_\alpha^0 A_0 + \pi_\alpha^\beta A_\beta. \tag{6.40}$$

Consider the normal bundle associated with a submanifold V^m normalized by the first and second normals. The forms ω^i are the base forms of V^m, and the forms $\theta_\alpha^\beta = \omega_\alpha^\beta - \delta_\alpha^\beta \omega_0^0$ and ω_α^0 are its fiber forms.

Equations (6.3) and (6.4) hold in the normal bundle, and equations (6.5) do not hold in this bundle. This implies the following form for equations (6.9) and (6.10):

$$(\nabla c_{\alpha j}^i - \delta_i^j \omega_\alpha^0) \wedge \omega^j = 0, \tag{6.41}$$

$$(\nabla l_{ij} + b_{ij}^\alpha \omega_\alpha^0) \wedge \omega^j = 0. \tag{6.42}$$

It follows from these equations that now the quantities $c^i_{\alpha j}$ form a quasitensor. This quasitensor can be always represented in the form:

$$c^i_{\alpha j} = \overset{\circ}{c}{}^i_{\alpha j} + \frac{1}{m}\delta^j_i c_\alpha, \tag{6.43}$$

where $c_\alpha = c^i_{\alpha i}$ and $\overset{\circ}{c}{}^i_{\alpha i} = 0$, i.e. $\overset{\circ}{c}{}^i_{\alpha j}$ is trace-free part of this quasitensor. Using equation (6.41), one can easily prove that this trace-free part is a tensor relative to all admissible transformations of the moving frames associated with a point A_0 of a normalized submanifold V^m. The tensor $\overset{\circ}{c}{}^i_{\alpha j}$ is connected with the congruence of first normals of V^m.

Similarly, equations (6.42) prove that now the quantities l_{ij} do not form a tensor but they form a tensor together with the quantities b^α_{ij}. Nevertheless, as the same equations (6.42) imply, the skew-symmetric part $l_{[ij]}$ of the quasitensor l_{ij} forms a tensor. The tensor $l_{[ij]}$ is connected with the pseudocongruence of second normals of V^m. As we noted earlier, the vanishing of this tensor is equivalent to the condition $d\omega^0_0 = 0$.

Note that when we defined the affine connection γ_t in the tangent bundle, we were forced to reduce the group of projective transformations (6.40) of the moving frames to (6.1) since the quantities l_{ij}, and consequently the quantities (6.25), defining the curvature tensor of the connection γ_t, do not form a tensor under transformations (6.40). When we consider an affine connection in the normal bundle, this reduction is not necessary, and we do not do it.

With the normal bundle of a normalized submanifold V^m, a distribution $\Delta^m \subset P^n$ is associated. This distribution can be constructed as follows. Let Y be an arbitrary point of the first normal n_x:

$$Y = y^0 A_0 + y^\alpha A_\alpha.$$

Consider the m-plane defined by this point and the second normal l_x. This m-plane and the first normal n_x have only one common point Y. A set of m-planar elements Δ^m_Y defined by a point Y and the m-plane $Y \wedge l_x$ is an m-dimensional distribution. We will call this distribution the *horizontal distribution* of the normal bundle. This distribution is invariant under transformations of the subgroup of the group of projective transformations preserving a point $x \in V^m$. If $\omega^i = 0$, the fiber forms θ^α_β and ω^0_α indicated above are invariant forms of this subgroup.

This implies that the distribution Δ^m defines a connection in the normal bundle. This connection is called the *normal connection* of a normalized submanifold V^m and is denoted by γ_n. The fiber forms θ^β_α and ω^0_α are called the *connection forms* of the connection γ_n.

The connection γ_n is not an affine connection since a fiber in the normal bundle in which this connection is defined is a centroprojective space n_x. By this reason, we can call this connection *centroprojective*. The operator $\nabla^\perp$ which is defined for a tensor t^α_β by the formula:

$$\nabla^\perp t^\alpha_\beta = dt^\alpha_\beta - t^\alpha_\gamma \theta^\gamma_\beta + t^\gamma_\beta \theta^\alpha_\gamma \tag{6.44}$$

does no longer lead to covariant derivatives that are invariant. However, if we make a specialization of admissible frames by placing the points A_α in the polar of the point A_0 relative to the focus variety F (see Section **6.1**), the forms ω_α^0 become principal (see formula (6.41)), and the connection γ_n becomes affine. After this specialization is completed, the operator $\nabla^\perp$ becomes the operator of covariant differentiation and leads to covariant derivatives of invariant nature. In our example, we have:

$$\nabla^\perp t_\beta^\alpha = t_{\beta k}^\alpha \omega^k$$

and the quantities $t_{\beta k}^\alpha$ themselves form a tensor.

In what follows, we will not assume that the moving frames are specialized as indicated above but we will use the notation $\nabla^\perp$.

The combined operator $\overline{\nabla} = (\nabla, \nabla^\perp)$, which is used for mixed tensors of type t_{ij}^α and has the form:

$$\overline{\nabla} t_{ij}^\alpha = dt_{ij}^\alpha - t_{kj}^\alpha \theta_i^k - t_{ik}^\alpha \theta_j^k + t_{ij}^\beta \theta_\beta^\alpha, \tag{6.45}$$

is called the *operator of van der Waerden–Bortolotti* (see, for example, [Ch 73]). Actually, we have already used this operator in Chapter 2 where we simply denoted it by ∇. However, since a submanifold V^m considered in Chapter 2 was not normalized, application of this operator did not lead to covariant derivatives of invariant nature.

In our further considerations, as we did in Chapter 2, we will denote each of these three operators by ∇—the practice shows that this will not bring us to an ambiguity since Latin and Greek indices take on different values.

2. Let us find the structure equations of the connection γ_n. To do this, we apply exterior differentiation to the fiber forms θ_β^α and ω_α^0:

$$\begin{aligned} d\theta_\beta^\alpha &= \theta_\beta^\gamma \wedge \theta_\gamma^\alpha + \omega_\beta^i \wedge \omega_i^\alpha - \delta_\beta^\alpha \omega^i \wedge \omega_i^0, \\ d\omega_\alpha^0 &= \theta_\alpha^\beta \wedge \omega_\beta^0 + \omega_\alpha^i \wedge \omega_i^0. \end{aligned} \tag{6.46}$$

Using equations (6.2), (6.3) and (6.4), we write equations (6.46) in the form:

$$\begin{aligned} d\theta_\beta^\alpha - \theta_\beta^\gamma \wedge \theta_\gamma^\alpha &= R_{\beta ij}^\alpha \omega^i \wedge \omega^j, \\ d\omega_\alpha^0 - \theta_\alpha^\beta \wedge \omega_\beta^0 &= R_{\alpha ij}^0 \omega^i \wedge \omega^j, \end{aligned} \tag{6.47}$$

where

$$R_{\beta ij}^\alpha = c_{\beta[i}^k b_{|k|j]}^\alpha - \delta_\beta^\alpha l_{[ij]}, \tag{6.48}$$

$$R_{\alpha ij}^0 = c_{\alpha[i}^k l_{|k|j]}. \tag{6.49}$$

The exterior quadratic forms (6.47) are called the *curvature forms* of the normal connection γ_n, and their coefficients (6.48) and (6.49) are called the *curvature tensor* of this connection.

Let us apply relations (6.43) to find another form for expressions (6.48) and (6.49). By symmetry of the tensor b^α_{ij}, the first of these expressions takes the form:

$$R^\alpha_{\beta ij} = \tilde{c}^k_{\beta[i} b^\alpha_{|k|j]} - \delta^\alpha_\beta l_{[ij]}, \tag{6.50}$$

and the second one can be written as follows:

$$R^0_{\alpha ij} = \tilde{c}^k_{\alpha[i} l_{|k|j]} + \frac{1}{m} c_\alpha l_{[ij]}. \tag{6.51}$$

Since all quantities in the right-hand side of expression (6.50) are tensors, the quantities $R^\alpha_{\beta ij}$ form a tensor relative to all admissible transformations of the moving frames associated with a point x of a normalized submanifold V^m. On the other hand, the quantities $R^0_{\alpha ij}$ taken separately do not form a tensor. However, the quantities $R^0_{\alpha ij}$ taken together with the quantities $R^\alpha_{\beta ij}$ form a tensor.

We will now write the expression for the curvature tensor of the normal connection γ_n for special normalizations introduced in Section **6.1**.

1. If a normalization of V^m is central, then, by relations (6.13), we have:

$$R^\alpha_{\beta ij} = -\delta^\alpha_\beta l_{[ij]}, \quad R^0_{\alpha ij} = 0. \tag{6.52}$$

2. If a normalization of V^m is affine, then, by relations (6.16), we have:

$$R^\alpha_{\beta ij} = \tilde{c}^k_{\beta[i} b^\alpha_{|k|j]}, \quad R^0_{\alpha ij} = 0. \tag{6.53}$$

3. If a normalization of V^m is quadratic, then, by relations (6.19), we have:

$$R^\alpha_{\beta ij} = -g^{kl} g_{\beta\gamma} b^\gamma_{l[i} b^\alpha_{j]k}, \quad R^0_{\alpha ij} = 0. \tag{6.54}$$

Consider the tensor:

$$\widetilde{R}_{ij} = R^\alpha_{\alpha ij} = \tilde{c}^k_{\alpha[i} b^\alpha_{|k|j]} - m l_{[ij]}. \tag{6.55}$$

This tensor is called the *Ricci tensor* of the normal connection γ_n. Since, by (6.36), the alternated Ricci tensor of the connection γ_n is expressed by the formula:

$$R_{[jk]} = \frac{1}{2}\left((m+1) l_{[jk]} - b^\alpha_{[j|l|} \tilde{c}^l_{|\alpha|k]}\right), \tag{6.56}$$

it is easy to find from equations (6.55) and (6.56) that

$$4R_{[jk]} - 2\widetilde{R}_{jk} = l_{[jk]}. \tag{6.57}$$

In the case of the harmonic normalization, we have:

$$\widetilde{R}_{jk} = 2R_{[jk]}. \tag{6.58}$$

Equations (6.56), (6.50) and (6.58) give the following result.

Theorem 6.1 *The connection γ_t is harmonic if and only if one of the following conditions holds:*

(i) *Its alternated Ricci tensor is completely defined by a congruence of first normals of a submanifold V^m.*

(ii) *The subtensor $\{R^\alpha_{\beta ij}\}$ of the curvature tensor $\{R^\alpha_{\beta ij}, R^0_{\beta ij}\}$ of the normal connection γ_n is completely defined by a congruence of first normals of a submanifold V^m.*

(iii) *The Ricci tensor of the normal connection γ_n is equal to the doubled alternated Ricci tensor of the connection γ_t.* ■

Relation (6.57) has an interesting geometric meaning. To clarify it, we recall the following definitions (cf. [N 76]). The affine connection γ_t is said to be *equiaffine* if its Ricci tensor is symmetric. The immediate verification proves that for such a connection the 1-form θ^i_i is a total differential, and the volume element $\omega^1 \wedge \omega^2 \wedge \ldots \omega^m$ is preserved under parallel displacement. If the tensor $\widetilde{R}_{ij}$ vanishes: $\widetilde{R}_{ij} = 0$, then the connection γ_n is said to be *Ricci-flat.*

Relations (6.57) imply the following theorem:

Theorem 6.2 *If any two of the following conditions hold:*

(i) *The connection γ_t is harmonic;*

(ii) *The connection γ_t is equiaffine;*

(iii) *The connection γ_n is Ricci-flat,*

then the third condition also holds. ■

3. We will now establish a geometric meaning for the curvature tensor of the normal connection γ_n of a normalized submanifold V^m. To do this, we first find the differential equations of the distribution Δ^m: Let

$$Y = A_0 + y^\alpha A_\alpha$$

be an arbitrary point of the subspace n_x. Its differential has the form:

$$dY = (\omega^0_0 + y^\alpha \omega^0_\alpha) Y + (\omega^i + y^\alpha \omega^i_\alpha) A_i + (\nabla y^\alpha - y^\alpha y^\beta \omega^0_\beta) A_\alpha, \tag{6.59}$$

where $\nabla y^\alpha = dy^\alpha + y^\beta \theta^\alpha_\beta$ is the covariant differential of y^α in the normal connection γ_n. If Y describes an integral curve of the distribution Δ^m, then the coordinates y^α of this point satisfy the condition:

$$\nabla y^\alpha - y^\alpha y^\beta \omega_\beta^0 = 0. \tag{6.60}$$

Equation (6.60) defines the parallel displacement of the point Y in the normal bundle of a submanifold V^m. This parallel displacement can be realized by integrating equation (6.60) along a curve l joining the points x and x' and belonging to V^m.

In general, this parallel displacement depends on a choice of a curve l. Let us find conditions under which this parallel displacement does not depend on a curve l. This condition is the condition of complete integrability of the system of equations (6.60). Applying exterior differentiation to equation (6.60) and excluding ∇y^α by means of the same equations (6.60), we arrive at the following exterior quadratic equations:

$$y^\beta(\omega_\beta^i \wedge \omega_i^\alpha - \delta_\beta^\alpha \omega^i \wedge \omega_i^0) - y^\alpha y^\beta \omega_\beta^i \wedge \omega_i^0 = 0. \tag{6.61}$$

Comparing these equations with equations (6.46), we see that the factors in y^β and $y^\alpha y^\beta$ are precisely the curvature forms of the normal connection γ_n. By the Frobenius theorem, condition (6.61) must be satisfied identically with respect to y^α. This implies that

$$\omega_\beta^i \wedge \omega_i^\alpha - \delta_\beta^\alpha \omega^i \wedge \omega_i^0 = 0, \quad \omega_\beta^i \wedge \omega_i^0 = 0. \tag{6.62}$$

Since the left-hand sides of these expressions are the curvature forms of the connection γ_n, condition (6.62) indicates that *the normal connection γ_n is flat.*

There is another geometric meaning of equations (6.62) giving the necessary and sufficient conditions for the system of equations (6.60) to be completely integrable.

To establish this meaning, we first give the following definition. A pair, consisting of a congruence $\{n_x\}$ of the first normals n_x and a pseudocongruence $\{l_x\}$ of the second normals l_x, is called *one-sided stratifiable in the direction from $\{n_x\}$ to $\{l_x\}$* if there exists a one-to-one correspondence between their elements, and there exists ∞^{n-m} of m-dimensional submanifolds (Y) which are transversal to the congruence $\{n_x\}$ and whose tangent $(n-m)$-subspaces taken at the points Y of their intersection with the normal n_x pass through the corresponding $(m-1)$-plane l_x of the pseudocongruence $\{l_x\}$.

If a point $Y = A_0 + y^\alpha A_\alpha$ describes such a submanifold, then, by the above definition, we have the condition

$$dY \wedge Y \wedge A_1 \wedge \ldots \wedge A_m = 0$$

which implies equations (6.60). Since relations (6.62) are the necessary and sufficient conditions for the system of equations (6.62) to be completely integrable, we arrive at the following result.

Theorem 6.3 *A parallel displacement of a point Y on a normalized submanifold V^m in the connection γ_n does not depend on a path of integration if and only if one of the following conditions holds:*

(i) *The curvature tensor of the connection* γ_n *vanishes.*

(ii) *A congruence* $\{n_x\}$ *of the first normals* n_x *and a pseudocongruence* $\{l_x\}$ *of the second normals* l_x *are one-sided stratifiable in the direction from* $\{n_x\}$ *to* $\{l_x\}$. ■

6.4 Submanifolds with Flat Normal Connection

1. In this section we will study submanifolds V^m admitting the harmonic normalization with flat normal connection. For such a submanifold equations (6.48) and (6.49) imply the following conditions:

$$l_{ij} = l_{ji}, \tag{6.63}$$

$$c^k_{\alpha i} b^\beta_{kj} = c^k_{\alpha j} b^\beta_{ki}, \tag{6.64}$$

$$c^k_{\alpha i} l_{kj} = c^k_{\alpha j} l_{ki}. \tag{6.65}$$

Suppose now that the focus variety F of the first normal does not have multiple components. We choose the point $A_{m+1} \in n_x$ in such a way that the straight line $A_0 A_{m+1}$ intersects the variety F at m distinct points. By equation (6.5), these points of intersection are determined by the equation:

$$\det(y^0 \delta^i_j + y^{m+1} c^i_{m+1,j}) = 0. \tag{6.66}$$

But this equation is the characteristic equation for the quasitensor $c^i_{m+1,j}$. Since the straight line $A_0 A_{m+1}$ contains m different points of intersection with the variety F, equation (6.66) must have m distinct roots. Thus, the quasitensor $c^i_{m+1,j}$ has m distinct eigenvalues. This proves that this tensor can be reduced to the diagonal form:

$$c^i_{m+1,j} = \delta^i_j c^i, \quad c^i \neq c^j \;\text{ if }\; i \neq j. \tag{6.67}$$

Consider now equations (6.64) and (6.65) for $\alpha = m+1$ and substitute the values of $c^i_{m+1,j}$ from (6.67) into these equations. This gives:

$$c^i b^\beta_{ij} = c^j b^\beta_{ji}, \quad c^i l_{ij} = c^j l_{ji}. \tag{6.68}$$

In this formula and in other formulas of this section there is no summation over the indices i and j. Since $c^i \neq c^j$ if $i \neq j$, it follows that

$$b^\alpha_{ij} = 0, \quad l_{ij} = 0 \;\text{ if }\; i \neq j. \tag{6.69}$$

The first of these relations proves that the submanifold V^m *carries a conjugate net* (see Section **3.1**). Moreover, by means of relations (6.69), equation (6.7) of the focus cone Φ takes the form:

$$\prod_{i=1}^{m}(\xi_0 l_{ii} + \xi_\alpha b^\alpha_{ii}) = 0, \tag{6.70}$$

i.e. *the focus cone Φ decomposes into m bundles of hyperplanes* whose centers are spanned by the $(m-1)$-plane l_x and one of the points

$$B_{ii} = l_{ii} A_0 + b^\alpha_{ii} A_\alpha. \tag{6.71}$$

Let us write equations (6.64) and (6.65) taking into account relations (6.69):

$$c^j_{\alpha i} b^\beta_{jj} = c^i_{\alpha j} b^\beta_{ii}, \quad c^j_{\alpha i} l_{jj} = c^i_{\alpha j} l_{ii}. \tag{6.72}$$

Contract first of these equations with the point A_β and add the obtained sum to the second of equations (6.72) multiplied by A_0. This gives:

$$c^j_{\alpha i} B_{jj} = c^i_{\alpha j} B_{ii}. \tag{6.73}$$

It follows from equation (6.73) that, if the points B_{ii} are mutually distinct, then we have:

$$c^j_{\alpha i} = 0, \quad i \neq j. \tag{6.74}$$

This equation implies that equation (6.6) of the focus variety F has the form:

$$\prod_{i=1}^{m}(y^0 + y^\alpha c^i_{\alpha i}) = 0. \tag{6.75}$$

This proves that the focus variety F decomposes into m planes of dimension $n-m-1$. Since, by our assumption, the variety F does not have multiple components, all these $(n-m-1)$-planes are distinct. Moreover, the equation

$$(y^0 \delta^i_j + y^\alpha c^i_{\alpha j})\omega^j = 0,$$

defining the focal directions in the congruence of first normals n_x (see Section **4.3**), proves that to each of these $(n-m-1)$-planes there corresponds the focal displacement along the curve C_i defined by the equations $\omega^j = 0, j \neq i$. In this case, the congruence of first normals n_x is said to be *conjugate* to the submanifold V^m, and the corresponding normalization of V^m is also called *conjugate*.

All these considerations prove the following theorem.

Theorem 6.4 *Suppose that a submanifold V^m has the harmonic normalization and the flat normal connection γ_n. If the focus variety F of each of its first normals n_x does not have multiple components, then the submanifold V^m carries a net of conjugate lines, and the focus cone Φ of each of its second normals*

l_x decomposes into m bundles of hyperplanes. If in addition, the points B_{ii}, which together with l_x define the centers of these bundles, are mutually distinct, then the focus variety F decomposes into m planes of dimension $n-m-1$, and the congruence of first normals n_x is conjugate to the submanifold V^m. Moreover, the pseudocongruence of second normals l_x is foliated by m families of torses corresponding to the lines of the conjugate net of V^m. ■

The last statement of Theorem 6.4 follows from formulas (6.4) and (6.69).

2. Consider now a submanifold V^m which is normalized by means of first and second normals n_x and l_x and which carries a conjugate net Σ. If the straight lines A_0A_i are tangent to the conjugate lines passing through the point A_0, then, as was established in Chapter 3, on V^m we have:

$$\omega^\alpha = 0, \quad \omega_i^\alpha = b_{ii}^\alpha \omega^i \tag{6.76}$$

and

$$\omega_j^i = l_{jk}^i \omega^k. \tag{6.77}$$

(see (3.5) and (3.14)). Since the submanifold V^m is normalized by means of first and second normals, we have:

$$\omega_\alpha^i = c_{\alpha j}^i \omega^j, \quad \omega_i^0 = l_{ij}\omega^j \tag{6.78}$$

(see (6.3) and (6.4)).

Let us find a geometric meaning for the condition:

$$l_{ij} = 0, \quad i \neq j, \tag{6.79}$$

which arose in the previous subsection (see (6.69)). Since

$$dA_i = \omega_i^0 A_0 + \omega_i^j A_j + \omega_i^\alpha A_\alpha, \tag{6.80}$$

by (6.76), (6.79) and (6.80), we have

$$dA_i - \omega_i^j A_j = (l_{ii}A_0 + b_{ii}^\alpha A_\alpha)\omega_i. \tag{6.81}$$

It follows that if the point A_0 moves in any direction conjugate to the direction A_0A_i, i.e. $\omega^i = 0$, the point A_i describes a curve which is tangent to the subspace $T_x^{(1)}(V^m)$. We will call such a point A_i the *strong pseudofocus* of the straight line A_0A_i. (For the general definition of a pseudofocus see [Cas 50] and [Ba 65b].) Thus all the points A_i are strong pseudofoci of the straight lines A_0A_i.

Now it is not so difficult to prove the following result.

Theorem 6.5 *If a submanifold V^m of a space P^n carries a conjugate net and is normalized in such a way that its congruence of first normals n_x is conjugate to V^m and the points of intersection of the second normals l_x with the tangents to the conjugate lines are strong pseudofoci of these tangent lines, then the normal connection γ_n induced on V^m by this normalization is flat.*

Proof. If we take a subbundle of moving frames associated with V^m as indicated above, then the conditions of the theorem can be written as

$$b^\alpha_{ij} = 0, \;\; c^i_{\alpha j} = 0, \;\; l_{ij} = 0, \;\; i \neq j. \tag{6.82}$$

But these relations imply that the tensors of normal curvature, $R^\alpha_{\beta ij}$ and $R^0_{\alpha ij}$, defined by formulas (6.48) and (6.49) vanish. ■

3. Let us prove the existence of the conjugate–harmonic normalization of a submanifold V^m carrying a conjugate net Σ and admitting the flat normal connection γ_n.

By (6.69) and (6.74), equations (6.3) and (6.4) take the form:

$$\omega^i_\alpha = c^i_{\alpha i}\omega^i, \;\; \omega^0_i = l_{ii}\omega^i. \tag{6.83}$$

Exterior differentiation of (6.83) by means of (6.41), (6.42), (6.76) and (6.77) leads to the following conditions:

$$(c^j_{\alpha j} - c^i_{\alpha i})l^i_{jk} - (c^k_{\alpha k} - c^i_{\alpha i})l^i_{kj} = 0, \tag{6.84}$$

$$l_{ii}(l^i_{jk} - l^i_{kj}) + l_{jj}l^j_{ik} - l_{kk}l^k_{ij} = 0, \tag{6.85}$$

where $i \neq j, k, \;\; j \neq k$. Let us recall that equations (6.76) also imply the conditions:

$$b^\alpha_{ii}(l^i_{jk} - l^i_{kj}) + b^\alpha_{jj}l^j_{ik} - b^\alpha_{kk}l^k_{ij} = 0. \tag{6.86}$$

Relations (6.84), (6.85) and (6.86) form a homogeneous system of equations with respect to the quantities l^i_{jk}. In general, this system has only the trivial solution:

$$l^i_{jk} = 0, \;\; i \neq j, k, \;\; j \neq k. \tag{6.87}$$

Geometrically, this solution signifies that the net Σ on V^m is holonomic and that the lines of this net also form a conjugate net on any submanifold defined by the equation $\omega^i = 0$.

The system of equations defining the conjugate-harmonic normalization of a submanifold V^m consists of equations (6.83) and quadratic equations that are obtained by exterior differentiation of (6.83). By (6.87), these quadratic equations have the form:

$$(\nabla c^i_{\alpha i} - \omega^0_\alpha) \wedge \omega^i = 0, \;\; (\nabla l_{ii} + b^\alpha_{ii}\omega^0_\alpha) \wedge \omega^i = 0. \tag{6.88}$$

Applying Cartan's test to the system of equations (6.87) and (6.88), we find that $q = s_1 = m(n - m + 1)$, $s_2 = \ldots = s_m = 0$, $N = Q = m(n - m + 1)$.

Therefore, *the system, defining the indicated above conjugate-harmonic normalization of a submanifold V^m, is in involution, and a general integral manifold, defined by this system, depends on $m(n - m + 1)$ arbitrary functions of one variable.*

6.5 Intrinsic Normalization of Submanifolds

1. So far we have assigned the first and second normals, n_x and l_x, of a submanifold V^m by the points A_0, A_α and A_i of a moving frame. However, this is not always convenient. On many occasions, the first normal is assigned by the point A_0 and the points

$$Y_\alpha = A_\alpha + y^i_\alpha A_i, \tag{6.89}$$

and the second normal by the points

$$Z_i = A_i + z_i A_0. \tag{6.90}$$

As earlier, in this case, the normals n_x and l_x must be invariantly connected with the point $x = A_0$. Thus, if a point x is fixed, i.e. $\omega^i = 0$, the points Y_α must satisfy the conditions:

$$\delta Y_\alpha = \sigma^0_\alpha A_0 + \sigma^\beta_\alpha Y_\beta, \tag{6.91}$$

and the points Z_i the conditions:

$$\delta Z_i = \sigma^j_i Z_j. \tag{6.92}$$

Differentiating equations (6.89) and (6.90) by means of the operator δ (i.e. by setting $\omega^i = 0$) and using the equations of infinitesimal displacement of projective frames and equations (2.5) and (6.2), we find that

$$\delta Y_\alpha = \pi^\beta_\alpha Y_\beta + (\pi^0_\alpha + y^i_\alpha \pi^0_i) A_0 + (\delta y^i_\alpha + y^j_\alpha \pi^i_j - y^i_\beta \pi^\beta_\alpha + \pi^i_\alpha) A_i, \tag{6.93}$$

$$\delta Z_i = \pi^j_i Z_j + (\delta z_i + z_i \pi^0_0 - z_j \pi^j_i + \pi^0_i) A_0. \tag{6.94}$$

Comparing these equations with equations (6.91) and (6.92), we find that the conditions for the normals n_x and l_x to be invariant have the forms

$$\delta y^i_\alpha + y^j_\alpha \pi^i_j - y^i_\beta \pi^\beta_\alpha + \pi^i_\alpha = 0$$

and

$$\delta z_i + z_i \pi^0_0 - z_j \pi^j_i + \pi^0_i = 0$$

respectively, or

$$\nabla_\delta y^i_\alpha + \pi^i_\alpha = 0 \tag{6.95}$$

and

$$\nabla_\delta z_i + \pi^0_i = 0. \tag{6.96}$$

Note that if $y^i_\alpha = 0$ and $z_i = 0$, the last two equations are reduced to the form $\pi^i_\alpha = 0$ and $\pi^0_i = 0$. This corresponds to equations (6.3) and (6.4) and implies that the normal n_x is defined by the points A_0 and A_α and the normal l_x by the points A_i.

The quantities y^i_α satisfying equations (6.95) are called the *normalizing objects of the first kind* and the quantities z_i satisfying equations (6.96) are called the *normalizing objects of the second kind.* In Section **6.1** we assumed these quantities to be equal to zero and obtained a normalization of a submanifold V^m which however was not intrinsically connected with the geometry of a submanifold V^m.

The problem of construction of an invariant normalization that is intrinsically connected with the structure of a submanifold V^m is more difficult. To construct such a normalization, we must express the normalizing objects y^i_α and z_i in terms of the fundamental objects generated by the submanifold V^m, i.e. the objects which are obtained by differential prolongations of the initial equation $\omega^\alpha = 0$ of the submanifold V^m and those equations that are connected with a specific structure of this submanifold.

As we noted earlier, to define an affine connection γ_t in the tangent bundle, in addition to the first and second normals, n_x and l_x, we additionally have to find an $(n - m - 1)$-dimensional subnormal $m_x \subset n_x$ which is also invariantly connected with the point $x = A_0$. The latter subnormal can be given by the points:

$$Z_\alpha = Y_\alpha + z_\alpha A_0. \tag{6.97}$$

The condition for this subnormal to be invariant has the form:

$$\delta Z_\alpha = \tau^\beta_\alpha Z_\beta. \tag{6.98}$$

Differentiating equations (6.97) by means of the operator δ, we obtain

$$\delta Z_\alpha = \pi^\beta_\alpha Z_\beta + (\delta z_\alpha + z_\alpha \pi^0_0 - z_\beta \pi^\beta_\alpha + y^i_\alpha \pi^0_i + \pi^0_\alpha) A_0, \tag{6.99}$$

where y^i_α is the normalizing object of the first kind satisfying equations (6.95). Comparing these equations with equations (6.98), we obtain the condition for the subnormal m_x to be invariant in the form:

$$\delta z_\alpha + z_\alpha \pi^0_0 - z_\beta \pi^\beta_\alpha + y^i_\alpha \pi^0_i + \pi^0_\alpha = 0. \tag{6.100}$$

Note that if $y^i_\alpha = 0$ and $z_\alpha = 0$, these equations are reduced to the form $\pi^0_\alpha = 0$. This corresponds to equation (6.5) and proves that the subspace m_x is spanned by the points A_α.

Let us prove that if the geometric object y^i_α defining the first normal n_x is constructed, then, by its prolongation, we can easily construct an object z_α determining the subnormal m_x.

It follows from equations (6.95), which the object y^i_α satisfies, that

$$\nabla y^i_\alpha + \omega^i_\alpha = y^i_{\alpha k} \omega^k. \tag{6.101}$$

Exterior differentiation of these equations leads to the exterior quadratic equations:

$$\{\nabla y^i_{\alpha j} - \delta^i_j(\omega^0_\alpha + y^k_\alpha \omega^0_k) + y^k_\alpha b^\beta_{kj}\omega^i_\beta + y^i_\beta b^\beta_{kj}\omega^k_\alpha\} \wedge \omega^j = 0. \qquad (6.102)$$

Applying Cartan's lemma to these quadratic equations and setting $\omega^i = 0$, we find that

$$\nabla_\delta y^i_{\alpha j} - \delta^i_j(\pi^0_\alpha + y^k_\alpha \pi^0_k) + b^\beta_{kj}(y^k_\alpha \pi^i_\beta + y^i_\beta \pi^k_\alpha) = 0. \qquad (6.103)$$

Using equations (6.95) and the fact that $\nabla_\delta b^\beta_{kj} = 0$, we eliminate the forms π^i_α from equations (6.103). This gives:

$$\nabla_\delta(y^i_{\alpha j} - b^\beta_{kj} y^i_\beta y^k_\alpha) - \delta^i_j(\pi^0_\alpha + y^k_\alpha \pi^0_k) = 0. \qquad (6.104)$$

Contracting these equations with respect to the indices i and j, we obtain the equation:

$$\nabla_\delta(y^l_{\alpha l} - b^\beta_{kl} y^l_\beta y^k_\alpha) - m(\pi^0_\alpha + y^k_\alpha \pi^0_k) = 0. \qquad (6.105)$$

Comparing these equations with equations (6.100), we see that the geometric object z_α, defining the location of the subnormal m_x, can be determined by the formula:

$$z_\alpha = -\frac{1}{m}(y^l_{\alpha l} - b^\beta_{kl} y^l_\beta y^k_\alpha). \qquad (6.106)$$

One can easily prove that the subnormal m_x, defined by means of this geometric object z_α, is *the polar of the point A_0 with respect to the focus surface F of the first normal n_x.*

2. As an example of construction of the invariant normalization of a submanifold V^m which is intrinsically connected with its geometry, we will perform such a construction for a submanifold V^m belonging to its osculating subspace $T^{(2)}_x$ of dimension $m + m_1$, where $m_1 \leq \frac{1}{2}m(m+1)$.

On such a manifold, the following equations hold:

$$\omega^\alpha = 0, \qquad (6.107)$$

$$\omega^\alpha_i = b^\alpha_{ij}\omega^j, \;\; b^\alpha_{ij} = b^\alpha_{ji} \qquad (6.108)$$

(see (2.5) and (2.21)), and all second fundamental forms

$$\Phi^\alpha_{(2)} = b^\alpha_{ij}\omega^i\omega^j \qquad (6.109)$$

are linearly independent. The coefficients of these forms satisfy the following system of equations:

$$\nabla b^\alpha_{ij} = b^\alpha_{ijk}\omega^k, \tag{6.110}$$

where the coefficients b^α_{ijk} are symmetric in all lower indices (see formulas (2.26)), and as was indicated in Chapter 2, these coefficients do not form a tensor.

We assume that a relative invariant

$$I = I(b^\alpha_{ij}) \neq 0 \tag{6.111}$$

can be constructed from the components of the tensor b^α_{ij}, and that this invariant is a homogeneous algebraic polynomial of degree p in b^α_{ij}. Consider the tensors

$$b^{ij}_\alpha = \frac{\partial \ln I}{\partial b^\alpha_{ij}}. \tag{6.112}$$

By Euler's theorem on homogeneous functions, we have

$$\frac{\partial I}{\partial b^\alpha_{ij}} b^\alpha_{ij} = pI. \tag{6.113}$$

Using relation (6.112), we write the latter equation in the form:

$$b^{ij}_\alpha b^\alpha_{ij} = p. \tag{6.114}$$

On the other hand, relation (6.112) implies that

$$d \ln I = b^{ij}_\alpha \, db^\alpha_{ij}. \tag{6.115}$$

Substituting expressions db^α_{ij} from (6.110) into (6.115), we obtain

$$d \ln I = 2b^{ij}_\alpha b^\alpha_{ik}\theta^k_j - b^{ij}_\alpha b^\beta_{ij}\theta^\alpha_\beta + b^{ij}_\alpha b^\alpha_{ijk}\omega^k, \tag{6.116}$$

where $\theta^k_j = \omega^k_j - \delta^k_j\omega^0_0$ and $\theta^\beta_\alpha = \omega^\beta_\alpha - \delta^\beta_\alpha\omega^0_0$.

But the relative invariant I satisfies an equation of the type

$$dI = I(\theta + I_k\omega^k),$$

which can be written as

$$d \ln I = \theta + I_k\omega^k. \tag{6.117}$$

Comparing formulas (6.116) and (6.117), we see that the form θ must be a linear combination of the forms θ^i_j and θ^α_β. But since the form θ does not have free indices, it is of the form:

$$\theta = \lambda\theta^i_i - \mu\theta^\alpha_\alpha.$$

i.e.

$$d \ln I = \lambda \theta_i^i - \mu \theta_\alpha^\alpha + I_k \omega^k. \tag{6.118}$$

Comparing formulas (6.118) and (6.116), we find that

$$2b_\alpha^{ij} b_{ik}^\alpha = \lambda \delta_k^j, \quad b_\alpha^{ij} b_{ij}^\beta = \mu \delta_\alpha^\beta \tag{6.119}$$

and

$$I_k = b_\alpha^{ij} b_{ijk}^\alpha. \tag{6.120}$$

Contracting the first of equations (6.119) with respect to j and k and the second one with respect to α and β and taking into account formula (6.114), we find the values of λ and μ:

$$\lambda = \frac{2p}{m}, \quad \mu = \frac{p}{n-m}. \tag{6.121}$$

By (6.121), formula (6.118) becomes:

$$d \ln I = \frac{2p}{m} \theta_i^i - \frac{p}{n-m} \theta_\alpha^\alpha + I_k \omega^k. \tag{6.122}$$

Since the quantities b_α^{ij} form a tensor, they satisfy a system of differential equations of the form:

$$\nabla b_\alpha^{ij} = b_{\alpha k}^{ij} \omega^k. \tag{6.123}$$

If we take exterior derivatives of system (2.26), apply the Cartan lemma and set $\omega^i = 0$, we obtain

$$\nabla_\delta b_{ijk}^\alpha = 3 b_{(ij}^\beta \sigma_{k)\beta}^\alpha, \tag{6.124}$$

where

$$\sigma_{k\beta}^\alpha = b_{lk}^\alpha \pi_\beta^l - \delta_\beta^\alpha \pi_k^0. \tag{6.125}$$

In the same way, we find from equations (6.123) that

$$\nabla_\delta b_{\alpha k}^{ij} = -3 b_\alpha^{(jl} \sigma_{kl}^{i)}, \tag{6.126}$$

where

$$\sigma_{kl}^i = b_{kl}^\beta \pi_\beta^i - \delta_k^i \pi_l^0.$$

Consider the contraction $b_{ijk}^\alpha b_\beta^{jk}$. Equations (6.124) and (6.123) imply that

$$\nabla_\delta (b_{ijk}^\alpha b_\beta^{jk}) = W_{\beta i}^{\gamma l} \sigma_{l\gamma}^\alpha, \tag{6.127}$$

where

$$W^{\gamma l}_{\beta i} = \frac{p}{n-m}\delta^{\gamma}_{\beta}\delta^{l}_{i} + 2b^{jl}_{\beta}b^{\gamma}_{ij}. \tag{6.128}$$

The quantities $W^{\gamma l}_{\beta i}$ form a tensor, and therefore

$$\nabla_\delta W^{\gamma l}_{\beta i} = 0. \tag{6.129}$$

The number of equations in system (6.127) is the same as the number of forms $\sigma^{\alpha}_{l\gamma}$. This system can be solved with respect to the forms $\sigma^{\alpha}_{l\gamma}$ if the tensor $W^{\gamma l}_{\beta i}$ is nonsingular. Assuming that this is the case, we introduce the tensor $\widetilde{W}^{\beta i}_{\epsilon j}$ such that

$$W^{\gamma l}_{\beta i}\widetilde{W}^{\beta i}_{\epsilon j} = \delta^{\gamma}_{\epsilon}\delta^{l}_{j}. \tag{6.130}$$

Contracting relation (6.127) with the tensor $\widetilde{W}^{\beta i}_{\epsilon j}$, we obtain

$$\nabla_\delta(b^{\alpha}_{ijk}b^{jk}_{\beta}\widetilde{W}^{\beta i}_{\epsilon h}) = \sigma^{\alpha}_{h\epsilon}. \tag{6.131}$$

Set

$$b^{\alpha}_{ijk}b^{jk}_{\beta}\widetilde{W}^{\beta i}_{\epsilon h} = -A^{\alpha}_{h\epsilon}. \tag{6.132}$$

Then relations (6.131), (6.132) and (6.125) imply

$$\nabla_\delta A^{\alpha}_{i\beta} = -b^{\alpha}_{ik}\pi^{k}_{\beta} + \delta^{\alpha}_{\beta}\pi^{0}_{i}. \tag{6.133}$$

Contracting the constructed quantities $A^{\alpha}_{i\beta}$ with the tensor b^{ki}_{α}, we obtain a new object

$$l^{k}_{\beta} = A^{\alpha}_{i\beta}b^{ki}_{\alpha}. \tag{6.134}$$

Applying (6.119) and (6.121), we can find from equations (6.133) and (6.126) that this new object satisfies the following system of differential equations:

$$\nabla_\delta l^{i}_{\alpha} = -\frac{p}{m}\pi^{i}_{\alpha} + b^{ik}_{\alpha}\pi^{0}_{k}. \tag{6.135}$$

In order to construct the normalizing objects y^{i}_{α} and z_i satisfying equations (6.95) and (6.96), we must separate the forms in formula (6.135). For this we need to construct one more object of the same type as l^{i}_{α}.

It is possible to prove that such an object can be constructed, and it has the following form:

$$h^{i}_{\alpha} = \frac{p(2n-m)}{2m(n-m)}b^{il}_{\alpha l} + b^{ik}_{\alpha}b^{\beta}_{kjh}b^{jh}_{\beta}. \tag{6.136}$$

Applying (6.126), (6.123) and (6.124), we find that the object h^{i}_{α} satisfy the equations

$$\nabla_\delta h^i_\alpha = \frac{p}{2m(n-m)}\Big[-\frac{p(2n-m)}{n-m}\pi^i_\alpha + m(m+2)b^{ij}_\alpha \pi^0_j\Big]. \qquad (6.137)$$

Eliminate now the forms π^0_k from differential equations (6.135) and (6.137). As a result, we obtain

$$\nabla_\delta \Big[h^i_\alpha - \frac{p(m+2)}{2(n-m)} l^i_\alpha\Big] = \frac{p^2(n-m-1)}{2(n-m)^2}\pi^i_\alpha. \qquad (6.138)$$

Suppose that $n > m+1$, i.e. the submanifold V^m under consideration is not a hypersurface. Then we can set

$$\rho^i_\alpha = \frac{n-m}{p^2(n-m-1)}\Big[p(m+2)l^i_\alpha - 2(n-m)h^i_\alpha\Big]. \qquad (6.139)$$

Relations (6.138) and (6.139) imply

$$\nabla_\delta \rho^i_\alpha + \pi^i_\alpha = 0. \qquad (6.140)$$

Comparing equations (6.140) and (6.95), we find that *the constructed quantities* ρ^i_α *form a normalizing object of the first kind.*

To construct a normalizing object of the second kind, we eliminate the forms π^i_α from equations (6.140) and (6.135). As a result, we obtain

$$\nabla_\delta \big(\frac{p}{m}\rho^i_\alpha - l^i_\alpha\big) = -b^{ik}_\alpha \pi^0_k. \qquad (6.141)$$

If we contract equation (6.141) with the tensor $\frac{m}{p}b^\alpha_{ij}$, then we find that

$$\nabla_\delta \big(b^\alpha_{ij}(\rho^j_\alpha - \frac{m}{p}l^j_\alpha)\big) = -\pi^0_i. \qquad (6.142)$$

Introduce the notation:

$$\tau_i = b^\alpha_{ij}(\rho^j_\alpha - \frac{m}{p}l^j_\alpha). \qquad (6.143)$$

Then equations (6.142) and (6.143) imply

$$\nabla_\delta \tau_i + \pi^0_i = 0. \qquad (6.144)$$

Comparing equations (6.144) and (6.96), we find that *the constructed quantities* τ_i *form a normalizing object of the second kind.*

Formulas (6.134) and (6.136) imply that if $n > m+1$, the normalizing objects of the first and second kinds can be expressed in terms of the tensors b^α_{ij} and b^α_{ijk} connected with a third order differential neighborhood of the submanifold V^m. Thus, if $n > m+1$, there exists an invariant normalization of a submanifold $V^m \subset P^n$ defined by a third order differential neighborhood of V^m.

As we will see in Chapter 7, if $n = m+1$, i.e. for a hypersurface, the construction of invariant normalization of $V^m \subset P^{m+1}$ requires the use of objects connected with a fourth order differential neighborhood of the submanifold V^m.

As was noted in subsection **6.5.1**, if a family of first normals n_x is found for a submanifold $V^m \subset P^n$, there is a standard procedure to construct a family of subnormals $m_x \subset n_x$ but this requires the objects connected with a fourth order differential neighborhood of the submanifold V^m.

In our construction of invariant normalization for a submanifold $V^m \subset P^n$, we mostly followed the paper [Os 66] by Ostianu.

The assumption for existence of a nonvanishing invariant I, which is necessary for construction of the inverse tensor b^{ij}_α, was already substantiated in the papers of Weise (see [We 38]). The possibility to construct the inverse tensor for the tensor $W^{\gamma l}_{\beta i}$, which is necessary in our construction, was proved in the paper [Os 66] mentioned above.

6.6 Normalization of Submanifolds Carrying a Conjugate Net of Lines

1. We will construct now an invariant normalization of a submanifold V^m carrying a conjugate net. Suppose that $m > 2$ and that a conjugate net on V^m is not conic (see Section **3.6**). On such a submanifold the following equations hold:

$$\omega^\alpha = 0, \tag{6.145}$$

$$\omega_i^\alpha = b_i^\alpha \omega^i, \tag{6.146}$$

$$\omega_j^i = l^i_{ji}\omega^i + l^i_{jj}\omega^j + \sum_{k\neq i,j} l^i_{jk}\omega^k,\ i \neq j \tag{6.147}$$

(cf. (2.5), (3.5) and (3.14)). Exterior differentiation of equations (6.145) leads to an identity by means of relations (6.146). The exterior differentials of equations (6.146) have the form (3.21), and by relations (3.23), they can be written in the form:

$$\nabla b_i^\alpha \wedge \omega^i = 0; \tag{6.148}$$

in these equations as well as in (6.146) and (6.147) there is no summation in $i, j, \ldots$ unless there is the summation sign, and

$$\nabla b_i^\alpha = db_i^\alpha + b_i^\beta \omega_\beta^\alpha - b_i^\alpha(2\omega_i^i - \omega_0^0). \tag{6.149}$$

Differentiating equations (6.147), we obtain an exterior quadratic equation of the type

$$\sum_k \Delta l^i_{jk} \wedge \omega^k + \Omega^i_j = 0, \tag{6.150}$$

where the quadratic form Ω^i_j is a linear combination of exterior products of the basis forms ω^i and

$$\begin{aligned}\Delta l^i_{jj} &= \nabla l^i_{jj} + b^\alpha_j \omega^i_\alpha, \\ \Delta l^j_{ij} &= \nabla l^j_{ij} - \omega^0_i, \\ \Delta l^i_{jk} &= \nabla l^i_{jk}, \quad i \neq j \neq k \neq i.\end{aligned}$$

In these expressions, the operator ∇ is constructed in the same manner as in Chapter 2 with the additional restriction following from the fact that on a submanifold V^m carrying a conjugate net, the group of admissible transformations of moving frames is reduced by the equations $\pi^i_j = 0,\ i \neq j$. Because of this, we have:

$$\begin{aligned}\nabla l^i_{jj} &= dl^i_{jj} + l^i_{jj}(\omega^0_0 + \omega^i_i - 2\omega^j_j), \\ \nabla l^j_{ij} &= dl^j_{ij} + l^j_{ij}(\omega^0_0 - \omega^i_i), \\ \nabla l^i_{jk} &= dl^i_{jk} + l^i_{jk}(\omega^0_0 + \omega^i_i - \omega^j_j - \omega^k_k).\end{aligned}$$

It follows from equations (6.148) and (6.150) that

$$\nabla_\delta b^\alpha_i = 0, \tag{6.151}$$

$$\nabla_\delta l^i_{jj} + b^\alpha_j \pi^i_\alpha = 0, \tag{6.152}$$

$$\nabla_\delta l^j_{ij} - \pi^0_i = 0, \tag{6.153}$$

$$\nabla_\delta l^i_{jk} = 0. \tag{6.154}$$

Let us denote by l_i the expression:

$$l_i = \frac{1}{m-1} \sum_{j \neq i} l^j_{ij}. \tag{6.155}$$

Then, using relations (6.153), we find that the quantities l_i satisfy the equations:

$$\nabla_\delta l_i - \pi^0_i = 0. \tag{6.156}$$

Comparing equations (6.156) and (6.96), we see that the latter equations are satisfied if we set

$$z_i = -l_i, \tag{6.157}$$

i.e. the quantities $-l_i$ form a normalizing object of the second kind. This implies that the points $Z_i = A_i - l_i A_0$ define the second normal l_x intrinsically connected with the geometry of a submanifold V^m carrying a conjugate net.

We shall separate the construction of a normalizing object of the first kind into a few steps.

1. Construct the object:

$$\widetilde{l}^j_{ij} = l^j_{ij} - l_i, \tag{6.158}$$

which by (6.156) and (6.153) satisfies the equation

$$\nabla_\delta \widetilde{l}^j_{ij} = 0. \tag{6.159}$$

2. Consider the mean square of the quantities $\widetilde{l}^j_{ij}$:

$$\widetilde{l}_i = \sqrt{\frac{1}{m-1}\sum_{j\neq i}(\widetilde{l}^j_{ij})^2}, \tag{6.160}$$

which satisfies the equation

$$\nabla_\delta \widetilde{l}_i = 0. \tag{6.161}$$

We will assume that $\widetilde{l}_i \neq 0$ since otherwise $l^j_{ij} = l_i$ and as will be proved below, if $m > 2$, then this implies that a conjugate net on a submanifold V^m is conic.

3. Introduce the following quantities:

$$\widetilde{b}^\alpha_i = \frac{b^\alpha_i}{(\widetilde{l}^2_i)}, \quad \widetilde{l}^i_{jj} = \frac{l^i_{jj}}{(\widetilde{l}_j)^2}. \tag{6.162}$$

By relations (6.151), (6.152) and (6.161), these quantities satisfy the equations:

$$\begin{aligned} &\delta\widetilde{b}^\alpha_i - \widetilde{b}^\alpha_i\pi^0_0 + \widetilde{b}^\beta_i\pi^\alpha_\beta = 0, \\ &\delta\widetilde{l}^i_{jj} + \widetilde{l}^i_{jj}(\pi^i_i - \pi^0_0) + \widetilde{l}^\alpha_j\pi^i_\alpha = 0. \end{aligned} \tag{6.163}$$

Introduce also the object:

$$\widetilde{l}^{\alpha i} = \sum_{j\neq i}\widetilde{b}^\alpha_j\widetilde{l}^i_{jj}. \tag{6.164}$$

By equation (6.163), this object satisfies the equations:

$$\delta\widetilde{l}^{\alpha i} + \widetilde{l}^{\alpha i}(\pi^i_i - 2\pi^0_0) + \widetilde{l}^{\beta i}\pi^\alpha_\beta + \sum_{j\neq i}\widetilde{b}^\alpha_j\widetilde{b}^\beta_j\pi^i_\beta = 0. \tag{6.165}$$

4. Assume now that *the conditions of the generalized Segre theorem hold on a submanifold* V^m (see Section **3.3**), i.e. we have $m_1 < m$, and any subsystem of points $\widetilde{B}_i = b_i^\alpha \widetilde{A}_\alpha$, consisting of $m-1$ points, is of rank m_1. For the matrix $B = (b_i^\alpha)$ of coordinates of the points $\widetilde{B}_i$ this assumption means that its rank is m_1 and it will not be reduced if we delete any column. The matrix $\widetilde{B} = (\widetilde{b}_i^\alpha)$ satisfies the same conditions as the matrix B.

   The matrix

$$\Big(\sum_{j\neq i} \widetilde{b}_j^\alpha \widetilde{b}_j^\beta\Big) = (\widetilde{b}_i^{\alpha\beta}) \tag{6.166}$$

   in equations (6.165) is the Gramm matrix for the nonsingular matrix obtained from the matrix $\widetilde{B}$ by deleting the ith column. Thus, this matrix has the inverse matrix $(\widetilde{b}^i_{\alpha\beta})$.

   By (6.163), the entries of the matrices $(\widetilde{b}_i^{\alpha\beta})$ and $(\widetilde{b}^i_{\alpha\beta})$ satisfy the equations:

$$\begin{aligned} &\delta\widetilde{b}_i^{\alpha\beta} - 2\widetilde{b}_i^{\alpha\beta}\pi_0^0 + \widetilde{b}_i^{\alpha\gamma}\pi_\gamma^\beta + \widetilde{b}_i^{\gamma\beta}\pi_\gamma^\alpha = 0, \\ &\delta\widetilde{b}^i_{\alpha\beta} + 2\widetilde{b}^i_{\alpha\beta}\pi_0^0 - \widetilde{b}^i_{\alpha\gamma}\pi_\beta^\gamma - \widetilde{b}^i_{\gamma\beta}\pi_\alpha^\gamma = 0. \end{aligned} \tag{6.167}$$

   These formulas prove that the matrices $(\widetilde{b}_i^{\alpha\beta})$ and $(\widetilde{b}^i_{\alpha\beta})$ are the matrices of components of mutually inverse symmetric tensors that are connected only with transformations of the points A_α.

5. Consider the system of the following quantities:

$$\widetilde{l}^i_\alpha = \widetilde{b}^i_{\alpha\beta}\widetilde{l}^{\beta i}. \tag{6.168}$$

   By (6.167) and (6.165), these quantities satisfy the equations:

$$\nabla_\delta \widetilde{l}^i_\alpha + \pi^i_\alpha = 0. \tag{6.169}$$

   Comparing these equations with equations (6.147) defining the normalized object of the first kind we see that the latter equations are satisfied if we set

$$y^i_\alpha = \widetilde{l}^i_\alpha. \tag{6.170}$$

   Thus, the points $Y_\alpha = A_\alpha + \widetilde{l}^i_\alpha A_i$ and the point A_0 determine the first normal n_x intrinsically connected with a submanifold V^m.

We will now clarify the geometric meaning of the first normal we just constructed. To do this, we find the osculating two-plane E_i^2 of the curve C_i of the conjugate net Σ on the manifold V^m. This curve is defined by the equations:

$$\omega^j = 0, \quad j \neq i. \tag{6.171}$$

For the curve C_i we have:

$$\begin{aligned} dA_0 &= \omega_0^0 A_0 + \omega^i A_i, \\ d^2 A_0 &\equiv \Big(\sum_{j \neq i} l_{ii}^j A_j + b_i^\alpha A_\alpha\Big)(\omega^i)^2 \pmod{A_0, A_i}. \end{aligned} \tag{6.172}$$

By setting

$$L_i = \sum_{j \neq i} l_{ii}^j A_j + b_i^\alpha A_\alpha,$$

we conclude that the osculating two-plane E_i^2 of the curve C_i is defined by the points A_0, A_i and L_i. Let us change the normalization of the point L_i by setting

$$\widetilde{L}_i = \frac{L_i}{(\widetilde{l}_i)^2} = \sum_{j \neq i} \widetilde{l}_{ii}^j A_j + \widetilde{b}_i^\alpha A_\alpha. \tag{6.173}$$

Consider the first normal defined by the points A_0 and Y_α whose form is given by relations (6.89). Eliminating the points A_α from relations (6.173) and (6.91), we find that

$$\widetilde{L}_i = \sum_{j \neq i} (\widetilde{l}_{ii}^j - \widetilde{b}_i^\alpha y_\alpha^j) A_j + \widetilde{b}_i^\alpha Y_\alpha. \tag{6.174}$$

The quantities $\widetilde{l}_{ii}^j - \widetilde{b}_i^\alpha y_\alpha^j$ give a deviation of the osculating two-plane E_i^2 from the normal defined by the points Y_α. The quantity

$$S = \sum_{\substack{i,j \\ j \neq i}} (\widetilde{l}_{ii}^j - \widetilde{b}_i^\alpha y_\alpha^j)^2$$

is the mean square of these deviations.

We will try to find the quantities y_α^i in such a way that this mean square is a minimum. To do this, we find $\frac{\partial S}{\partial y_\alpha^i}$:

$$\frac{\partial S}{\partial y_\alpha^i} = -2 \sum_{j \neq i} (\widetilde{l}_{ii}^j - \widetilde{b}_i^\beta y_\beta^j) \widetilde{b}_i^\alpha = -2(\widetilde{l}^{\alpha j} - \widetilde{b}_j^{\alpha\beta} y_\beta^j).$$

At the point where the minimum occurs, all these derivatives must vanish. This implies that

$$\widetilde{l}^{\alpha j} = \widetilde{b}_j^{\alpha\beta} y_\beta^j.$$

Solving these equations with respect to y_β^j and using (6.168), we find that

$$y_\alpha^j = \widetilde{b}_{\alpha\beta}^j \widetilde{l}^{\beta j} = \widetilde{l}_\alpha^j. \tag{6.175}$$

Thus, the first normal we have constructed minimizes the mean square S of above mentioned deviations.

We will next clarify the geometric meaning of the second normal we have constructed in this section. Consider a family of lines (6.171) of the conjugate net Σ on the submanifold V^m. Consider also the straight line $A_0 A_i$ tangent to the curve C_i and passing through the point A_0, and consider the displacements of this straight line along the integral curves of the equation $\omega^i = 0$ where i is fixed. For such displacements we have:

$$\begin{aligned} dA_0 &= \omega_0^0 A_0 + \sum_{j \neq i} \omega^j A_j, \\ dA_i &= \omega_i^0 A_0 + \omega_i^i A_i + \sum_{j,k \neq i} l_{ik}^j \omega^k A_j. \end{aligned} \tag{6.176}$$

We define the *foci* of the straight line $A_0 A_i$ as the points

$$X = A_i + x A_0$$

satisfying the condition:

$$dX \wedge A_0 \wedge A_i = 0 \tag{6.177}$$

for some integral curve of the equation $\omega^i = 0$ passing through the point A_0 (cf. Section **3.5**). Since

$$dX \equiv (l_{ik}^j + x\delta_k^j)\omega^k A_j \pmod{A_0, A_i}, \tag{6.178}$$

it follows from condition (6.177) that

$$(l_{ik}^j + x\delta_k^j)\omega^k = 0. \tag{6.179}$$

This system has a nontrivial solution defining a focal displacement of the straight line $A_0 A_i$ if and only if

$$\det(l_{ik}^j + x\delta_k^j) = 0. \tag{6.180}$$

Equation (6.180) is an algebraic equation of degree $m-1$ with respect to x. This equation can be written in the form:

$$x^{m-1} + \sum_{j \neq i} l_{ij}^j x^{m-2} + \ldots = 0. \tag{6.181}$$

By the Vièta theorem, we have

$$x_1 + \ldots + x_{m-1} = -\sum_{j \neq i} l_{ij}^j, \tag{6.182}$$

where $x_1, \ldots, x_{m-1}$ are roots of equation (6.180). Thus the point

$$F_i = A_i - \frac{1}{m-1} \sum_{j \neq i} l_{ij}^j A_0 \tag{6.183}$$

is the *harmonic pole* of the point A_0 with respect to the foci $X_j = A_i + x_j A_0$, $j \neq i$. But by (6.155), the points F_i coincide with the points $Z_i = A_i - l_i A_0$ determining the location of the second normal l_x.

Note that if the conjugate net Σ on a submanifold V^m is the conjugate system, i.e. if $l_{jk}^i = 0$, $i, j \neq k$, $j \neq k$, then equations (6.179) take the form:

$$(l_{ij}^j + x)\omega^j = 0.$$

It follows that the foci X_j of the straight line $A_0 A_i$ are determined by the formulas: $X_j = A_i - l_{ij}^j A_0$ and correspond to the focal displacement along the curve C_j.

We can now justify our assumption $\widetilde{l}_i \neq 0$ used in formulas (6.162). If $\widetilde{l}_i = 0$, then formulas (6.160) and (6.158) imply that

$$\widetilde{l}_{ij}^j = 0 \text{ and } l_{ij}^j = l_i.$$

But this implies that all foci of the straight line $A_0 A_i$ coincide, and if $m > 2$, then the latter fact means that a conjugate system on V^m is conic (cf. Section **3.6**). We excluded this case from consideration in this section.

NOTES

6.1. In the recently published book [Sto 92b], Stolyarov considers the topics closed to that of this chapter.

6.2. Schouten and Haantjes in [SH 36] considered the geometry of the general normalized submanifolds V^m in spaces with a projective connection, and as an application of their studies, they investigated a hypersurface in P^n. The general normalized submanifolds V^m in P^n were considered by Grove in [Gro 39], Norden (see his papers [N 35], [N 37], [N 45a], [N 47], [N 48] and the book [N 76]) and Atanasyan in [At 58b]. The central normalization for submanifolds in the affine space was considered by Atanasyan in [At 52] who called this normalization axial. Ivlev in [Iv 90] showed that a torsion-free projective connection is invariantly associated with a codimension two normalized submanifold $V^m \subset P^n$ and a tangentially degenerate normalized submanifold $V^m \subset P^n$ and found the fields of geometric images defined by the curvature tensor of this connection.

6.3. The normal connection in a Euclidean space was actually considered by É. Cartan in his lectures [Ca 60]. The curvature tensor of this connection was called by Cartan the geodesic torsion of a submanifold. For a normalized submanifold in a space of constant curvature, the normal connection was investigated by Chen in [Ch 73], Chakmazyan in [Cha 77a], [Cha 77b] and [Cha 78b] and Lumiste and Chakmazyan in [LC 81]. For a normalized submanifold in an affine space, the normal connection was investigated by Akivis and Chakmazyan in [AC 75] and by Chakmazyan in [Cha 77c], [Cha 84b], [Cha 87] and [Cha 89] (see also the book [Cha 90]). Submanifolds V^m admitting a parallel p-dimensional subbundle in the normal bundle were considered by Chakmazyan in [Cha 76], [Cha 77a], [Cha 77b], [Cha 77d], [Cha 77e], [Cha 78a], [Cha 78b], [Cha 80], [Cha 84b] and [Cha 89] and Lumiste and Chakmazyan in [LC 74] (see also the book [Cha 90]). Normalized submanifolds in a projective space P^n were considered by Norden in the book [N 76]. Apparently, the normal connection on such submanifolds was first considered in this book.

6.4. Submanifolds with flat normal connection were considered in [Ch 73], [Cha 77c], [Cha 77d], [Cha 78b], [Cha 83], [Cha 87] [Cha 89] and Lumiste and Chakmazyan in [LC 81] (see also the book [Cha 90]). Theorems 6.1 and 6.2 are proved in this book for the first time. Theorem 6.3 can be found in [Cha 77d] and [Cha 78b] (see Theorems 3 and 4 in [Cha 90], pp. 26–27). Note that the relation between the flatness of a connection and the one-sided stratifiability of some pair of subspaces was noticed by Lumiste in [Lu 65] and [Lu 67]. For the conjugate-harmonic normalization of m-conjugate system this fact was proved in [G 66] by Goldberg.

The conjugate, harmonic and conjugate-harmonic normalizations of m-conjugate systems were first studied by Goldberg in [G 66]. He also studied pairs of m-conjugate systems with a common conjugate-harmonic normalizations (see [G 69]).

The notion of a pseudofocus was introduced by Bazylev in [Ba 65b] for a planar multidimensional nets. Later in [Ba 66] he considered the pseudofoci on the tangents to conjugate lines of $V^m \subset P^n$. However, in the latter case the location of pseudofoci depends on the choice of the normal subspace n_x.

Theorem 6.4 was proved by Chakmazyan in [Cha 77d] and [Cha 78b] (see Theorem 1, p. 37 and Lemma, p. 31 in [Cha 90]). Theorem 6.5 is proved in this book for the first time.

6.5. The problem of construction of an invariant normalization of an m-dimensional submanifold V^m in an affine space A^n and a projective space P^n for the general case and different special cases was considered by many authors. The affine normal for a surface V^2 in an affine space A_3 was known for a long time (see, for example, the book [Bl 23]). Weise in [We 38] suggested a scheme for a construction of an invariant normalization of V^m in A^n, where $n \leq \frac{1}{2}m(m+3)$. His scheme is based on the existence of a nonvanishing differential invariant. Atanasyan in [At 58a] used Weise's scheme to construct the inner geometry of V^m in A^n. In 1951, in [Kli 52], Klingenberg gave a scheme which is valid for any m and n. Another scheme which is valid in a centroaffine, affine and projective space was suggested by Liber in 1952–1956, first in the case $n \leq \frac{1}{2}m(m+3)$ and later for any m and n (see [Lib 52], [Lib 53], [Lib 55], [Lib 56], and [Lib 66]). Using the normal objects introduced by Hlavatý in [H 49] (who also proved their existence), Shveykin in [Shv 55], [Shv 56] and [Shv 58], constructed these objects and applied them to find an invariant normalization of a submanifold V^m in an affine space A^n. In [Lo 50a] and [Lo 50b], Lopshits studied the problem of an invariant normalization of a hypersurface and a submanifold of codimension two in an equiaffine n-space. Pyasetsky found an invariant normalization of a hypersurface and a submanifold of codimension two in a complex affine n-space (see [Pya 53a], [Pya 60] and [Pya 61]) and of a hypersurface in a centroaffine n-space (see [Pya 53b]). Izmailov considered the problem of an invariant normalization for a two-dimensional submanifold in an affine n-space (see [Iz 54] and [Iz 57b]). The survey paper [Shv 66] contains the most important results on the problem of normalization of a submanifold V^m in an affine space A^n.

In [Ud 63], Udalov constructed an invariant normalization of a submanifold V^m in a projective n-space for the case when $n = \binom{m+p}{p} - 1$. In another paper [Ud 68], Udalov considered the same problem for general m and n provided that the submanifold V^m possesses two nonvanishing relative invariants. In the case $m + 1 < n < \frac{1}{2}m(m+3)$, an invariant normalization of V^m in P^n was constructed by Berezina (see [Ber 64a] and [Ber 64b]), and in the cases $1 < n - m < \frac{1}{2}m(m+1)$ and $\frac{1}{2}m(m+1) < n - m < \frac{1}{2}m(m+1) + \frac{1}{6}m(m+1)(m+2)$ by Ostianu (see [Os 64] and [Os 66]). Berezina and Ostianu followed Weise's scheme in their constructions. Ermakov constructed an invariant normalization of V^m in P^n with some restriction on the dimensions of osculating subspaces of higher orders of V^m (see [E 65] and [E 66]).

Laptev in [Lap 59], constructed an invariant normalization of a submanifold V^m in a space with an affine connection where $n \leq \frac{1}{2}m(m+3)$.

If $\frac{1}{2}m(m+3) < n \leq \frac{1}{6}m(m+1)(m+2)$ and some additional conditions are satisfied, an invariant normalization of a submanifold V^m in a space with an affine connection was found by Laptev and Zaits in [LZ 63] (see also [Lap 64]). Izmailov constructed an invariant normalization of a two-dimensional submanifold in a space with an affine connection (see [Iz 61]).

Invariant normalizations of a family of multidimensional planes in P^n, of a distribution of m-dimensional linear elements in a space with a projective connection, of a hyperplanar distribution in P^n, of manifolds imbedded in a space with a projective structure was considered in the series of papers authored or co-authored by Ostianu (see [Os 69], [LO 71], [Os 71], [Os 73], and [OB 78]).

For a more detailed description of the developments in the construction of an invariant normalization of submanifolds in different spaces see the survey papers [Lap 65] and [Lu 75] by Laptev and Lumiste.

The described construction of an invariant normalization of $V^m \subset P^n$, $n < \frac{1}{2}m(m+3)$, which is intrinsically connected with its geometry is due to Ostianu (see [Os 64] and [Os 66]).

6.6. Invariant normalizations of submanifolds V^m of some special types in the space P^n were constructed by Atanasyan and Vorontsova for cones V^m with a p-dimensional vertex and $(p+1)$-dimensional generators (see [AV 63], [AV 65] and [Vo 64]), by Goldberg for the Cartan submanifold (see [G 70]), by Ostianu for a submanifold V^m with a given net on it (see [Os 70]), by Akivis for a submanifold V^m, $m > 2$, $n < 2m$, carrying a conjugate net (see [A 70b]), and by Ivlev and Luchinin for a submanifold V^m of codimension two (see [IL 67])—in the latter case a conjugate net on V_m always exists.

In [Lib 64], Liber indicated a new scheme for construction of an affine connection and subsequently of an invariant normalization of a two-dimensional submanifold V^2 in P^n. His scheme is based on the construction of a differential p-form and the corresponding p-web defined in a differential neighborhood of order r, where $(r+1)(r+2) < 2n \leq (r+2)(r+3)$.

For a more detailed description of the developments in the construction of an invariant normalization of submanifolds of special kinds in different spaces see the survey papers [Lap 65] and [Lu 75] by Laptev and Lumiste.

In this section, following the paper [A 70b], we construct a normalization of submanifolds $V^m \subset P^n$ carrying a conjugate net of lines.

Chapter 7

Projective Differential Geometry of Hypersurfaces

The theory of hypersurfaces V^m in a projective space P^{m+1} occupies a special place in the projective differential geometry of submanifolds. First of all, this is explained by the fact that this theory is the closest generalization of the theory of surfaces in a three-dimensional projective space. The projective differential geometry of two-dimensional surfaces in a three-dimensional projective space was studied in many books and papers in the 1920's – 1930's. The monographs of Fubini and Čech [FČ 26] and [FČ 31]), Finikov ([Fi 37]), Bol ([Bo 50a]) and Lane ([La 32] and [La 42]) were devoted to this theory. In these monographs, the investigations in this field conducted in the years preceding monographs' publication were summarized.

As to the proper theory of hypersurfaces (i.e. the case when $m > 2$), this theory was also studied in great detail. The deepest results in this theory are due to Laptev (see his papers [Lap 53] and [Lap 65]).

In this present chapter we will present the basic results of this theory and introduce some new concepts.

7.1 Basic Equations of the Theory of Hypersurfaces

1. As in the case of an arbitrary submanifold V^m in a projective space P^n, we associate a family of moving frames with any point x of a hypersurface V^m in a projective space P^{m+1} in such a way that the point A_0 of these frames coincides with the point x, the points A_i, $i = 1, \ldots, m$, belong to the tangent subspace $T_x^{(1)}(V^m)$ which is a hyperplane of P^{m+1} if V^m is a hypersurface.

In such a frame, the initial equation of the hypersurface V^m has the form:

$$\omega_0^n = 0, \quad n = m + 1, \tag{7.1}$$

and

$$dA_0 = \omega_0^0 A_0 + \omega_0^i A_i. \tag{7.2}$$

The forms ω_0^i are basis forms of the hypersurface V^m. As we did in the previous chapters, we set $\omega_0^i = \omega^i$. Exterior differentiation of equation (7.1) leads to the single exterior quadratic equation:

$$\omega^i \wedge \omega_i^n = 0. \tag{7.3}$$

Applying Cartan's lemma to this equation, we find that

$$\omega_i^n = b_{ij}\omega^j, \quad b_{ij} = b_{ji} \tag{7.4}$$

(cf. (2.21), (2.94) or (6.2)).

The forms ω_i^n define the Gauss mapping $\gamma(V^m)$ of the hypersurface V^m into the Grassmannian $G(m, m+1)$ which is the dual projective space P^{n*}:

$$\gamma : V^m \to G(m, m+1) = P^{n*}. \tag{7.5}$$

The hypersurface V^m is tangentially nondegenerate if its Gauss mapping γ is nondegenerate, i.e. if

$$b = \det(b_{ij}) \neq 0. \tag{7.6}$$

The determinant b is called the *discriminant of second order* of the hypersurface V^m. The Gauss mapping $\gamma(V^m)$ defines a hypersurface V^{m*} in the space P^{n*}. In what follows in this chapter, we will assume that the hypersurface V^m is tangentially nondegenerate, i.e. condition (7.6) holds, and we will study the hypersurface V^m along with its dual image V^{m*}.

By relation (7.6), the 1-forms ω_i^n are linearly independent, and they are basis forms on the hypersurface V^{m*}. The generating element of the hypersurface V^{m*} is a hyperplane $\alpha^n = A_0 \wedge A_1 \wedge \ldots \wedge A_m$. The differential of the latter hyperplane has the form:

$$d\alpha^n = -\omega_n^n \alpha^n - \omega_i^n \alpha^i \tag{7.7}$$

(cf. (1.72)).

The single second fundamental form $\Phi_{(2)}$ of the hypersurface V^m is written in the form:

$$\Phi_{(2)} = \omega^i \omega_i^n = b_{ij}\omega^i\omega^j. \tag{7.8}$$

This form defines the asymptotic cone of second order

$$b_{ij}\omega^i\omega^j = 0 \tag{7.9}$$

in the tangent hyperplane $T_x^{(1)}(V^m)$. Since

$$d^2 A_0 \equiv \Phi_{(2)} A_n \pmod{T_x^{(1)}},$$

equations (7.9) define directions along which the tangent hyperplane $T_x^{(1)}(V^m)$ has a tangency of second order with the hypersurface V^m.

The second fundamental form $\Phi^*_{(2)}$ of the hypersurface V^{m*}, which is defined by the formula:

$$d^2 \alpha^n \equiv -\omega_i^n \omega^i \alpha^0 \pmod{\alpha^n, \alpha^i},$$

differs from the form $\Phi_{(2)}$ only in sign. By (7.4), this form can be written as follows:

$$\Phi^*_{(2)} = -b^{ij} \omega_i^n \omega_j^n = -b_{ij} \omega_0^i \omega_0^j = -\Phi_{(2)},$$

where b^{ij} is the inverse tensor of the tensor b_{ij}. Since the asymptotic lines on the hypersurface V^m are determined by the equation $\Phi_{(2)} = 0$, and on hypersurface V^{m*} they are determined by the equation $\Phi^*_{(2)} = 0$, the Gauss mapping γ preserves asymptotic lines. The same is true for conjugate directions.

To study further the geometry of the hypersurface V^m, we take the exterior differentials of equations (7.4):

$$[db_{ij} - b_{ik}\omega_j^k - b_{kj}\omega_i^k + b_{ij}(\omega_0^0 + \omega_n^n)] \wedge \omega^j = 0. \tag{7.10}$$

As earlier, in order to make our notation shorter, we will use the operator ∇:

$$\nabla b_{ij} = db_{ij} - b_{ik}\theta_j^k - b_{kj}\theta_i^k, \tag{7.11}$$

where $\theta_i^j = \omega_i^j - \delta_j^i \omega_0^0$.

Application of Cartan's lemma to equation (7.10) gives:

$$\nabla b_{ij} - b_{ij}(\omega_0^0 - \omega_n^n) = b_{ijk}\omega^k, \tag{7.12}$$

where the quantities b_{ijk} are symmetric in all lower indices. These quantities are connected with the third order neighborhood of a point x of the hypersurface V^m.

If the point A_0 is fixed, i.e. if $\omega^i = 0$, equations (7.12) give

$$\nabla_\delta b_{ij} = b_{ij}(\pi_0^0 - \pi_n^n). \tag{7.13}$$

Since $b_{ik} b^{kj} = \delta_i^j$, we find from relations (7.12) that

$$\nabla b^{ij} + b^{ij}(\omega_0^0 - \omega_n^n) = -b_k^{ij}\omega^k, \tag{7.14}$$

where

$$\nabla b^{ij} = db^{ij} + b^{kj}\theta_k^i + b^{ik}\theta_k^j, \tag{7.15}$$

and

$$b_k^{ij} = b^{ip} b^{jq} b_{pqk}. \tag{7.16}$$

If $\omega^i = 0$, relation (7.14) gives

$$\nabla_\delta b^{ij} = -b^{ij}(\pi_0^0 - \pi_n^n). \tag{7.17}$$

Formulas (7.13) and (7.17) confirm that the quantities b_{ij} and b^{ij} are relative tensors.

In the same manner as in Section **2.1**, we can prove that

$$\delta\omega^i = -(\pi_j^i - \delta_j^i \pi_0^0)\omega^j, \tag{7.18}$$

or

$$\nabla_\delta \omega^i = 0. \tag{7.19}$$

Using formulas (7.13) and (7.19), we obtain:

$$\delta\Phi_{(2)} = (\pi_0^0 - \pi_n^n)\Phi_{(2)}. \tag{7.20}$$

In the same manner, we find that

$$\delta\Phi_{(2)}^* = (\pi_0^0 - \pi_n^n)\Phi_{(2)}^*. \tag{7.21}$$

Relations (7.20) and (7.21) prove that the forms $\Phi_{(2)}$ and $\Phi_{(2)}^*$ are relatively invariant, and this must be expected.

Note that on a tangentially degenerate hypersurface, the relatively invariant second fundamental form $\Phi_{(2)}$ defines a conformal structure or a pseudoconformal structure. Such a structure on a hypersurface of the space P^n was studied by Sasaki in [Sas 88] and by Akivis and Konnov in [AK 93].

2. Exterior differentiation of equations (7.12) and application of Cartan's lemma to the exterior quadratic equations obtained as a result of this differentiation leads to the following equations:

$$\nabla b_{ijk} + b_{ijk}(\omega_n^n - \omega_0^0) + 3b_{(ij}\omega_{k)}^0 - 3b_{(ij}b_{k)l}\omega_n^l = b_{ijkl}\omega^l, \tag{7.22}$$

where the parentheses mean the cycling, for example,

$$b_{(ij}\omega_{k)}^0 = \frac{1}{3}(b_{ij}\omega_k^0 + b_{jk}\omega_i^0 + b_{ki}\omega_j^0),$$

and the quantities b_{ijkl} are symmetric in all lower indices. These quantities are connected with the fourth order neighborhood of a point x of the hypersurface V^m.

Let us find how the quantities b_{ijk} are changed under admissible transformations of moving frames associated with a point $x \in V^m$. To do this, we set $\omega^i = 0$ in equations (7.22). This gives:

$$\nabla_\delta b_{ijk} = b_{ijk}(\pi_0^0 - \pi_n^n) - 3b_{(ij}(\pi_{k)}^0 - b_{k)l}\pi_n^l). \tag{7.23}$$

It follows from these equations that the quantities $\{b_{ij}, b_{ijk}\}$ form a differential-geometric object of third order which is associated with the hypersurface V^m. In addition, equations (7.23) show that the quantities b_{ijk} do not form even a relative tensor since expressions (7.23) contain the forms π_k^0 and π_n^l which determine the displacements of the first and second normals associated with the point A_0.

However, the quantities b_{ijk} allow us to construct a relative tensor associated with the third order neighborhood of a point x of the hypersurface V^m. To do this, we contract the quantities b_{ijk} with the relative tensor b^{jk} and define

$$b_i = \frac{1}{m+2} b^{jk} b_{ijk}. \tag{7.24}$$

Using relations (7.14) and (7.22), we find that

$$\nabla b_i = -\omega_i^0 + b_{ij}\omega_n^j + l_{ij}\omega^j, \tag{7.25}$$

where the quantities l_{ij} are expressed in terms of the object b_{ijkl} (associated with the fourth order neighborhood) as follows:

$$l_{ij} = -b_{ikl}b_j^{kl} + b^{kl}b_{klij}. \tag{7.26}$$

If $\omega^i = 0$, equation (7.25) implies

$$\nabla_\delta b_k = -\pi_k^0 + b_{kl}\pi_n^l. \tag{7.27}$$

Note that formulas (7.23) and (7.27) contain the same forms $\pi_k^0 - a_{kl}\pi_n^l$. In order to eliminate these forms, we introduce the object:

$$B_{ijk} = b_{ijk} - 3b_{(ij}b_{k)}. \tag{7.28}$$

Applying the operator ∇_δ to this object, we find that

$$\nabla_\delta B_{ijk} = B_{ijk}(\pi_0^0 - \pi_n^n). \tag{7.29}$$

It follows from these equations that the quantities B_{ijk} form a tensor. This tensor is called the *Darboux tensor* of the hypersurface V^m.

It follows from relations (7.28) that the tensor B_{ijk} is connected with the tensor b_{ij} by the relations:

$$B_{ijk}b^{ij} = 0. \tag{7.30}$$

which are called the *conditions of apolarity* of the tensor B_{ijk} to the tensor b_{ij}.

The Darboux tensor determines the cubic form:

$$\Psi_{(3)} = B_{ijk}\omega^i\omega^j\omega^k. \tag{7.31}$$

This form is defined in the third order neighborhood. By relations (7.29) and (7.19), we have

$$\delta\Psi_{(3)} = (\pi_0^0 - \pi_n^n)\Psi_{(3)}. \tag{7.32}$$

This proves that the form $\Psi_{(3)}$ is relatively invariant.

Comparing formulas (7.21) and (7.32), we see that they contain the same factor $\pi_0^0 - \pi_n^n$. Thus, the ratio of these forms:

$$\sigma = \frac{\Psi_{(3)}}{\Phi_{(2)}} = \frac{B_{ijk}\omega^i\omega^j\omega^k}{b_{ij}\omega^i\omega^j} \tag{7.33}$$

is an absolute differential invariant. This invariant is called the *Fubini linear element* since in the case of a surface V^2 in P^3, this invariant is a projective differential invariant introduced by Fubini (see [FČ 31], p. 66).

The cubic form (7.31) defines the cone

$$B_{ijk}\omega^i\omega^j\omega^k = 0 \tag{7.34}$$

of third order in the tangent hyperplane $T_x^{(1)}(V^m)$. This cone is called the *Darboux cone* of the hypersurface V^m.

3. In this section we encountered the condition (7.30) of apolarity of the tensors b_{ij} and B_{ijk}. We will also encounter in this chapter the apolarity condition of two (0, 2)-tensors. Since in the literature we could not find a description of a geometric meaning of this type apolarity conditions for $n > 2$ (for $n = 2$ such a description can be found in the book [SS 59]), we will fill this gap in this subsection.

Let g_{ij} and h_{ij} be two symmetric (0, 2)-tensors and let g_{ij} be nondegenerate. The tensor h_{ij} is said to be *apolar* to the tensor g_{ij} if

$$g^{ij}h_{ij} = 0. \tag{$*$}$$

Here $i, j = 1, \ldots, n$, and g^{ij} is the inverse tensor of the tensor g_{ij}. Next, in a vector space L^n, we consider two cones:

$$g_{ij}x^ix^j = 0 \tag{g}$$

and

$$h_{ij}x^ix^j = 0. \tag{h}$$

Condition ($*$) means that *there exists a frame* $\{e_1, \ldots, e_n\}$ *which is conjugate with respect to the cone* (g) *and inscribed into the cone* (h).

We will prove this by induction with respect to n. If $n = 2$, then the cone (h) has two different zero directions. Taking the vectors e_1 and e_2 of a frame along these directions, we reduce the equation of the cone (h) to the form

$$2h_{12}x^1x^2 = 0,$$

where $h_{12} \neq 0$. Because of this, the apolarity condition ($*$) takes the following form:

$$g^{12}h_{12} = 0.$$

This implies $g^{12} = 0$, i.e. the vectors e_1 and e_2 are conjugate with respect to the cone (g).

Suppose now that our statement is true for any k less than n. We will prove that it is valid for $k = n$. To prove this, we take the vector e_n of a frame in such a way that e_n belongs to the cone (h) but not the cone (g). Then $h_{nn} = 0$ and $g_{nn} \neq 0$. Consider a subspace $L^{n-1} \subset L^n$ which is conjugate to the vector e_n with respect to the cone (g), and take the remaining vectors $e_1, \ldots, e_{n-1}$ of the frame in L^{n-1}. In this frame, the matrix g of the quadratic form on the left-hand side of (g) takes the form

$$g = \begin{pmatrix} g_{uv} & 0 \\ 0 & g_{nn} \end{pmatrix}, \quad u, v = 1, \ldots, n-1; \quad g_{nn} \neq 0,$$

and thus the inverse matrix g^{-1} takes the form:

$$g^{-1} = \begin{pmatrix} g^{uv} & 0 \\ 0 & g^{nn} \end{pmatrix}, \quad g^{nn} = \frac{1}{g_{nn}}.$$

This implies that the apolarity condition (*) reduces to the relations:

$$h_{uv} g^{uv} = 0.$$

The latter condition is the restriction of the condition (*) to the subspace L^{n-1}. Under the induction assumption, there exists a frame $\{e_1, \ldots, e_{n-1}\}$ in L^{n-1} which is conjugate with respect to the cone $g_{uv}x^u x^v = 0$ and inscribed into the cone $h_{uv}x^u x^v = 0$. The vectors $\{e_1, \ldots, e_{n-1}\}$ together with the vector e_n form a basis in the space L^n satisfying the required condition.

Note that if $n > 2$, the apolarity condition for tensors g_{ij} and h_{ij} is not symmetric, i.e. in general, the condition (*) does not imply the condition

$$g_{ij} h^{ij} = 0.$$

Consider now the apolarity condition for (0, 3)-tensor h_{ijk} and (0, 2)-tensor g_{ij}. Suppose that these tensors are both symmetric and, as earlier, the tensor g_{ij} is nondegenerate.

Let $y = y^i e_i$ be an arbitrary vector in L^n. Then this vector defines (0, 2)-tensor $h_{ij}(y) = h_{ijk} y^k$, and the latter tensor defines the second order cone

$$h_{ijk} x^i x^j y^k = 0, \qquad (**)$$

called the *second polar* of the vector y with respect to the cubic hypercone $h_{ijk} x^i x^j x^k = 0$. Since the apolarity condition

$$g^{ij} h_{ijk} = 0$$

implies that

$$g^{ij} h_{ij}(y) = 0$$

for any vector $y \in L^n$, the geometric meaning of apolarity of two (0, 2)-tensors proves that *it is possible to inscribe a frame into any cone* (**) *in such a way that this frame is conjugate with respect to the cone* (g).

For the case $n = 2$, in the book [SS 59] (see §23), the following geometric meaning of the apolarity condition $g^{ij} h_{ijk} = 0, \; i, j, k = 0, 1, 2$, is given: *if the equation* $h_{ijk} x^i x^j x^k = 0$ *has three mutually distinct solutions* e_1, e_2, e_3, *and each of the vectors* u_a, $a, b, c = 1, 2, 3$, *which is harmonic conjugate of the vector* e_a *with respect to* *the vectors e_b and e_c (the indices a, b and c are mutually distinct), is also conjugate to the vectors e_a with respect to directions defined by the equation $g_{ij}x^i x^j = 0$*, i.e. $g(e_a, u_a) = 0$.

7.2 Osculating Hyperquadrics of a Hypersurface

1. The tangent hyperplane $T_x^{(1)}(V^m)$ has a first order tangency with the hypersurface V^m at the point x. The order of their tangency along the asymptotic lines defined by equations (7.9) is equal to two.

It is natural to try to find a hypersurface Q of second order that has a second order tangency with the hypersurface V^m in all directions emanating from the point A_0.

We will write the equation of a hyperquadric Q relative to a first order frame associated with the point $A_0 \in V^m$ in the form:

$$Q(x, x) = A_{uv}x^u x^v = 0, \quad u, v = 0, 1, \ldots, m+1. \tag{7.35}$$

Since the hyperquadric Q must pass through the point A_0, we have

$$Q(A_0, A_0) = A_{00} = 0. \tag{7.36}$$

Since the hyperquadric Q is tangent to V^m at A_0, the following condition must be satisfied:

$$Q(A_0 + dA_0, A_0 + dA_0) = 0,$$

i.e.

$$Q(A_0 + \omega_0^0 A_0 + \omega^i A_i, A_0 + \omega_0^0 A_0 + \omega^i A_i) = 0.$$

Taking into account relations (7.36) and neglecting infinitesimals of order higher than one in the latter equation, we obtain

$$A_{0i} = 0. \tag{7.37}$$

As a result, the equation of the hyperquadric Q takes the form:

$$2A_{0n}x^0 x^n + A_{ij}x^i x^j + 2A_{in}x^i x^n + A_{nn}(x^n)^2 = 0. \tag{7.38}$$

Moreover, we have $A_{0n} \neq 0$ since otherwise the hyperquadric Q degenerates into a hypercone. We normalize equation (7.38) by imposing the condition:

$$A_{0n} = -1. \tag{7.39}$$

Finally, requesting that the hyperquadric Q has a second order tangency with the hypersurface V^m, we find that

$$Q(A_0 + dA_0 + \frac{1}{2}d^2A_0, A_0 + dA_0 + \frac{1}{2}d^2A_0) = 0.$$

Taking into account relations (7.36), (7.37) and (7.39) and neglecting infinitesimals of order higher than two, we obtain

$$A_{ij} = b_{ij}. \tag{7.40}$$

After these simplifications, the equation of the family of osculating hyperquadrics Q takes the form:

$$2x^0x^n = b_{ij}x^ix^j + 2A_{in}x^ix^n + A_{nn}(x^n)^2. \tag{7.41}$$

We see from this equation that the family of osculating hyperquadrics depends on $m+1$ parameters.

Let us find directions along which a hyperquadric (7.41) has a third order tangency with the hypersurface V^m. Such directions are called *characteristic directions*. To find them, we should substitute coordinates of the point

$$x = A_0 + dA_0 + \frac{1}{2}d^2A_0 + \frac{1}{6}d^3A_0$$

into equation (7.41) of a hyperquadric Q. We write the differentials d^2A_0 and d^3A_0 in the form:

$$\begin{aligned} d^2A_0 &= \Omega_0^0 A_0 + \Omega_0^i A_i + \Omega_0^n A_n, \\ d^3A_0 &= \Phi_0^0 A_0 + \Phi_0^i A_i + \Phi_0^n A_n. \end{aligned} \tag{7.42}$$

If we substitute coordinates of the point x into equation (7.41), then the terms of orders 0, 1 and 2 cancel out since the hyperquadric has a second order tangency with the hypersurface V^m, and the third order terms give

$$\frac{1}{3}\Phi_0^n + \Omega_0^n\omega_0^0 = b_{ij}\omega^i\Omega_0^j + A_{in}\Omega_0^n\omega^i. \tag{7.43}$$

By (7.42), the forms Ω_0^i, Ω_0^n and Φ_0^n have the form:

$$\begin{aligned} \Omega_0^i &= d\omega^i + \omega_0^0\omega^i + \omega^j\omega_j^i, \\ \Omega_0^n &= \omega^i\omega_i^n = b_{ij}\omega^i\omega^j, \\ \Phi_0^n &= d\Omega_0^n + \Omega_0^j\omega_j^n + \Omega_0^n\omega_n^n. \end{aligned} \tag{7.44}$$

Hence, equation (7.43) takes the form:

$$\frac{1}{3}(d\Omega_0^n + \Omega_0^j\omega_j^n + \Omega_0^n\omega_n^n) + \Omega_0^n\omega_0^0 = b_{ij}\Omega_0^i\omega^j + A_{in}\Omega_0^n\omega^i. \tag{7.45}$$

To complete our calculations, using (7.44) and (7.12), we find $d\Omega_0^n$:

$$\begin{aligned} d\Omega_0^n = {} & db_{ij}\omega^i\omega^j + 2b_{ij}\omega^i d\omega^j \\ = {} & (b_{ik}\omega_j^k + b_{kj}\omega_i^k - b_{ij}(\omega_0^0 + \omega_n^n) + b_{ijk}\omega^k)\omega^i\omega^j \\ & +2b_{ij}\omega^i(\Omega_0^j - \omega_0^0\omega^j - \omega^k\omega_k^j) \\ = {} & (-b_{ij}(3\omega_0^0 + \omega_n^n) + b_{ijk}\omega^k)\omega^i\omega^j + 2b_{ij}\omega^i\Omega_0^j. \end{aligned}$$

Substituting this expression into equation (7.45), after uncomplicated calculations we obtain

$$(b_{ijk} - 3A_{n(i}b_{jk)})\omega^i\omega^j\omega^k = 0. \tag{7.46}$$

Thus, the characteristic directions, along which the hyperquadric Q has a third order tangency with the hypersurface V^m at the point A_0, are defined by equation (7.46). These directions form a *cubic cone* for each osculating hyperquadric (7.37).

A hyperquadric (7.41) is called the *Darboux hyperquadric* if the cone (7.46) coincides with the Darboux cone (7.34). Since the Darboux tensor B_{ijk} is defined by formula (7.28), the cone (7.46) coincides with the Darboux cone if and only if

$$A_{ni} = b_i. \tag{7.47}$$

Because of this, the equation of the family of Darboux osculating hyperquadrics has the form:

$$2x^0x^n = b_{ij}x^ix^j + 2b_ix^ix^n + A_{nn}(x^n)^2. \tag{7.48}$$

We remind that in formulas (7.47) and (7.48) $n = m + 1$. Since this family depends on one parameter, it is called the *pencil of Darboux osculating hyperquadrics*.

We will prove the following property of hyperquadrics of this pencil.

Theorem 7.1 *If one of osculating hyperquadrics has a third order tangency with a hypersurface V^m at a point $x \in V^m$ in any direction emanating from the point x, then this hyperquadric is a Darboux hyperquadric, and the Darboux tensor of the hypersurface V^m vanishes at this point.*

Proof. In fact, if a hyperquadric of the family (7.41) has a third order tangency with a hypersurface V^m at a point $x \in V^m$, then equation (7.46) has to become an identity at this point, that is we must have

$$b_{ijk} - 3A_{n(i}b_{jk)} = 0. \tag{7.49}$$

Contracting this relation with the tensor b^{ij}, we find that

$$A_{ni} = b_i. \tag{7.50}$$

Substituting these values of A_{ni} into equation (7.41), we see that this equation coincides with equation (7.48) of the family of Darboux hyperquadrics. On the other hand, if we substitute these values of A_{ni} into equation (7.49), we find that $B_{ijk} = 0$. ■

2. It is easy to see that the second fundamental tensor b_{ij} of a hypersurface V^m vanishes at all points of V^m if and only if the hypersurface V^m is a part of a hyperplane. The Darboux tensor possesses a similar property. We can now prove the following result.

Theorem 7.2 *A hypersurface V^m of a projective space P^{m+1} is a hyperquadric or a part of a hyperquadric if and only if the Darboux tensor of this hypersurface is identically equal to* 0.

Proof. *Necessity.* Consider a hyperquadric Q which is defined in a frame $\{A_u\}$ by the equation:

$$Q(x, x) = A_{uv} x^u x^v = 0, \quad u, v = 0, 1, \ldots, n. \tag{7.51}$$

Since the frame $\{A_u\}$ is a moving frame, the coefficients A_{uv} of equation (7.51) depend on the location of this frame. Let us find the differential equations which the coefficients A_{uv} of equation (7.51) must satisfy in order that the hyperquadric Q be fixed in the space P^n. To do this, first of all, we find the conditions for a point

$$X = x^u A_u \tag{7.52}$$

to be invariant. Since the equations of infinitesimal displacement of a moving frame have the form

$$dA_u = \omega_u^v A_v, \tag{7.53}$$

we have:

$$dX = (dx^u + x^v \omega_v^u) A_u. \tag{7.54}$$

Since the coordinates of the point X admit multiplication by an arbitrary factor, the condition for this point to be invariant has the form $dX = \sigma X$. By equation (7.54), this implies

$$dx^u + x^v \omega_v^u = \sigma x^u. \tag{7.55}$$

Similarly, the condition for the hyperquadric Q to be invariant has the form

$$dQ = \rho Q. \tag{7.56}$$

Differentiating equation (7.51) and using formulas (7.55), we find that

$$dA_{uv} - A_{uw} \omega_v^w - A_{wv} \omega_u^w = \theta A_{uv}, \tag{7.57}$$

where $\theta = \rho + 2\sigma$.

We now assume that a hypersurface V^m is a hyperquadric Q. We associate a frame of first order with Q placing the point A_0 into a point $x \in Q$ and placing the points A_i into the tangent hyperplane to Q at the point $x = A_0$. As a result, we obtain the following relations:

$$A_{00} = A_{0i} = 0. \tag{7.58}$$

In addition, we will normalize the equation of the hyperquadric Q by the condition

$$A_{0n} = -1. \tag{7.59}$$

Let us write equation (7.57) for different pairs of values of the indices u and v. If $u = v = 0$, we obtain an identity by means of relations (7.58) and (7.1). If $u = 0$ and $v = i$, by means of relations (7.58), (7.59) and (7.1), we find easily that

$$\omega_i^n = A_{ij}\omega^j. \tag{7.60}$$

Comparing these equations with equations (7.4), we find that

$$A_{ij} = b_{ij}. \tag{7.61}$$

Next, if $u = 0$ and $v = n$, we obtain

$$\theta = A_{ni}\omega^i - \omega_0^0 - \omega_n^n. \tag{7.62}$$

If $u = i$ and $v = j$, using all relations obtained earlier, we find that

$$\nabla b_{ij} = b_{ij}(\omega_0^0 - \omega_n^n) + 3A_{n(k}b_{ij)}\omega^k. \tag{7.63}$$

Comparing this relation with equation (7.12), we arrive at the following equation:

$$b_{ijk} = 3A_{n(k}b_{ij)}. \tag{7.64}$$

Substituting these expressions into formula (7.24), we obtain:

$$b_k = A_{nk}, \text{ where } n = m + 1. \tag{7.65}$$

If we use this equation, we find from formula (7.28) that

$$B_{ijk} = 0, \tag{7.66}$$

i.e. the Darboux tensor of the hyperquadric Q is identically equal to 0.

Sufficiency. For $u = i$ and $v = n$, from relations (7.57) we find that

$$\nabla A_{in} = -\omega_i^0 + b_{ij}\omega_n^j + (A_{nn}b_{ij} + A_{ni}A_{nj})\omega_n^j, \tag{7.67}$$

and if $u = v = n$, we have

$$dA_{nn} = A_{nn}(\omega_n^n - \omega_0^0) - 2\omega_n^0 + A_{ni}\omega_n^i + A_{nn}A_{nj}\omega^i. \tag{7.68}$$

Assume that condition (7.66) holds at any point of a hypersurface V^m. Then, at any point of V^m its Darboux hyperquadrics has a third order tangency with V^m. We prove that there is one fixed hyperquadric among the Darboux hyperquadrics. This means that for some value of A_{nn}, all conditions of invariance, considered above, will be satisfied provided that relation (7.68) holds. In fact, equations (7.57) of invariance of hyperquadric (7.48) are identically satisfied for the following pairs of u and v: $u = 0, v = 0$; $u = 0, v = i$ and $u = 0, v = n$. If $u = i, v = j$, condition (7.57) is satisfied by means of equation (7.63). Suppose that $u = i$ and $v = n$. Substituting the values A_{ni} from equations (7.65) into equations (7.67), we obtain:

$$\nabla b_i = -\omega_i^0 + b_{ij}\omega_n^j + (A_{nn}b_{ij} + b_i b_j)\omega^j. \tag{7.69}$$

Comparing this equation with relation (7.25), we find that

$$l_{ij} = A_{nn}b_{ij} + b_i b_j. \tag{7.70}$$

Contracting this relation with b^{ij}, we find that

$$A_{nn} = \frac{1}{m} b^{ij}(l_{ij} - b_i b_j). \tag{7.71}$$

One can easily check that the quantities A_{nn} satisfy relations (7.68). Hence, the hyperquadric Q is fixed, and the hypersurface V^m is a part of this hyperquadric. ■

3. Theorem 7.2 is closely connected with the problem of existence of a tangentially nondegenerate hypersurface with a parallel second fundamental form. In an Euclidean and affine geometries this problem was considered by many authors (for the Euclidean space, see, for example, §3 of the book [Ch 81] by Chen and the joint papers [CV 80] and [CV 81] by Chen and Vanhecke). However, there is a projective interpretation of this problem.

To study the problem of a parallelism of the second fundamental form, we must naturally define an affine connection on the hypersurface in question. As we saw in Chapter 6, such an affine connection is defined if the hypersurface is normalized by means of first and second normals.

Suppose that such a normalization of a hypersurface $V^m \subset P^{m+1}$ is given. Suppose also that the vertex A_{m+1} of a frame associated with V^m lies on the first normal n_x, and the points $A_1, \ldots, A_m$ belong to the second normal l_x. Then in formula (7.12), ∇b_{ij} is the covariant differential of the second fundamental tensor b_{ij}, and the quantities b_{ijk} form a tensor since the fact that first and second normals are fixed implies the vanishing the forms π_k^0 and π_n^l in equations (7.23) and the following form of these equations:

$$\nabla_\delta b_{ijk} = b_{ijk}(\pi_0^0 - \pi_n^n).$$

In this formula the tensor b_{ijk} is the covariant derivative of the second fundamental tensor b_{ij}.

The condition for the second fundamental form to be parallel in the affine connection indicated above can be written as

$$\nabla b_{ij} = 0$$

(see [Vil 72]). By (7.12), this implies

$$-b_{ij}(\omega_0^0 - \omega_n^n) = b_{ijk}\omega^k.$$

It follows

$$\omega_0^0 - \omega_n^n = -\alpha_k \omega^k$$

and

$$b_{ijk} = 3b_{(ij}\alpha_{k)}.$$

Let us find the Darboux tensor B_{ijk} of the hypersurface V^m. From equation (7.24) we find that

$$b_i = \frac{1}{m+2} b^{jk}(b_{ij}\alpha_k + b_{jk}\alpha_i + b_{ki}\alpha_j),$$

and thus, by (7.28), we have $B_{ijk} = 0$.

This proves the following result.

Theorem 7.3 *The second fundamental form of a tangentially nondegenerate hypersurface $V^m \subset P^{m+1}$ is parallel with respect to an affine connection induced by a normalization of V^m if and only if the hypersurface V^m is a hyperquadric or a part of a hyperquadric.* ■

We have proved the necessity of the condition of this theorem. Its sufficiency follows from the fact the Darboux tensor B_{ijk} of a hyperquadric is identically equal to 0.

7.3 Invariant Normalizations of a Hypersurface

1. In this section we will construct an invariant normalization of a hypersurface V^m in a projective space V^{m+1}, i.e. we will find a congruence of one-dimensional first normals n_x and a pseudocongruence of $(m-1)$-dimensional second normals l_x which are intrinsically connected with a hypersurface V^m.

The first normal n_x at a point $A_0 = x$ is spanned by the point A_0 and a point

$$Y_n = A_n + y^i A_i, \tag{7.72}$$

and the second normal l_x at a point $A_0 = x$ is spanned by points

$$Z_i = A_i + z_i A_0 \tag{7.73}$$

(cf. formulas (6.91) and (6.92)). The normalizing objects of the first and second kinds, y^i and z_i, must satisfy the equations:

$$\nabla_\delta y^i = y^i(\pi_n^n - \pi_0^0) - \pi_n^i \tag{7.74}$$

and

$$\nabla_\delta z_i = -\pi_i^0, \tag{7.75}$$

which are obtained from equations (6.95) and (6.96) by setting $\alpha = n$.

We can establish a connection between the first and second normals of a hypersurface V^m at a point A_0 by requesting that these normals are polar-conjugate to one another with respect to the pencil of Darboux osculating hyperquadrics. To do this, we consider the quantities:

$$y_i = b_{ij} y^j. \tag{7.76}$$

It follows from equations (7.74) and (7.13) that these quantities y_i satisfy the equations:

$$\nabla_\delta y_i = -b_{ij} \pi_n^j. \tag{7.77}$$

We now set

$$u_i = z_i - y_i. \tag{7.78}$$

Using (7.75) and (7.77), we obtain

$$\nabla_\delta u_i = -(\pi_i^0 - b_{ij} \pi_n^j). \tag{7.79}$$

Comparing this equation with equation (7.27), we see that by setting

$$u_i = b_i, \tag{7.80}$$

we have equation (7.79) be satisfied. By this reason, we set

$$z_i = y_i + b_i. \tag{7.81}$$

We will now prove that *the first and second normals, which are defined by the objects y_i and z_i connected by relation* (7.81) *are polar-conjugate to one another with respect to the pencil* (7.48) *of Darboux osculating hyperquadrics.* To do this, we write the bilinear equation which is polar to the quadratic equation (7.48):

$$x^0 \xi^n + x^n \xi^0 = b_{ij} x^i \xi^j + b_i (x^i \xi^n + x^n \xi^i) + A_{nn} x^n \xi^n. \tag{7.82}$$

This equation implies that the point A_0 with coordinates $1, 0, \ldots, 0$ is polar-conjugate to the hyperplane $\xi^n = 0$, and the point $Y_n = A_n + y^i A_i$ is polar-conjugate to the hyperplane

$$\xi^0 = b_{ij} y^i \xi^j + b_i (y^i \xi^n + \xi^i) + A_{nn} \xi^n.$$

Because of this, the straight line $A_0 Y_n$ is polar-conjugate to the $(m-1)$-plane defined by the equations:

$$\begin{cases} \xi^n = 0, \\ \xi^0 = (b_{ij} y^j + b_i)\xi^i. \end{cases} \tag{7.83}$$

But, by relation (7.76), this $(m-1)$-plane contains the points

$$Z_i = A_i + (y_i + b_i)A_0. \tag{7.84}$$

Hence, this $(m-1)$-plane coincides with the second normal l_x defined by these points.

Note that if a hypersurface V^m belongs to an affine space A^{m+1}, then the second normal is defined as the intersection of the tangent space $T_x^{(1)}(V^m)$ with the hyperplane at infinity of the space A^{m+1}. The normal n_x which is polar-conjugate to l_x with respect to the pencil of Darboux osculating hyperquadrics is the *Blaschke affine normal* (see [Bl 23]).

2. To construct the normalizing objects of the first and second kinds, we must use the fourth order differential neighborhood of a point $x \in V^m$. Exterior differentiation of equations (7.25), application of Cartan's lemma to the exterior equations obtained and setting $\omega^i = 0$ in the resulting relations lead to

$$\nabla_\delta l_{ij} = (b_{ijk} + b_{ij} b_k)\pi_n^k - (b_i \pi_j^0 + b_j \pi_i^0) - 2b_{ij}\pi_n^0. \tag{7.85}$$

The right-hand side of this equation contains the forms π_i^0 and π_n^k. This allows us, after rather lengthy calculations, in which the quantities l_{ij} and the quantities b_i satisfying equations (7.27) are used, to construct a few pairs of normalizing objects of the first and second kinds, related by relation (7.81), and thus to find the first and second normals intrinsically connected with a hypersurface V^m.

We will perform this construction following Laptev (see [Lap 53]). Our formulas will be slightly different from the similar formulas in [Lap 53] because Laptev did not use the factor $\frac{1}{m+2}$ in formula (7.24).

First let us recall that we consider only nondegenerate hypersurfaces for which inequality (7.6) holds. The differential of the discriminant b of second order of V^m has the form:

$$db = bb^{ij} db_{ij}, \tag{7.86}$$

where bb^{ij} is the cofactor of the element b_{ij} of the determinant b. If we substitute the expressions of db_{ij} from (7.12) into (7.86), we obtain

$$d \ln |b| = 2\omega_i^i - m(\omega_0^0 + \omega_n^n) + (m+2)b_i\omega^i, \tag{7.87}$$

where b_i is expressed by formula (7.24). Thus, the discriminant b of second order of V^m is a relative invariant.

Next, we construct several quasitensors of third order. We already have the object b_i expressed by formula (7.24). This object satisfies differential equations (7.25) which prove that the system $\{b_i, b_{ij}\}$ of two objects, b_i and b_{ij}, forms a quasitensor of third order. Note that equation (7.25), from which equation

(7.26) was derived, can also be obtained by exterior differentiation of equation (7.87) and application of Cartan's lemma. Note also that the quasitensor b_k was introduced by A.P. Norden (see [N 45b] or [N 76], p. 175) who called it the Chebyshev covector.

Further we define

$$b^i = b^{ik} b_k. \tag{7.88}$$

Differentiating this relation by means of (7.12) and (7.87), we obtain a differential equation which the object b^i satisfies. We will write these and further differential equations assuming that a point $x \in V^m$ is fixed, i.e. that $\omega^i = 0$:

$$\nabla_\delta b^i = -b^i(\pi_0^0 - \pi_n^n) - b^{ik}\pi_k^0 + \pi_n^i. \tag{7.89}$$

Thus, the system $\{b^i, b_{ij}\}$ of two objects, b_i and b_{ij}, forms a quasitensor of third order.

If we contract the objects b_k and b^k, we obtain a new object

$$\hat{b} = b_k b^k = b^{ij} b_i b_j. \tag{7.90}$$

It follows from equations (7.89) and (7.16) that

$$\delta\hat{b} = \hat{b}(\pi_n^n - \pi_0^0) - 2(b^k \pi_k^0 - b_k \pi_n^k). \tag{7.91}$$

Equations (7.91), (7.27), (7.89), (7.13), and (7.16) imply that the systems $\{\hat{b}, b_i, b^{ij}\}$ and $\{\hat{b}, b^i, b_{ij}\}$ form quasitensors of third order.

3. Now we construct two tensors B_{ij} and B^{ij} and two relative invariants B_0 and B associated with the third order neighborhood of a hypersurface V^m.

Define the system of quantities B_{ij} by

$$B_{ij} = b^{kl} b^{pq} B_{kpi} B_{lqj}, \tag{7.92}$$

where b^{ij} is the inverse tensor of the tensor b_{ij} and B_{ijk} is the Darboux tensor of the hypersurface V^m satisfying differential equations (7.17) and (7.29), respectively. Using these two differential equations, we find from equation (7.92) that

$$\nabla_\delta B_{ij} = 0. \tag{7.93}$$

It follows from equations (7.93) that the quantities B_{ij} form a tensor.

If we contract this tensor with the tensor b^{ij}, we obtain a new quantity

$$B_0 = b^{ij} B_{ij} = b^{ij} b^{kl} b^{pq} B_{kpi} B_{lqj}, \tag{7.94}$$

which is called the *trace* of the tensor B_{ij}. By equations (7.93) and (7.16), the differential of the trace has the form:

$$dB_0 = B_0(\omega_n^n - \omega_0^0) + B_0 c_k \omega^k. \tag{7.95}$$

This equation implies that the trace B_0 is a relative invariant.

We will now prove that in general, the trace B_0 of the the tensor B_{ij} is nonvanishing. Suppose first that the second fundamental tensor b_{ij} of a hypersurface V^m is positive definite. Then the matrix of this tensor can be reduced to the identity matrix: $b_{ij} = \delta_{ij}$. If the Darboux tensor B_{ijk} is not identically equal to 0, then this implies that

$$B_0 = \sum_{i,j,k=1}^{m} (B_{ijk})^2 > 0. \tag{7.96}$$

But if the tensor b_{ij} is indefinite, then the expression (7.96) has both positive and negative terms. However, in this case the vanishing of the trace B_0 is also connected with a special structure of the tensor B_{ijk}.

This is the reason that in what follows we will assume that the trace B_0 of the the tensor B_{ij} is nonvanishing on a hypersurface V^m. This assumption allows us to write relation (7.95) in the form:

$$d \ln B_0 = \omega_n^n - \omega_0^0 + c_k \omega^k. \tag{7.97}$$

The determinant

$$B = \det(B_{ij}). \tag{7.98}$$

is called the *discriminant* of the hypersurface V^m.

As we did for the differential of the second order discriminant b, we write the differential of B as

$$dB = BB^{ij} dB_{ij}, \tag{7.99}$$

where BB^{ij} is the cofactor of the element B_{ij} of the discriminant B. It follows from equation (7.99) that

$$\delta B = 2B(\pi_k^k - m\pi_0^0), \tag{7.100}$$

i.e. the discriminant B of the tensor B_{ij} is a relative invariant.

Definition 7.4 The Darboux tensor B_{ijk} of a hypersurface V^m is said to be *nondegenerate* if the cubic form

$$\Psi_{(3)} = B_{ijk} \omega^i \omega^j \omega^k,$$

defined by this tensor, cannot be expressed in terms of less than m linearly independent 1-forms.

Lemma 7.5 *The discriminant B of the tensor B_{ij} is nonvanishing if and only if the Darboux tensor B_{ijk} is nondegenerate.*

Proof. In fact, consider a vector space L^N of dimension $N = \frac{1}{2}m(m+1)$ with a basis $\{E^{kl}\}$, $E^{kl} = E^{lk}$, and take the vectors $B_i = B_{ikl}E^{kl}$ in this vector space. Then expression (7.92) for the tensor B_{ij} can be represented in the form:

$$B_{ij} = < B_i, B_j >,$$

where $< B_i, B_j >$ is the scalar product of the vectors B_i and B_j with respect to the metric defined in the space L^N by the nondegenerate tensor $g^{klpq} = b^{kl}b^{pq}$. Therefore, the invariant B is the Gramm determinant of the system of vectors B_i. It is well-known that this determinant vanishes if and only if the vectors B_i are linearly dependent, and this is equivalent to the degeneracy of the Darboux tensor. ■

In what follows we will assume that the Darboux tensor B_{ijk} of the hypersurface V^m is nondegenerate. This assumption implies the following inequality:

$$B = \det(B_{ij}) \neq 0. \tag{7.101}$$

This and relation (7.100) imply that

$$\delta \ln B = 2(\pi_k^k - m\pi_0^0). \tag{7.102}$$

Inequality (7.101) means that the tensor B_{ij} is also nondegenerate and has the inverse tensor B^{ij} such that $B^{ik}B_{jk} = \delta_j^i$. Applying the operator ∇_δ to this relation, by (7.93), we find that

$$\nabla_\delta B^{ij} = 0. \tag{7.103}$$

Note also that the relative invariants b, B_0 and B allow us to construct an absolute invariant

$$C = \frac{B}{b(B_0)^m} \tag{7.104}$$

In fact, it follows from equations (7.87), (7.97) and (7.100) that

$$\delta C = 0. \tag{7.105}$$

Since the tensors B_{ij}, B^{ij} and the invariants b, B_0 and B were constructed by means of the Darboux tensor B_{ijk} connected with a third order neighborhood of a point $x \in V^m$, they are also connected with a third order neighborhood of this point.

4. Now we are able to find quasitensors which are the normalizing objects y^i and z_i of the first and second kinds, i.e. quasitensors satisfying equations (7.74) and (7.75).

To do this, we first recall that the object l_{ij} of fourth order in equations (7.25) satisfies the differential equations (7.85). The quantities l_{ij} and the quasitensors and tensors of the second and third orders constructed above, allow us to find new objects $l, \hat{l}, \hat{l}_{ij}, \hat{l}_k, l^i, l_j$ and k_j. We define the object l by

$$l = b^{ij} l_{ij}. \tag{7.106}$$

By (7.17) and (7.85), this object satisfies the differential equation:

$$\delta l = l(\pi_n^n - \pi_0^0) + 2(m+1)b_k \pi_n^k - 2b^i \pi_i^0 - 2m\pi_n^0. \tag{7.107}$$

To eliminate the form π_i^0 from this equation, we define the object $\hat{l}$ by setting

$$\hat{l} = l - \hat{b}, \tag{7.108}$$

where $\hat{b}$ is defined by equation (7.90). Using (7.107) and (7.91), we find that this object satisfies the differential equation:

$$\delta \hat{l} = \hat{l}(\pi_n^n - \pi_0^0) + 2mb_k \pi_n^k - 2m\pi_n^0. \tag{7.109}$$

To eliminate the forms π_n^0 from equations (7.107), we introduce the quantities $\hat{l}_{ij}$ by setting

$$\hat{l}_{ij} = l_{ij} - b_i b_j - \frac{1}{m}\hat{l} b_{ij}. \tag{7.110}$$

It is easy to check that the quantities $\hat{l}_{ij}$ and b_{ij} are apolar: $b^{ij}\hat{l}_{ij} = 0$. However, the quantities $\hat{l}_{ij}$ do not form a tensor since by (7.85), (7.27), (7.109) and (7.13), they satisfy the differential equations:

$$\nabla_\delta \hat{l}_{ij} = B_{ijk}\omega_n^k, \tag{7.111}$$

where B_{ijk} is the Darboux tensor defined by equation (7.28).

Next, we define the quantities $\hat{l}_k$:

$$\hat{l}_k = b^{ip} b^{jq} \hat{l}_{ij} B_{pqk}. \tag{7.112}$$

By (7.16), (7.111) and (7.29), these quantities satisfy the differential equations:

$$\nabla_\delta \hat{l}_k = \hat{l}_k(\pi_n^n - \pi_0^0) + B_{kl}\pi_n^l, \tag{7.113}$$

where B_{kl} is a nondegenerate tensor defined by relations (7.92). If, using the tensor B_{ij}, we raise the index k in $\hat{l}_k$:

$$l^i = B^{ik}\hat{l}_k, \tag{7.114}$$

then, by equations (7.103) and (7.113), we obtain the following differential equations for the object l^i:

$$\nabla_\delta l^i = l^i(\pi_n^n - \pi_0^0) + \pi_n^i. \tag{7.115}$$

If, using the tensor b_{ij}, we lower the index i in l^i:

$$l_j = b_{ji} l^i \tag{7.116}$$

and apply equations (7.13) and (7.115), we find that the object l^i satisfies the differential equations:

$$\nabla_\delta l_j = b_{ji}\pi^i_n. \tag{7.117}$$

Finally, using the objects b_j and l_j, we define another object

$$k_j = b_j - l_j, \tag{7.118}$$

which, by equations (7.27) and (7.118), satisfies the differential equations:

$$\nabla_\delta k_j = -\pi^0_j. \tag{7.119}$$

It follows from differential equations (7.107), (7.109), (7.113), (7.115) and (7.119) that we constructed the following quasitensors of the fourth order: $\{l, b_i, b^i\}, \{\hat{l}, b_i\}$,
$\{\hat{l}_{ij}, B_{ijk}\}, \{\hat{l}_i, B_{ij}\}, \{l^i\}, \{l_i\}$ and $\{k_i\}$. The quasitensors $\{l^i\}$ and $\{k_i\}$ are called the *first pair of normal quasitensors.*

To construct the second pair of normal quasitensors, we first find differential equations for the quantities c_k introduced in equations (7.95), which the trace B_0 satisfies. The prolongation of equation (7.95) leads to the following differential equation for c_k:

$$\nabla_\delta c_k = -b_{lk}\pi^l_n - \pi^0_k. \tag{7.120}$$

Using the quantities c_k, b_k and b^{ik}, we define the following quantities:

$$j_k = \frac{1}{2}(c_k - b_k), \tag{7.121}$$

$$j^i = b^{ik} j_k, \tag{7.122}$$

$$h_k = \frac{1}{2}(c_k + b_k), \tag{7.123}$$

and

$$h^i = b^{ik} h_k. \tag{7.124}$$

By equations (7.13), (7.27) and (7.120), these quantities satisfy the following differential equations:

$$\nabla_\delta j_k = -b_{kl}\pi^l_n, \tag{7.125}$$

$$\nabla_\delta j^i = j^i(\pi^n_n - \pi^0_0) - \pi^i_n, \tag{7.126}$$

$$\nabla_\delta h_k = -\pi^0_k \tag{7.127}$$

and

$$\nabla_\delta h^i = h^i(\pi_n^n - \pi_0^0) - b^{ik}\pi_k^0. \tag{7.128}$$

Equations (7.120) and (7.125)–(7.128) prove that the quantities $\{c_i, b_{ij}\}$, $\{j_k, b_{ij}\}$, $\{j^i\}$, $\{h_i\}$ and $\{h^i, b^{ik}\}$ form quasitensors of the fourth order. The quasitensors $\{j^i\}$ and $\{h_i\}$ are called the *second pair of normal quasitensors.*

Using the quantities c_i, l_i and k_i we can define the object:

$$\hat{c}_i = c_i + b_{ij}l^j - k_i, \tag{7.129}$$

which by relations (7.120), (7.116) and (7.119) satisfies the differential equations:

$$\nabla_\delta \hat{c}_i = 0, \tag{7.130}$$

i.e. the quantities $\hat{c}_i$ form a $(0, 1)$-tensor. This tensor is called the *first canonical tensor* of the hypersurface V^m. It was introduced by Bushmanova and Norden in their joint paper [BN 48].

Let us construct another relative invariant:

$$\hat{c} = b^{ij}\hat{c}_i\hat{c}_j. \tag{7.131}$$

By (7.130) and (7.17), the quantity $\hat{c}$ satisfy the equation

$$\delta\hat{c} = \hat{c}(\pi_n^n - \pi_0^0). \tag{7.132}$$

This invariant vanishes if and only if the covector $\hat{c}_i$ vanishes or if the vector $\hat{c}^i = b^{ij}\hat{c}_j$ (associated with the covector $\hat{c}_i$) defines an asymptotic direction on the hypersurface V^m. Assuming that the covector $\hat{c}_i$ is not 0 or asymptotic, we can consider the relative invariant $\hat{c}$ to be nonvanishing:

$$\hat{c} \neq 0. \tag{7.133}$$

Now we can easily verify that the quasitensors $-l^i$ and j^i are solutions of equations (7.74), and the quasitensors k_i and h_i are solutions of equations (7.75), i.e. we found two solutions of each of equations (7.74) and (7.75). This means that the straight line A_0Y_n, where

$$Y_n = A_n - l^iA_i \quad (\text{or } Y_n = A_n + j^iA_i),$$

is *an invariant first normal of the hypersurface* V^m, and the points

$$Z_i = A_i + k_iA_0 \quad (\text{or } Z_i = A_i + h_iA_0)$$

span *an invariant second normal of the hypersurface* V^m. Moreover, it is easy to see that each of the pairs of normal quasitensors, $\{l^i, k_i\}$ and $\{j^i, h_i\}$, satisfies relation (7.81), i.e. *each of the two pairs of first and second normals* we have constructed *are polar-conjugate with respect to the pencil* (7.48) *of Darboux osculating hyperquadrics.*

5. The two pairs of first and second normals, which we have constructed, allow us to construct the pencil of invariant normalizations of the hypersurface V^m. In fact, it is obvious that equations (7.74) and (7.75) have not only the solutions

$$y^i_{(1)} = -l^i, \ y^i_{(2)} = j^i, \tag{7.134}$$

and

$$z^{(1)}_i = k_i, \ z^{(2)}_i = h_i \tag{7.135}$$

indicated above, but also the solutions

$$y^i = -l^i + \tau(j^i + l^i), \tag{7.136}$$

and

$$z_i = k_i + \tau(h_i - k_i), \tag{7.137}$$

where τ is an absolute invariant, i.e. $\delta\tau = 0$. Moreover, these two pencils of solutions are still connected by relation (7.81):

$$k_i + \tau(h_i - k_i) = -l^i + \tau(j^i + l^i) + b_i. \tag{7.138}$$

Therefore, *we obtain a pencil of first normals and the corresponding pencil of second normals defined by the points:*

$$A_0 \quad \text{and} \quad Y_n(\tau) = A_n + [(\tau - 1)l^i + \tau j^i]A_i, \tag{7.139}$$

and by the points

$$Y_i(\tau) = A_i + [(1 - \tau)k_i + \tau h_i]A_0, \tag{7.140}$$

respectively. The normals of these two pencils corresponding to the same value of τ are polar-conjugate with respect to the pencil (7.48) *of Darboux osculating hyperquadrics.*

We will see later that the pencil of first normals defined by the points (7.139) is a generalization of the classical canonical pencil of projective normals of a surface $V^2 \subset P^3$.

Note that using the objects $\hat{l}$ and B_0, defined by formulas (7.108) and (7.94), we can represent the parameter A_{nn} of the pencil of Darboux osculating hyperquadrics in the form:

$$A_{nn} = \frac{1}{m+2}\hat{l} + \mu B_0, \tag{7.141}$$

where μ is an arbitrary absolute invariant, i.e. $\delta\mu = 0$. In particular, we can set $\mu = 0$. By equations (7.109),(7.95), (7.39) and (7.47), we find that if a point $x \in V^m$ is fixed, then the quantity A_{nn}, defined by relation (7.141), satisfies equation (7.68). By (7.65), the latter relation takes the form:

$$\delta A_{nn} = A_{nn}(\pi^n_n - \pi^0_0) + \frac{m+2}{m} b_i \pi^i_n - 2\pi^0_n. \tag{7.142}$$

Note also that using the object $\hat{l}$ defined by equation (7.107), we can write the expression (7.71) of the coefficient A_{nn} of the fixed hyperquadric found in Theorem 7.2 in a shorter form:

$$A_{nn} = \frac{1}{m+2}\hat{l}, \tag{7.143}$$

i.e. that the fixed hyperquadric, obtained in Theorem 7.2, is a hyperquadric from the pencil (7.48) of Darboux osculating hyperquadrics corresponding to the value $\mu = 0$ of its parameter.

6. Two pairs of quasitensors, which we have constructed in subsection **4**, allow us to find two symmetric (0, 2)-tensors associated with the fifth order differential neighborhood of a point $x \in V^m$.

To construct the first of these (0, 2)-tensors, we write differential equations (7.115) and (7.119) without making the assumption that a point $x \in V^m$ is fixed, i.e. without assuming that $\omega^i = 0$:

$$\nabla l^i = l^i(\omega_n^n - \omega_0^0) + \omega_n^i + q_k^i \omega^k, \tag{7.144}$$

$$\nabla k_j = -\omega_j^0 + p_{ji}\omega^i. \tag{7.145}$$

Note that the quantities q_k^i and p_{ji} are associated with the fifth order differential neighborhood of a point $x \in V^m$.

Exterior differentiation of equations (7.144) and (7.145), application of Cartan's lemma and setting $\omega^i = 0$ in the resulting equations leads to the following differential equations for the objects q_k^i and p_{ji}:

$$\nabla_\delta q_k^i = q_k^i(\pi_n^n - \pi_0^0) + \delta_k^i l^j \pi_j^0 - l^i b_{kj}\pi_n^j - l^j b_{kj}\pi_n^i - \delta_k^i \pi_n^0, \tag{7.146}$$

and

$$\nabla_\delta p_{ji} = -k_i \pi_j^0 - k_j \pi_i^0 + b_{ji}(k_l \pi_n^l - \pi_n^0). \tag{7.147}$$

Using these new objects q_k^i and p_{ji}, we construct the following quantities:

$$\hat{p}_{ij} = p_{ij} - k_i k_j, \tag{7.148}$$

$$q_{ij} = b_{ik} q_j^k, \tag{7.149}$$

$$\hat{q}_{ij} = q_{ij} + b_{ip} b_{jq} l^p l^q = b_{ik} q_j^k + b_{ip} b_{jq} l^p l^q, \tag{7.150}$$

$$c_{ij} = \hat{p}_{ij} - \hat{q}_{ij} - b_{ij} k_p l^p, \tag{7.151}$$

$$c = b^{ij} c_{ij}, \tag{7.152}$$

$$\hat{c}_{ij} = c_{ij} + c_{ji} - \frac{2}{m} b_{ij} c. \tag{7.153}$$

Using (7.146), (7.147), (7.119), (7.114), (7.115) and (7.117), we find that the quantities defined in (7.148)–(7.153) satisfy the following differential equations:

$$\nabla_\delta \hat{p}_{ij} = b_{ij} k_l \pi_n^l - b_{ij} \pi_n^0, \tag{7.154}$$

$$\nabla_\delta q_{ij} = b_{ij}(l^k \pi_k^0 - \pi_n^0) - b_{ik} b_{jl} l^k \pi_n^l - b_{jk} b_{il} l^k \pi_n^l, \tag{7.155}$$

$$\nabla_\delta \hat{q}_{ij} = b_{ij} l^k \pi_k^0 - b_{ij} \pi_n^0, \tag{7.156}$$

$$\nabla_\delta c_{ij} = 0, \tag{7.157}$$

$$\delta c = c(\pi_n^n - \pi_0^0), \tag{7.158}$$

$$\nabla_\delta \hat{c}_{ij} = 0. \tag{7.159}$$

It follows from equations (7.146)–(7.147) and (7.154)–(7.158) that the objects $\{p_{ij}, k_i, b_{ij}\}$, $\{q_j^i, l^k, b_{ij}\}$, $\{\hat{p}_{ij}, k_i, b_{ij}\}$, $\{q_{ij}, l^i, b_{ij}\}$ and $\{\hat{q}_{ij}, l^i, b_{ij}\}$ are quasitensors, the object $\{c\}$ is a relative invariant, and the objects $\{c_{ij}\}$ and $\{\hat{c}_{ij}\}$ are (0, 2)-tensors, and the latter tensor is symmetric. Equations (7.152) and (7.153) imply that the tensor $\hat{c}_{ij}$ is apolar to the tensor b^{ij}:

$$b^{ij} \hat{c}_{ij} = 0. \tag{7.160}$$

7. In Section **7.1** we found the relatively invariant forms $\Phi_{(2)}$ and $\Psi_{(3)}$ (see equations (7.8) and (7.31)) and the absolute differential invariant $\sigma = \Psi_{(3)}/\Phi_{(2)}$ (see equation (7.33)). These forms and the invariant were connected with a second and third differential neighborhood of a point $x \in V^m$. The tensors $\hat{c}_i$ and $\hat{c}_{ij}$, which we constructed in this section, allow us to introduce the 1-form:

$$\Psi_{(4)} = \hat{c}_i \omega^i \tag{7.161}$$

and the 2-form

$$\Psi_{(5)} = \hat{c}_{ij} \omega^i \omega^j, \tag{7.162}$$

connected with the fourth and fifth differential neighborhoods of V^m, respectively.

Using equations (7.130), (7.159) and (7.20), we find that

$$\nabla_\delta \Psi_{(4)} = 0, \quad \nabla_\delta \Psi_{(5)} = 0, \tag{7.163}$$

i.e. the forms $\Psi_{(4)}$ and $\Psi_{(5)}$ are absolutely invariant.

7.4 The Rigidity Problem in a Projective Space

1. As in the differential geometry of the Euclidean space, in the projective differential geometry of a hypersurface the following problem arises: to determine a system of differential forms or a system of tensors that defines a hypersurface up to a projective transformation. In this section we present two approaches to the solution of this problem.

Fubini, in his works on projective differential geometry, has already considered the rigidity theorem for a hypersurface. However, in his papers and books, the final solution of this problem has not been given (see his papers [Fu 18a], [Fu 18b], [Fu 20] and the book [FČ 26]). É. Cartan also considered this problem while studying the problem of projective deformation of a hypersurface (see [Ca 20c]). Griffiths and Harris mentioned this problem in Appendix B of their joint paper [GH 79]. Lastly, Jensen and Musso presented the final solution of the problem in their preprint [JM 92].

We believe that the apparatus we have developed in this chapter allows us to present the solution of this problem in a more orderly manner. Further, we do not reduce the second fundamental form of a hypersurface V^m to the sum of squares as was done in [JM 92], and thus our proof is valid in both complex and real domains.

We will now prove the following result.

Theorem 7.6 *Suppose that hypersurfaces V^m and $\overline{V}^m$ are given in a projective space P^{m+1}, $m > 2$, and these hypersurfaces are both tangentially nondegenerate and not parts of hyperquadrics. Suppose also that there is a one-to-one correspondence between points of these hypersurfaces, and for corresponding elements in the tangent bundles $T(V^m)$ and $T(\overline{V}^m)$, the Fubini projective linear elements σ and $\overline{\sigma}$ defined by equation (7.33) coincide. Then the hypersurfaces V^m and $\overline{V}^m$ are projectively equivalent.*

Proof. The proof will consist of a few stages.

a) Let A_0 and $\overline{A}_0$ be corresponding points of the hypersurfaces V^m and $\overline{V}^m$. By the given correspondence of the tangent bundles $T(V^m)$ and $T(\overline{V}^m)$, the basis forms ω^i and $\overline{\omega}^i$ on these hypersurfaces can be made coincide:

$$\overline{\omega}^i = \omega^i. \tag{7.164}$$

The coincidence of the Fubini projective linear elements σ and $\overline{\sigma}$ implies that

$$\frac{\overline{\Psi}_{(3)}}{\overline{\Phi}_{(2)}} = \frac{\Psi_{(3)}}{\Phi_{(2)}}. \tag{7.165}$$

First note that if $m > 2$, the nominators and the denominators of these two fractions have no common factors. In fact, since the rank of the form $\Phi_{(2)}$ is greater than two, then this form cannot be decomposed into linear factors, and if the form $\Psi_{(3)}$ has a common factor with the form $\Phi_{(2)}$, then the form $\Phi_{(2)}$ must divide $\Psi_{(3)}$, and thus

$$\Psi_{(3)} = \theta\Phi_{(2)}.$$

Substituting the values of the forms $\Phi_{(2)}$ and $\Psi_{(3)}$ from formulas (7.8) and (7.31) into the above equation and representing the form θ as $\theta = 3\theta_i\omega^i$, we find that

$$B_{ijk} = 3\theta_{(i}b_{jk)}.$$

Contracting the latter relation with the tensor b^{ij} and applying equation (7.30), we obtain the equation

$$(m+2)\theta_i = 0,$$

which implies $\theta_i = 0$ and $B_{ijk} = 0$, i.e. $\Psi_{(3)} = 0$, and the hypersurface V^m is a hyperquadric or a part of a hyperquadric. This contradicts the theorem hypothesis.

Note that if $m = 2$, then the forms $\Phi_{(2)}$ and $\Psi_{(3)}$ can have a common factor even if $\Psi_{(3)} \neq 0$. If this is the case, these two forms can be reduced to

$$\Phi_{(2)} = 2\omega^1\omega^2, \quad \Psi_{(3)} = \beta(\omega^1)^3,$$

and the surface $V^2 \subset P^3$ is a ruled surface.

Since the nominators and the denominators in (7.165) have no common factors, it follows that

$$\overline{\Phi}_{(2)} = t\Phi_{(2)}, \quad \overline{\Psi}_{(3)} = t\Psi_{(3)}. \tag{7.166}$$

In fact, relations (7.165) can be written as

$$\overline{\Phi}_{(2)}\Psi_{(3)} = \Phi_{(2)}\overline{\Psi}_{(3)}.$$

As $\overline{\Phi}_{(2)}$ and $\overline{\Psi}_{(3)}$ have no common factors, it follows that relation (7.166) takes place. By (7.164), relation (7.166) implies that

$$\overline{b}_{ij} = tb_{ij}, \quad \overline{B}_{ijk} = tB_{ijk}. \tag{7.167}$$

b) Consider the third order frame bundle on a hypersurface V^m. Third order frames are characterized by the fact that the first normal $n_x = A_0 \wedge A_n$ is polar-conjugate to the second normal $l_x = A_1 \wedge \ldots \wedge A_{n-1}$ relative to the pencil (7.48) of osculating Darboux hyperquadrics. This implies $b_i = 0$, and equation (7.12) can be written as

$$db_{ij} - b_{kj}\omega_i^k - b_{ik}\omega_j^k + b_{ij}(\omega_0^0 + \omega_n^n) = B_{ijk}\omega^k. \tag{7.168}$$

Contract equations (7.168) with the tensor b^{ij}. By (7.87), this leads to

$$d\ln|b| = 2\omega_i^i - m(\omega_0^0 + \omega_n^n), \tag{7.169}$$

where $b = \det(b_{ij})$. Moreover, the condition of normalization of a projective frame (see (1.64)) implies that

$$\omega_0^0 + \omega_i^i + \omega_n^n = 0$$

(see (1.66)). Eliminating the form $\omega_0^0 + \omega_n^n$ from relation (7.169), we find that

$$d\ln|b| = (m+2)\omega_i^i. \tag{7.170}$$

This equation implies that the quantity $|b|$ can be reduced to 1 on the hypersurface V^m: $|b| = 1$, and this leads to the equations

$$\omega_i^i = 0, \quad \omega_0^0 + \omega_n^n = 0. \tag{7.171}$$

By (7.171), relations (7.168) can be written as

$$db_{ij} - b_{kj}\omega_i^k - b_{ik}\omega_j^k = B_{ijk}\omega^k. \tag{7.172}$$

Applying similar considerations to a hypersurface $\overline{V}^m$, we arrive at the equations $\overline{b}_i = 0$, $|\overline{b}| = 1$, $\overline{\omega}_i^i = 0$, $\overline{\omega}_0^0 + \overline{\omega}_n^n = 0$, and

$$d\overline{b}_{ij} - \overline{b}_{kj}\overline{\omega}_i^k - \overline{b}_{ik}\overline{\omega}_j^k = \overline{B}_{ijk}\omega^k. \tag{7.173}$$

Compare now the determinants of the tensors b_{ij} and $\overline{b}_{ij}$. Relation (7.167) implies $\overline{b} = t^m b$. Since $|\overline{b}| = |b| = 1$, it follows that $t^m = 1$, and thus $|t| = 1$. Without loss of generality, we can assume that $t = 1$. As a result, equations (7.167) take the form

$$\overline{b}_{ij} = b_{ij}, \quad \overline{B}_{ijk} = B_{ijk}. \tag{7.174}$$

It follows that

$$\overline{\Phi}_{(2)} = \Phi_{(2)}, \quad \overline{\Psi}_{(3)} = \Psi_{(3)}. \tag{7.175}$$

Note also that in a third order frame, equations (7.25) on a hypersurface V^m can be written as

$$\omega_i^0 - b_{ij}\omega_n^j = l_{ij}\omega^j. \tag{7.176}$$

Prolongation of these equations leads to the relations which can be derived from equations (7.85) by setting $b_i = 0$ in them:

$$\nabla_\delta l_{ij} = B_{ijk}\pi_n^k - 2b_{ij}\pi_n^0.$$

Contracting the latter equations with the tensor b^{ij}, we arrive at

$$\delta l = l(\pi_n^n - \pi_0^0) - 2m\pi_n^0, \tag{7.177}$$

where l is defined by relation (7.106). Equation (7.177) implies that the quantity l can be reduced to 0: $l = 0$, and as a result, the form ω_n^0 becomes a linear combination of the basis forms ω^i:

$$\omega_n^0 = r_i \omega^i. \tag{7.178}$$

By (7.71) the condition $l = 0$ means that in equation (7.48) of the pencil of osculating Darboux hyperquadrics, the coefficient A_{nn} vanishes, and as a result, the point A_n of the constructed moving frame lies on an osculating Darboux hyperquadric.

The frame specialization which we have constructed is a part of specialization which separates the fourth order frame subbundle from the third order frame subbundle. The complete transfer to the fourth order frames will be used in the next subsection.

On a hypersurface $\overline{V}^m$, we obtain similar formulas:

$$\overline{\omega}_i^0 - \overline{b}_{ij}\overline{\omega}_n^j = \overline{l}_{ij}\omega^j, \quad \overline{l} = 0, \quad \overline{\omega}_n^0 = \overline{r}_i \omega^i. \tag{7.179}$$

c) Exterior differentiation of equations (7.164) gives the following exterior quadratic equations:

$$\omega^j \wedge [\overline{\omega}_j^i - \omega_j^i - \delta_j^i(\overline{\omega}_0^0 - \omega_0^0)] = 0. \tag{7.180}$$

Applying to these equations the Cartan lemma, we obtain

$$\overline{\omega}_j^i - \omega_j^i - \delta_j^i(\overline{\omega}_0^0 - \omega_0^0) = A_{jk}^i \omega^k, \tag{7.181}$$

where $A_{jk}^i = A_{kj}^i$. Contracting equations (7.181) with respect to the indices i and j and applying relations (7.171), we arrive at

$$\overline{\omega}_0^0 - \omega_0^0 = -A_k \omega^k, \tag{7.182}$$

where $A_k = \frac{1}{m} A_{ik}^i$. Eliminating the form $\overline{\omega}_0^0 - \omega_0^0$ by means of (7.182), we find that

$$\overline{\omega}_j^i - \omega_j^i = \widetilde{A}_{jk}^i \omega^k, \tag{7.183}$$

where

$$\widetilde{A}_{jk}^i = A_{jk}^i - \delta_j^i A_k. \tag{7.184}$$

Next, applying (7.172) and (7.173), we differentiate the first of equations (7.174). As a result, we have

$$b_{ik}(\overline{\omega}_j^k - \omega_j^k) + b_{kj}(\overline{\omega}_i^k - \omega_i^k) = 0. \tag{7.185}$$

Substituting expressions (7.183) into equations (7.185), we find that

$$b_{ik}\widetilde{A}_{jl}^k + b_{jk}\widetilde{A}_{il}^k = 0. \tag{7.186}$$

By (7.184), equations (7.186) can be written as

$$b_{ik}A^k_{jl} + b_{jk}A^k_{il} = 2b_{ij}A_l. \tag{7.187}$$

Cycling these relations with respect to the indices i, j and l, we find two more relations:

$$b_{jk}A^k_{li} + b_{lk}A^k_{ji} = 2b_{jl}A_i, \quad b_{lk}A^k_{ij} + b_{ik}A^k_{lj} = 2b_{li}A_j. \tag{7.188}$$

Subtracting equation (7.187) from the sum of equations (7.188), we obtain:

$$b_{kl}A^l_{ij} = -b_{ij}A_k + b_{jk}A_i + b_{ki}A^j.+ \tag{7.189}$$

Solving these equations with respect to A^k_{ij}, we find that

$$A^k_{ij} = \delta^k_i A_j + \delta^k_j A_i - b_{ij}b^{kl}A_l. \tag{7.190}$$

Equations (7.184) and (7.190) allow us to write equations (7.183) as

$$\overline{\omega}^i_j - \omega^i_j = (\delta^i_k A_j - b_{jk}A^i)\omega^k, \tag{7.191}$$

where $A_i = b^{ik}A_k$.

If we now apply apply exterior differentiation to equation (7.182), we arrive at

$$(\nabla A_k - (\overline{\omega}^0_k - \omega^0_k)) \wedge \omega^k = 0, \tag{7.192}$$

where as usually $\nabla A_k = dA_k - A_l(\omega^l_k - \delta^l_k\omega^0_0)$. Applying the Cartan lemma to equations (7.192) and setting $\omega^i = 0$ in the obtained relations, we find that

$$\nabla_\delta A_k = \overline{\pi}^0_k - \pi^0_k. \tag{7.193}$$

Equations (7.193) imply that the quantities A_k can be reduced to 0: $A_k = 0$. This specialization gives agreement of the third order frames associated with the hypersurfaces V^m and $\overline{V}^m$.

As a result of the latter specialization, equations (7.182) and (7.193) take the form

$$\overline{\omega}^0_0 = \omega^0_0, \quad \overline{\omega}^i_j = \omega^i_j, \tag{7.194}$$

and by the Cartan lemma, equations (7.192) give

$$\overline{\omega}^0_k - \omega^0_k = P_{kl}\omega^l, \tag{7.195}$$

where $P_{kl} = P_{lk}$.

Exterior differentiation of equations (7.194) leads to

$$(\overline{\omega}^0_j - \omega^0_j) \wedge \omega^i + b_{jk}\omega^k \wedge (\overline{\omega}^i_n - \omega^i_n) = 0. \tag{7.196}$$

These exterior quadratic equations imply that the forms $\overline{\omega}_n^i - \omega_n^i$ can be expressed in terms of the basis forms ω^j, i.e.

$$\overline{\omega}_n^i - \omega_n^i = Q_j^i \omega^j. \tag{7.197}$$

Substituting these expressions and expressions (7.195) into quadratic equations (7.196), we find that

$$P_{jk}\delta_l^i - P_{jl}\delta_k^i + b_{jk}Q_l^i - b_{jl}Q_k^i = 0. \tag{7.198}$$

Contracting these relations with respect to the indices i and k, we obtain

$$(m-1)P_{jl} - Q_{jl} + b_{jl}Q = 0, \tag{7.199}$$

where $Q_{jl} = b_{jk}Q_l^k$ and $Q = Q_k^k$. It follows that

$$Q_{jl} = Q_{lj}. \tag{7.200}$$

On the other hand, contracting equations (7.198) with respect to the indices i and k, we obtain

$$(m-1)Q_l^i - P_l^i + P\delta_l^i = 0, \tag{7.201}$$

where $P_l^k = b^{ik}P_{kl}$ and $P = P_k^k$.

If we contract relations (7.201) with respect to the indices i and l, we find that

$$P + Q = 0. \tag{7.202}$$

The latter relation can also be obtained if we contract equations (7.199) with the tensor b^{jl}.

Next, let us contract equations (7.201) with the tensor b_{ij}. This gives

$$(m-1)Q_{jl} - P_{jl} + b_{jl}P = 0. \tag{7.203}$$

Adding equations (7.199) and applying (7.202), we find that

$$(m-2)(P_{ij} + Q_{ij}) = 0. \tag{7.204}$$

Since, by the theorem hypothesis, $m > 2$, equations (7.204) imply that

$$P_{ij} + Q_{ij} = 0. \tag{7.205}$$

This and equations (7.199) and (7.203) imply that

$$P_{ij} = \frac{P}{m}b_{ij}, \quad Q_{ij} = \frac{Q}{m}b_{ij}. \tag{7.206}$$

Thus, equations (7.195) and (7.197) can be written as

$$\overline{\omega}_i^0 - \omega_i^0 = \frac{P}{m} b_{ij}\omega^j, \quad \overline{\omega}_n^i - \omega_n^i = \frac{Q}{m}\omega^i. \tag{7.207}$$

Applying formulas (7.176) and (7.179), we find that

$$\overline{\omega}_k^0 - \omega_k^0 - b_{ij}(\overline{\omega}_n^i - \omega_n^i) = (\bar{l}_{ij} - l_{ij})\omega^j. \tag{7.208}$$

Substituting expressions (7.207) into equation (7.208), we arrive at

$$\frac{1}{m}(P - Q)b_{ij} = (\bar{l}_{ij} - l_{ij}). \tag{7.209}$$

Let us now turn to the fourth order frame bundle, where $l = \bar{l} = 0$, and contract equations (7.208) with the tensor b^{ij}. This gives the equation

$$P - Q = 0. \tag{7.210}$$

Comparing this with equation (7.202), we find that

$$P = Q = 0.$$

By (7.207), it follows that

$$\overline{\omega}_i^0 = \omega_i^0, \quad \overline{\omega}_n^i = \omega_n^i. \tag{7.211}$$

Applying exterior differentiation to equations (7.211), we obtain

$$(\overline{\omega}_n^0 - \omega_n^0) \wedge \omega^i = 0.$$

Since $m > 1$, it follows that

$$\overline{\omega}_n^0 = \omega_n^0. \tag{7.212}$$

Equations (7.164), (7.1), (7.4), (7.174), (7.194), (7.211) and (7.212) imply that all components of the infinitesimal displacement of fourth order moving frames associated with the hypersurfaces V^m and $\overline{V}^m$ coincide. Therefore, there exists a projective transformation which transforms the hypersurface V^m with the fourth order frames we have constructed, into the hypersurface $\overline{V}^m$ with the same kind of fourth order moving frames.

This concludes our proof. ■

Note that if $m = 2$, the relation (7.204) becomes an identity, and the rest of the proof is not valid. This is related with the fact that for two-dimensional surfaces $V^2 \subset P^3$, the relation (7.165) does not imply that the surfaces V^2 and $\overline{V}^2$ are projectively equivalent. As was already proved by Fubini in [Fu 16] and later by É. Cartan in [Ca 20c], in P^3, there exist two surfaces V^2 and $\overline{V}^2$ admitting a nontrivial mapping one to another under which the projective linear element is preserved: $\overline{\sigma} = \sigma$. This mapping is called a *projective deformation.*

2. Let us now find a system of tensors defining a hypersurface V^m in the space P^{m+1} up to a projective transformation. To this end, at each point

$x \in V^m$ we will pass to a fourth order moving frame associated with this point and defined by one of the pairs of invariant normals constructed in Section **7.3**. In contrast to what we did in subsection **7.4.1**, we will now completely fix the first and second normals by setting

$$\widetilde{A}_n = A_n - l^i A_i, \quad \widetilde{A}_i = A_i + k_i A_0, \tag{7.213}$$

where l^i and k_i are defined by equations (7.114) and (7.118). As we know, in this new frame $b_i = 0$. This implies that the normalizing objects l^i and k_i vanish, and equations (7.144) and (7.145), which they satisfy, take the form:

$$\omega_n^i = -q_k^i \omega^k, \quad \omega_i^0 = p_{ik}\omega^k. \tag{7.214}$$

As we did in subsection **7.4.1**, we also fix an invariant point $Y = \widetilde{A}_n + yA_0$ on the first normal $A_0\widetilde{A}_n$. If $\omega^i = 0$, the differentiation of the point Y gives

$$\delta Y = \pi_n^n Y + (\delta y - y(\pi_n^n - \pi_0^0) + \pi_n^0)A_0.$$

Thus the condition for the point Y to be invariant has the following form:

$$\delta y = y(\pi_n^n - \pi_0^0) - \pi_n^0. \tag{7.215}$$

Consider now equation (7.109) which the object $\hat{l}$, defined by formula (7.108), satisfies. Since in our new moving frame equations (7.214) hold, we have $\pi_n^i = 0$, and equation (7.109) becomes

$$\delta \hat{l} = \hat{l}(\pi_n^n - \pi_0^0) - 2m\pi_n^0. \tag{7.216}$$

This equation coincides with equation (7.177) since even in a third order frame we have $\hat{l} = l$. Comparing equations (7.215) and (7.216), we find that a solution of the first of them can be represented in the form:

$$y = \frac{1}{2m}\hat{l}.$$

Therefore the point

$$\hat{A}_n = \widetilde{A}_n + \frac{1}{2m}\hat{l}A_0 \tag{7.217}$$

is an invariant point of the normal $A_0\widetilde{A}_n$. If we place the point $\widetilde{A}_n$ into the point $\hat{A}_n$, we obtain $\hat{l} = 0$. This and equation (7.216) imply that $\pi_n^0 = 0$ and

$$\omega_n^0 = r_i \omega^i. \tag{7.218}$$

This equation we already obtained in subsection **7.4.1** (cf. equation (7.178)). Having this specialization of a moving frame made, we see that the quantities q_k^i, p_{ik} and r_i become tensors. These tensors are defined in a fifth order neighborhood of a point $x \in V^m$. In an arbitrary frame, these tensors can be expressed in terms of the fifth order fundamental objects which arise in the differential prolongation of the fourth order fundamental objects defined by formulas (7.106)–(7.128).

Next, consider the following matrix of 1-forms:

$$\omega = \begin{pmatrix} \omega_0^0 & \omega_0^i & 0 \\ \omega_i^0 & \omega_j^i & \omega_i^n \\ \omega_n^0 & \omega_n^j & \omega_n^n \end{pmatrix}, \tag{7.219}$$

defining the infinitesimal displacement of a fourth order moving frame of the hypersurface V^m. The entries $\omega_0^i = \omega^i$ of this matrix are basis forms on V^m, and the 1-forms ω_i^n, ω_n^i, ω_i^0 and ω_n^0 are expressed in terms of the basis forms by formulas (7.4), (7.214) and (7.218).

In order to find the remaining 1-forms in the matrix (7.219), we will write equations following from equations (7.12), (7.93) and (7.132) relative to a fourth order moving frame. As a result, we obtain the following equations for the tensors b_{ij}, B_{ij} and the relative invariant $\hat{c}$ defined by formula(7.131):

$$db_{ij} - b_{ik}\omega_j^k - b_{kj}\omega_i^k + b_{ij}(\omega_0^0 + \omega_n^n) = B_{ijk}\omega^k, \tag{7.220}$$

$$dB_{ij} - B_{ik}\omega_j^k - B_{kj}\omega_i^k + 2B_{ij}\omega_0^0 = C_{ijk}\omega^k, \tag{7.221}$$

$$d\hat{c} + \hat{c}(\omega_0^0 - \omega_n^n) = \hat{c}C_k\omega^k. \tag{7.222}$$

The quantities in the right-hand sides of equations (7.221) and (7.222) are also relative tensors since after our specialization of the first and second normals and our choice $\widetilde{A}_n = \hat{A}_n$, only the secondary forms π_0^0, π_j^i and π_n^n (defining tensor transformations of all quantities under consideration) remain nonvanishing. Moreover, it follows from equations (7.221) and (7.222) that the tensor C_{ijk} is defined in a fourth order neighborhood, and the covector C_k in a fifth order neighborhood of a point $x \in V^m$.

If we assume that our fourth order moving frame of V^m is normalized by the condition $A_0 \wedge A_1 \wedge \ldots \wedge A_{m+1} = 1$, then, in addition to equations (7.220)–(7.222), we will have the relation

$$\omega_0^0 + \omega_k^k + \omega_n^n = 0 \tag{7.223}$$

(cf. equation (1.66)).

We will now prove the following lemma.

Lemma 7.7 *If the tensor b_{ij} is positive definite, the affinor $B_j^i = b^{ik}B_{jk}$ has distinct eigenvalues λ_i not connected by any functional relation, and also the inequality (7.232) (see below) holds, then equations (7.220)–(7.222) allows us to express all the forms $\omega_0^0, \omega_j^i, \omega_n^n$ and ω^i in terms of the tensors $b_{ij}, B_{ij}, B_{ijk}, C_{ijk}, \hat{c}, C_k$ and the differentials of the tensors b_{ij}, B_{ij} and $\hat{c}$.*

Proof. The lemma conditions allow us to reduce simultaneously the symmetric tensors b_{ij} and B_{ij} to the diagonal forms:

$$b_{ij} = \delta_{ij}, \quad B_{ij} = \lambda_i \delta_{ij} \tag{7.224}$$

by means of a frame transformation. Moreover, by inequality (7.101), all eigenvalues λ_i are different from 0, and by the lemma conditions, they are distinct. In such a moving frame, for $i \neq j$ equations (7.220) and (7.221) take the form:

$$-\omega_j^i - \omega_i^j = B_{ijk}\omega^k, \quad -\lambda_i\omega_j^i - \lambda_j\omega_i^j = C_{ijk}\omega^k, \tag{7.225}$$

where here and further in this subsection there is no summation over the indices i and j and there is summation over the index k. Since $\lambda_i \neq \lambda_j$ for $i \neq j$, it follows from system (7.225) that the forms ω_j^i, $i \neq j$, are expressed in terms of the basis forms ω^k:

$$\omega_j^i = l_{jk}^i \omega^k, \quad i \neq j. \tag{7.226}$$

If $i = j$, equations (7.220) and (7.221) take the form:

$$-2\omega_i^i + \omega_0^0 + \omega_n^n = B_{iik}\omega^k, \tag{7.227}$$

$$d\lambda_i - 2\lambda_i\omega_i^i + 2\lambda_i\omega_0^0 = C_{iik}\omega^k. \tag{7.228}$$

We have $2m+2$ equations (7.222), (7.223), (7.227) and (7.228) which are linear with respect to the forms $\omega_0^0, \omega_i^i, \omega_n^n$ and ω^i, and the number of these forms is also $2m+2$. From equations (7.227) we find that

$$\omega_i^i = \frac{1}{2}(\omega_0^0 + \omega_n^n - B_{iik}\omega^k). \tag{7.229}$$

Substituting these expressions into equations (7.228), we obtain

$$d\lambda_i + \lambda_i(\omega_0^0 - \omega_n^n) = (C_{iik} - \lambda_i B_{iik})\omega^k.$$

Eliminating the difference $\omega_0^0 - \omega_n^n$ by means of relation (7.222), we find that

$$d\ln|\lambda_i| - d\ln\hat{c} = \frac{1}{\lambda_i}(C_{iik} - \lambda_i(B_{iik} + C_k))\omega^k. \tag{7.230}$$

Let also suppose that the determinant of the matrix

$$\widetilde{C}_{ik} = \frac{1}{\lambda_i}(C_{iik} - \lambda_i(B_{iik} + C_k)) \tag{7.231}$$

is different from zero:

$$\det(\widetilde{C}_{ik}) \neq 0. \tag{7.232}$$

Then we find from equations (7.230) that

$$\omega^k = \widetilde{C}^{kl} d \ln \left| \frac{\lambda_l}{\hat{c}} \right|, \tag{7.233}$$

where $\widetilde{C}^{kl}$ is the inverse matrix of the matrix $\widetilde{C}_{ik}$. Since by the lemma hypotheses, the eigenvalues λ_k of the affinor B^i_j are functionally independent, the forms ω^k are linearly independent and can be taken as basis forms of V^m.

Finally, relations (7.222) and (7.223) allow us to express 1-forms ω_0^0 and ω_n^n in terms of the basis forms ω^k. As a result, all the forms $\omega_0^0, \omega_j^i, \omega_n^n$ and ω^i will be expressed in terms of the differentials $d \ln \left| \frac{\lambda_l}{\hat{c}} \right|$, and the coefficients of these expressions are functions of the quantities $\lambda_i, B_{iik}, C_{iik}$ and C_k.

Thus we found the expressions of the forms $\omega_0^0, \omega_j^i, \omega_n^n$ and ω^i in the moving frame where conditions (7.224) hold. But it follows that even in the case when the specialization (7.224) does not take place, these forms will also be expressed in terms of the tensors $b_{ij}, B_{ij}, B_{ijk}, C_{ijk}, \hat{c}, C_k$ and their differentials. ∎

Note if the tensor b_{ij} is not positive definite but the remaining lemma conditions are satisfied, the conclusion of Lemma 7.5 is still valid. However, relations (7.224) and (7.225) which were used in the proof will have a more complex form.

If the 1-forms composing matrix (7.219) satisfy the structure equations of a projective space (see equations (1.67) in Section **1.3**), then, up to a projective transformation, these forms define a hypersurface V^m together with the fourth order frame bundle associated with V^m. This proves the following result.

Theorem 7.8 *Suppose that in a domain D of an m-dimensional manifold M the quantities $b_{ij}, B_{ij}, B_{ijk}, C_{ijk}, \hat{c}$ and C_k are given, and the 1-forms of the matrix (7.219), which are defined as was indicated above, satisfy the structure equations of a projective space P^{m+1}. Then, up to a projective transformation, in P^{m+1} there exists a unique hypersurface $V^m = f(M)$ and a fourth order frame bundle associated with V^m such the tensor b_{ij} is the second fundamental tensor of V^m, the tensor B_{ijk} is the Darboux tensor of V^m, and the remaining tensors and the invariant $\hat{c}$ are connected with the hypersurface V^m in the way indicated above.* ∎

We will not write the integrability conditions which follow from the structure equations of the space P^{m+1} since these conditions are cumbersome. It turned out to be very difficult to analyze these compatibility conditions.

Note also that the tensors, in terms of which all the 1-forms of the matrix (7.219) can be expressed, are defined by a neighborhood of order not higher than five. This is the reason that fundamental object of order five is called *complete* (see [Lap 53]).

7.5 The Geometry of a Surface in Three-Dimensional Projective Space

1. In this section we apply all geometric objects, which we obtained in Section **7.3** for a hypersurface V^m in a projective space P^{m+1}, to a surface of a three-dimensional projective space P^3, i.e we will set everywhere $m = 2,\ n = 3$.

All formulas, which we derive in Section **7.3**, are invariant. They can be used for the calculation of the corresponding geometric objects with respect to any specific moving frame. Following the paper [Lap 53], we will apply them in the moving frame which was introduced and widely used by Finikov (see [Fi 37], p. 42). This frame has the following table of components:

	A_0	A_1	A_2	A_3
dA_0	$Pdu + Qdv$	du	dv	0
dA_1	$Cdu + Ldv$	$-Pdu + Qdv$	βdu	dv
dA_2	$Kdu + Adv$	γdv	$Pdu - Qdv$	du
dA_3	$A\beta du + C\gamma dv$	$Kdu + Adv$	$Cdu + Ldv$	$-Pdu - Qdv$

(7.234)

where

$$P = \frac{1}{2}\frac{\partial \ln \gamma}{\partial u}, \quad Q = \frac{1}{2}\frac{\partial \ln \beta}{\partial v} dv. \tag{7.235}$$

If we apply exterior differentiation to the Pfaffian equations corresponding to the entries of the table (7.234), we obtain only five independent exterior quadratic equations:

$$\begin{aligned}
&2dP \wedge du + (2L - \beta\gamma) du \wedge dv = 0,\\
&2dQ \wedge dv + (2K - \beta\gamma) du \wedge dv = 0,\\
&2dC \wedge du + dL \wedge dv + 2PL du \wedge dv = 0,\\
&2dK \wedge du + dA \wedge dv - 2QK du \wedge dv = 0,\\
&\beta dA \wedge du + \gamma dC \wedge dv + 4(PC\gamma - QA\beta) du \wedge dv = 0.
\end{aligned} \tag{7.236}$$

Equations (7.236) are convenient when one needs to prove the existence theorem for special classes of surfaces $V^2 \subset P^3$.

If in (7.236) we replace the differentials dP, dQ, dA, dC, dK and dL by their values, for example, $dP = \frac{\partial P}{\partial u} du + \frac{\partial P}{\partial v} dv$, we find that equations (7.236) are equivalent to the following equations connecting the surface invariants β, γ, K, L, A and C:

$$2K = \beta\gamma - \frac{\partial^2 \ln \beta}{\partial u \partial v}, \quad 2L = \beta\gamma - \frac{\partial^2 \ln \gamma}{\partial u \partial v}, \tag{7.237}$$

$$\frac{\partial A}{\partial u} = \frac{\partial K}{\partial v} + K\frac{\partial \ln \beta}{\partial v}, \quad \frac{\partial C}{\partial v} = \frac{\partial L}{\partial u} + L\frac{\partial \ln \gamma}{\partial u}, \tag{7.238}$$

$$\frac{\partial A}{\partial v}\beta + 2A\frac{\partial \beta}{\partial v} = \frac{\partial C}{\partial u}\gamma + 2C\frac{\partial \gamma}{\partial u}. \tag{7.239}$$

We see that in addition to the curvilinear coordinates u and v, equations (7.237)–(7.239) contains six relative invariants, and two of them, K and L, are expressed by formulas (7.237) in terms of β and γ. Thus a surface V^2 is defined by four relative invariants: β, γ, A and C which are connected by three equations (7.238) and (7.239).

2. We now compute the components of the relative tensors b_{ij} and b^{ij} and the discriminant b of second order of $V^2 \subset P^3$.

By formulas (7.4), we have

$$\omega_1^3 = b_{11}\omega^1 + b_{12}\omega^2, \quad \omega_2^3 = b_{12}\omega^1 + b_{22}\omega^2.$$

Comparing these two formulas with the table (7.234), we find that

$$b_{11} = b_{22} = 0, \; b_{12} = 1. \tag{7.240}$$

It follows that

$$b = \det(b_{ij}) = -1 \tag{7.241}$$

and

$$b^{11} = b^{22} = 0, \; b^{12} = 1. \tag{7.242}$$

Next, we compute the objects $b_{ijk}, b_k, b^k, B_{ij}, B, B^{ij}$ and B_0. By equations (7.240) and (7.12), we have

$$-2\omega_1^2 = b_{11k}\omega^k, \quad -\omega_1^1 - \omega_2^2 + \omega_0^0 + \omega_3^3 = b_{12k}\omega^k, \quad -2\omega_2^1 = b_{22k}\omega^k.$$

Taking the values of the forms from the table (7.234), we obtain from the first and the third of these equations:

$$-2\beta du = b_{111}du + b_{112}dv, \quad -2\gamma dv = b_{221}du + b_{222}dv.$$

It follows that

$$b_{111} = -2\beta, \quad b_{222} = -2\gamma, \quad b_{112} = b_{221} = 0. \tag{7.243}$$

By (7.24), (7.88) and (7.90), we now calculate b_k, b^i and $\hat{b}$:

$$b_k = \frac{1}{4}b^{ij}b_{ijk} = 0, \tag{7.244}$$

$$b^i = b^{ik}b_k = 0, \tag{7.245}$$

$$\hat{b} = b^{ij}b_ib_j = 0. \tag{7.246}$$

Formulas (7.28) and (7.244) imply that

$$B_{ijk} = b_{ijk}. \tag{7.247}$$

Thus, by (7.243) and (7.247), we find that

$$B_{111} = -2\beta, \quad B_{222} = -2\gamma, \quad B_{112} = B_{221} = 0. \tag{7.248}$$

From this and (7.242) and (7.243) it follows that on the surface V^2 the components of the tensor B_{ij}, which is defined by formula (7.92), have the following values:

$$B_{11} = B_{22} = 0, \quad B_{12} = 4\beta\gamma. \tag{7.249}$$

Equations (7.94), (7.97), (7.245) and (7.249) imply

$$B_0 = b^{ij} B_{ij} = 8\beta\gamma, \tag{7.250}$$

$$B = \det(B_{ij}) = -16\beta^2\gamma^2, \tag{7.251}$$

$$B^{11} = B^{22} = 0, \quad B^{12} = \frac{1}{4\beta\gamma}. \tag{7.252}$$

Using equations (7.240) and (7.244), from equation (7.25) we find that

$$-2(\omega_1^0 - \omega_3^2) + l_{11}du + l_{12}dv = 0, \quad -2(\omega_2^0 - \omega_3^1) + l_{12}du + l_{22}dv = 0. \tag{7.253}$$

Using the table (7.234), we find from this:

$$l_{ij} = 0. \tag{7.254}$$

By relation (7.254), formulas (7.108), (7.110), (7.112), (7.114), (7.116) and (7.118) give:

$$\hat{l} = l - \hat{b} = 0, \tag{7.255}$$

$$\hat{l}_{ij} = l_{ij} - b_i b_j - \frac{1}{2}\hat{l} b_{ij} = 0, \tag{7.256}$$

$$l^s = b^{ip} b^{jq} \hat{l}_{ij} B_{pqk} B^{ks} = 0, \tag{7.257}$$

$$l_j = b_{ij} l^j = 0, \tag{7.258}$$

$$k_j = b_j - l_j = 0. \tag{7.259}$$

Thus, the normals A_0A_3 and A_1A_2 of the Finikov frame correspond to the first and second normals defined by the points $\widetilde{A}_n$ and $\widetilde{A}_i$ (see relations (7.213)) on a hypersurface V^m. In addition, by (7.255), the point A_3 of the Finikov frame corresponds to the point $\hat{A}_n$ defined by formula (7.217) for a hypersurface V^m.

Next, in the Finikov frame, equation (7.97) has the form

$$\frac{\partial \ln(\beta\gamma)}{\partial u}du + \frac{\partial \ln(\beta\gamma)}{\partial v}dv = -\frac{\partial \ln \gamma}{\partial u}du - \frac{\partial \ln \beta}{\partial v}dv + c_1 du + c_2 dv. \tag{7.260}$$

Thus,

$$c_1 = \frac{\partial \ln(\beta\gamma^2)}{\partial u}, \quad c_2 = \frac{\partial \ln(\beta^2\gamma)}{\partial v}. \tag{7.261}$$

Equations (7.121) and (7.123) give

$$j_k = \frac{1}{2}(c_k - b_k) = \frac{1}{2}c_k, \quad h_k = \frac{1}{2}(c_k + b_k) = \frac{1}{2}c_k. \tag{7.262}$$

Hence,

$$j_1 = h_1 = \frac{1}{2}\frac{\partial \ln(\beta\gamma^2)}{\partial u}, \quad j_2 = h_2 = \frac{1}{2}\frac{\partial \ln(\beta^2\gamma)}{\partial v}. \tag{7.263}$$

Raising the indices:

$$j^i = b^{ik} j_k, \quad h^i = b^{ik} h_k,$$

we find that

$$j^1 = h^1 = \frac{1}{2}\frac{\partial \ln(\beta^2\gamma)}{\partial v}, \quad j^2 = h^2 = \frac{1}{2}\frac{\partial \ln(\beta\gamma^2)}{\partial u}. \tag{7.264}$$

By formulas (7.129), (7.257) and (7.259), we have

$$\hat{c}_i = c_i + l_i - k_i = c_i.$$

Therefore,

$$\hat{c}_1 = \frac{\partial \ln(\beta\gamma^2)}{\partial u}, \quad \hat{c}_2 = \frac{\partial \ln(\beta^2\gamma)}{\partial v}. \tag{7.265}$$

Finally, we will compute the objects c_{ij}, c and $\hat{c}_{ij}$. To this end, we will first calculate p_{ij} and q^i_j from equations (7.214), which in our moving frame have the form:

$$\begin{aligned} C du + L dv &= p_{11} du + p_{12} dv, \\ K du + A dv &= p_{21} du + p_{22} dv, \\ -K du - A dv &= q^1_1 du + q^1_2 dv, \\ -C du - L dv &= q^2_1 du + q^2_2 dv. \end{aligned}$$

Thus,

$$p_{11} = C, \quad p_{12} = L, \quad p_{21} = K, \quad p_{22} = A, \tag{7.266}$$

and

$$q_1^1 = -K, \quad q_2^1 = -A, \quad q_1^2 = -C, \quad q_2^2 = -L. \tag{7.267}$$

Now from equations (7.147), (7.148), (7.149) and (7.150) we compute the quantities c_{ij}, c and $\hat{c}_{ij}$:

$$c_{11} = 2C, \quad c_{12} = 2L, \quad c_{21} = 2K, \quad c_{22} = 2A, \tag{7.268}$$

$$c = 2(L + K), \tag{7.269}$$

$$\hat{c}_{11} = 4C, \quad \hat{c}_{12} = 0, \quad \hat{c}_{21} = 0, \quad \hat{c}_{22} = 4A. \tag{7.270}$$

4. Next, we will find the main geometric objects that are invariantly associated with a point $x \in V^2 \subset P^3$.

If $m = 2$, the equation of the pencil (7.48) of Darboux osculating quadrics, where A_{33} is defined by (7.161), has the form:

$$2x^0x^3 = b_{ij}x^ix^j + 2b_ix^ix^3 + (\frac{\hat{l}}{4} + \mu B_0)(x^3)^2. \tag{7.271}$$

Since in Finikov's moving frame $\hat{l} = b_i = b_{11} = b_{22} = 0, b_{12} = 1$ and $B_0 = 8\beta\gamma$, this equation becomes

$$XY - Z = \frac{1}{2}h\beta\gamma Z^2, \tag{7.272}$$

where we set

$$x^0 = 1, x^1 = X, \quad x^2 = Y, \quad x^3 = Z, \quad h = -8\mu. \tag{7.273}$$

Thus, the canonical pencil of Darboux osculating hyperquadrics corresponds to the pencil of Darboux quadrics of a surface V^2 in P^3 (cf. [Fi 37], pp. 85–86). This means that *equation* (7.271) *defines the classical pencil of Darboux quadrics in an arbitrary moving frame of the first order.* In particular,

- if $\mu = 0$, then we obtain the *Lie quadric*;
- if $\mu = \frac{1}{16}$, then we obtain the *Wilczynski–Bompiani quadric*;
- if $\mu = \frac{1}{48}$, then we obtain the *Fubini quadric*;
- if $\mu = \infty$, then we obtain the *double tangent plane*.

As we saw in Section **7.3**, the first normal from the canonical pencil is defined by a pair of points (7.139):

$$Y_0 = A_0, \ \ Y_3 = A_3 + [(\tau - 1)l^i + \tau j^i]A_i, \ i = 1, 2. \tag{7.274}$$

In Finikov's moving frame, $l^i = 0$ (see (7.257)) and j^i are defined by equations (7.264). By this, formulas (7.274) take the form:

$$Y_0 = A_0, \ \ Y_3 = A_3 + \frac{1}{2}\tau\Big\{\frac{\partial \ln(\beta\gamma^2)}{\partial u}A_2 + \frac{\partial \ln(\beta^2\gamma)}{\partial v}A_1\Big\}. \tag{7.275}$$

If we compare relations (7.275) with the formulas on pp. 106–107 of the book [Fi 37], we can again conclude that the *canonical pencil of the first normals which we found in Section* **7.3** *completely coincides with the classical canonical pencil.* In particular, this implies, that

- if $\tau = 0$, then we obtain the *Wilczynski directrix*;
- if $\tau = 1$, then we obtain the the *projective normal*;
- if $\tau = \frac{1}{2}$, then we obtain the *Green edge*;
- if $\tau = \frac{1}{3}$, then we obtain the *Čech axis*;
- if $\tau = \frac{1}{4}$, then we obtain the *Cartan line*;
- if $\tau = \infty$, then we obtain the *canonical tangent.*

If the parameter τ takes on the values indicated above, then *points* (7.275) *invariantly define the indicated straight lines of the canonical pencil.*

By (7.243), (7.247), (7.265) and (7.270), the relative and absolute invariant forms (7.8), (7.31), (7.158) and (7.159) and the absolute invariant (7.33), which we constructed in Sections **7.1** and **7.3**, in the Finikov moving frame have the forms:

$$\Phi_{(2)} = b_{ij}\omega^i\omega^j = 2du dv, \tag{7.276}$$

$$\Psi_{(3)} = B_{ijk}\omega^i\omega^j\omega^k = -2(\beta du^3 + \gamma dv^3), \tag{7.277}$$

$$\Psi_{(4)} = \frac{\partial \ln(\beta\gamma^2)}{\partial u}du + \frac{\partial \ln(\beta^2\gamma)}{\partial v}dv, \tag{7.278}$$

$$\Psi_{(5)} = \hat{c}_{ij}\omega^i\omega^j = 4(C du^2 + A dv^2), \tag{7.279}$$

$$\sigma = \frac{\Psi_3}{\Phi_2} = \frac{B_{ijk}\omega^i\omega^j\omega^k}{b_{ij}\omega^i\omega^j} = \frac{\beta du^3 + \gamma dv^3}{du dv}. \tag{7.280}$$

Therefore, the forms (7.276), (7.277) and (7.279) are *three fundamental forms*, the form (7.278) is *the linear form of the canonical tangent*, and the ratio (7.280) is the *projective (Fubini) linear element* of a surface $V^2 \subset P^3$.

This allows us to write the following invariant equations for the principal systems of lines of a surface $V^2 \subset P^3$:

- Equations of the *asymptotic lines*:

$$b_{ij}\omega^i\omega^j = 0; \tag{7.281}$$

- Equations of the *Darboux lines*:

$$B_{ijk}\omega^i\omega^j\omega^k = 0; \tag{7.282}$$

- Equations of the *Fubini net*:

$$\hat{c}_{ij}\omega^i\omega^j = 0; \tag{7.283}$$

- Equations of the lines of the *canonical tangents*:

$$\hat{c}_k\omega^k = 0. \tag{7.284}$$

5. We conclude this section by indicating the *invariant conditions of different kinds of degeneracy of a surface* $V^2 \subset P^3$:

(i) A surface $V^2 \subset P^3$ is a *plane* or a *part of plane* if and only if its second fundamental tensor vanishes:

$$b_{ij} = 0. \tag{7.285}$$

(ii) A surface $V^2 \subset P^3$ is a *quadric* or a *part of a quadric* if and only if its Darboux tensor vanishes:

$$B_{ijk} = 0 \tag{7.286}$$

since by (7.244), relation (7.286) is equivalent to

$$\beta = \gamma = 0$$

(cf. [Fi 37], p. 45).

(iii) A surface $V^2 \subset P^3$ is a *ruled surface* if and only if its invariant B_0 vanishes:

$$B_0 = 0 \tag{7.287}$$

since, by (7.250), equation (7.287) is equivalent to the condition:

$$\beta\gamma = 0$$

(cf. [Fi 37], p. 44).

(iv) *The surfaces V^2 and $\widetilde{V}^2$, described by the points A_0 and A_3, have common Lie quadrics at the corresponding points if and only if the tensor $\hat{c}_{ij}$ vanishes:*

$$\hat{c}_{ij} = 0. \tag{7.288}$$

To prove this, first note that by (7.270), equation (7.288) is equivalent to

$$A = C = 0 \tag{7.289}$$

(cf. [Fi 37], p. 50). From (7.271) we find that the equation of a Lie quadric at a point $x \in V^2$ is

$$x^0 x^3 - x^1 x^2 = 0. \tag{7.290}$$

By (7.289), equations (7.238) give:

$$K = \frac{U}{\beta}, \quad L = \frac{V}{\gamma},$$

where $U = U(u)$ and $V = V(v)$ are arbitrary functions of u and v respectively. If $U \neq 0$ and $V \neq 0$, then by changing parameters u and v the functions U and V can be reduced to 1. As a result, we have

$$K = \frac{1}{\beta}, \quad L = \frac{1}{\gamma}. \tag{7.291}$$

Let us associate a moving frame $\{\overline{A}_u\}, u = 0, 1, 2, 3$ with a point $A_3 \in (A_3)$ in such a way that

$$\overline{A}_0 = \sqrt{\beta\gamma} A_3, \quad \overline{A}_1 = \sqrt{\frac{\gamma}{\beta}} A_1, \quad \overline{A}_2 = \sqrt{\frac{\beta}{\gamma}} A_2, \quad \overline{A}_3 = \frac{1}{\sqrt{\beta\gamma}} A_0. \tag{7.292}$$

We also set:

$$\overline{\beta} = \gamma, \quad \overline{\gamma} = \beta. \tag{7.293}$$

Differentiating the points (7.292), by means of the table (7.234) and using relations (7.289), (7.291)–(7.293), we arrive at the following table:

	$\overline{A}_0$	$\overline{A}_1$	$\overline{A}_2$	$\overline{A}_3$
$d\overline{A}_0$	$\overline{P}du + \overline{Q}dv$	du	dv	0
$d\overline{A}_1$	$\frac{1}{\overline{\gamma}}dv$	$-\overline{P}du + \overline{Q}dv$	$\overline{\beta}du$	dv
$d\overline{A}_2$	$\frac{1}{\overline{\beta}}du$	$\overline{\gamma}dv$	$\overline{P}du - \overline{Q}dv$	du
$d\overline{A}_3$	0	$\frac{1}{\overline{\beta}}du$	$\frac{1}{\overline{\gamma}}dv$	$-\overline{P}du - \overline{Q}dv$

(7.294)

where

$$\overline{P} = \frac{1}{2}\frac{\partial \ln \overline{\gamma}}{\partial u}, \quad \overline{Q} = \frac{1}{2}\frac{\partial \ln \overline{\beta}}{\partial v} dv. \tag{7.295}$$

It is easy to see that the table (7.294) is completely analogous to the table (7.234) where the values of A, C, K and L are replaced according to formulas (7.289) and (7.291). Thus the surfaces $V^2 = (A_0)$ and $(\overline{A}_0) = (A_3)$ have the common moving frame at the corresponding points, and the tangents to the asymptotic coordinate lines u and v at the points A_0 and A_3 intersect one another at the points A_1 and A_2 respectively. Moreover, if $M = x^u A_u = \overline{x}^u \overline{A}_u$ are representations of a point $M \in P^3$ with respect to the frames $\{A_u\}$ and $\{\overline{A}_u\}$, then using relations (7.292), we find that the coordinates x^u and $\overline{x}^u$ of this point are connected as follows:

$$\overline{x}^0 = \frac{1}{\sqrt{\beta\gamma}} x^3, \quad \overline{x}^1 = \sqrt{\frac{\beta}{\gamma}} x^1, \quad \overline{x}^2 = \sqrt{\frac{\gamma}{\beta}} x^2, \quad \overline{x}^3 = \sqrt{\beta\gamma} x^0.$$

So, $\overline{x}^0\overline{x}^3 - \overline{x}^1\overline{x}^2 = x^0x^3 - x^1x^2$, and the surfaces $V^2 = (A_0)$ and $(\overline{A}_0) = (A_3)$ have the same Lie quadric at the corresponding points A_0 and A_3.

A pair of surfaces in a point correspondence having common Lie quadrics at the corresponding points is called *a pair of Godeaux surfaces* named after Godeaux who studied such pairs in detail (see for example his papers [Go 28]).

Equations (7.236) and (7.289) show that for a surface V^2 under consideration we have $q = 4, s_1 = 4, s_2 = 0$ and $Q = N = 4$. Thus, *the system defining such surfaces is in involution, and a two-dimensional integral manifold V^2, defined by this system, depends on four arbitrary functions of one variable.*

One can easily prove that if $U = 0$, the surface (A_3) degenerates into a curve and that the Lie quadrics of V^2, taken along one of asymptotic lines, pass through the point A_3. If $U = V = 0$, the surface (A_3) degenerates into a fixed point, and all Lie quadrics pass through the point A_3 at which they are tangent to the plane $A_1A_2A_3$ (see [Fi 37], pp. 51–52).

(v) *All first normals of the canonical pencil of a surface* $V^2 \subset P^3$ *coincide* if and only if the canonical covector $\hat{c}_k$ vanishes:

$$\hat{c}_k = 0 \tag{7.296}$$

since by (7.265), condition (7.285) is equivalent to

$$\frac{\partial \ln(\beta\gamma^2)}{\partial u} = 0, \quad \frac{\partial \ln(\beta^2\gamma)}{\partial v} = 0.$$

(cf. [Fi 37], p. 108). Integrating these equations, we find that $\beta^2\gamma = U(u)$ and $\beta\gamma^2 = V(v)$. Changing u and v, we can reduce these functions and subsequently β and γ to 1:

$$\beta = 1, \quad \gamma = 1. \tag{7.297}$$

By (7.297), equations (7.237)–(7.239) give

$$K = L = \frac{1}{2} \tag{7.298}$$

and

$$A_u = C_v = 0, \quad A_v = C_u. \tag{7.299}$$

It follows from equations (7.299) that $A = V_1(v)$ and $C = U_1(u)$ and $V_1' = U_1' = \text{const}$, i.e.

$$C = cu + c_1, \quad A = cv + c_2. \tag{7.300}$$

Therefore, *there is a three-parameter family of surfaces with all first normals of the canonical pencil coinciding.* (This matches the fact that, under conditions (7.297), (7.298) and (7.300), the system defined by the table (7.234) is completely integrable.)

If $c \neq 0$, then by introducing new parameters

$$u_1 = u + \frac{c_1}{2}, \quad v_1 = v + \frac{c_2}{2},$$

we (without changing β and γ) reduce the invariants A and C to the same value on all surfaces with the same $c \neq 0$. Thus, in this case *surfaces with all first normals of the canonical pencil coinciding are projectively identical.* When we change the constants c_1 and c_2, the surface V^2 undergoes a projective transformation.

If $c = 0$ and $c_1 \neq c_2$, the surfaces cannot be made projectively identical: the change of c_1 and c_2 implies a projective deformation of a surface.

For any $c = 0, c_1$ and c_2, the corresponding surface V^2 is a projective deformation of a surface for which $c = c_1 = c_2 = 0$. For the latter surface

$A = C = 0$ (see (7.298)). This, equations (7.297), (7.298) and the table (7.234) imply that

$$\begin{aligned} dA_0 &= duA_1 + dvA_2, \\ dA_1 &= \tfrac{1}{2}dvA_0 + duA_2 + dvA_3, \\ dA_2 &= \tfrac{1}{2}duA_0 + dvA_1 + duA_3, \\ dA_3 &= \tfrac{1}{2}duA_1 + \tfrac{1}{2}dvA_2. \end{aligned}$$

If we set $A_0 = e^{ru+sv}$, we immediately find from the above equations that

$$s = r^2, \quad r^4 = r.$$

If we denote by ϵ one of the cubic roots from 1, we find four linearly independent solutions of the above system: $1, e^{\epsilon u+\epsilon^2 v}, e^{\epsilon^2 u+\epsilon v}$ and e^{u+v}. These solutions can be taken as homogeneous coordinates x^0, x^1, x^2, x^3 of the point A_0. Then, since $1+\epsilon+\epsilon^2 = 0$, the nonhomogeneous coordinates $x = \frac{x^1}{x^0}$, $y = \frac{x^2}{x^0}$ and $z = \frac{x^1}{x^0}$ of this point satisfy the equation

$$xyz = 1.$$

Thus, in this case *surfaces with all first normals of the canonical pencil coinciding are projective deformations of the surface* $xyz = 1$.

7.6 The Geometry of Hyperbands

1. Under a correlative transformation $\mathcal{K}$, a tangentially nondegenerate hypersurface V^{n-1} is transferred again into a tangentially nondegenerate hypersurface $\mathcal{K}V^{n-1}$. Such submanifolds are said to be *self-dual* or *autodual* (cf. Section **4.1**).

If a submanifold V^m is nondegenerate and $m < n-1$, then, as we saw in Section **4.1**, under a correlative transformation $\mathcal{K}$, the submanifold V^m is transferred into a tangentially degenerate submanifold $\mathcal{K}V^m$ of rank m with $(n-m-1)$-dimensional plane generators. This implies that such a submanifold V^m is not autodual.

Consider another class of autodual manifolds —the *hyperbands*. The *element* of a hyperband is an autodual pair (x, ξ) in the space P^n consisting of a point x and a hyperplane ξ through x. An m-dimensional manifold of pairs (x, ξ) in the space P^n such that the point x describes an m-dimensional tangentially nondegenerate submanifold $V^m \subset P^n$, and the hyperplane ξ is tangent to V^m at the point x, is called an *m-dimensional hyperband* and is denoted by H^m. The submanifold V^m is called the *support submanifold* of the hyperband H^m. The hyperplanes ξ envelop a tangentially degenerate hypersurface V_m^{n-1} of rank m with $(n-m-1)$-dimensional plane generators E^{n-m-1}. Denote this generator by the symbol $\tau_\xi(V_m^{n-1})$. Thus we have $E^{n-m-1} = \tau_\xi(V_m^{n-1})$.

Under a correlative transformation $\mathcal{K}$, the support submanifold V^m of a hyperband H^m is transferred into a tangentially degenerate hypersurface $\mathcal{K}V^m = {}'V_m^{n-1}$ of rank m with $(n-m-1)$-dimensional plane generators, and a tangentially degenerate hypersurface V_m^{n-1} of rank m into an m-dimensional tangentially nondegenerate submanifold $\mathcal{K}V_m^{n-1} = {}'V^m$, which is the support submanifold of the new hyperband ${}'H^m$.

Thus, the correlation $\mathcal{K}$ transfers a hyperband H^m into a hyperband $\mathcal{K}H^m = {}'H^m$, and therefore, the hyperbands are autodual.

The theory of hyperbands, which we develop in this section, found application in the theory of submanifolds lying on hypersurfaces of a projective space P^n. If a tangentially nondegenerate submanifold V^m belongs to a hypersurface W^{n-1}, then in a natural way, with a submanifold V^m, there is associated a hyperband H^m for which the submanifold V^m is the support manifold and the hyperplanes ξ are tangent hyperplanes of the hypersurface W^{n-1} at the points $x \in V^m$. In particular, if the hypersurface W^{n-1} is a nondegenerate hyperquadric Q in the space P^n and $V^m \subset Q$, then the arising hyperband is called *quadratic*. Since the conformal or pseudoconformal geometry is realized on nondegenerate hyperquadrics of a projective space P^n, the geometry of quadratic hyperbands is closely connected with the geometry of submanifolds of a conformal space C^{n-1} or a pseudoconformal space C_l^{n-1} (see [Ro 66], p. 483–485 and p. 548–549).

The relationship between the geometry of quadratic hyperbands and the geometry of submanifolds of a conformal space was studied in detail by Vasilyan [Va 70b].

Another application of the theory of hyperbands can be obtained in the following way. As we noted in Section **4.6**, the manifold of hypercones of the projective space P^n is represented by a tangentially degenerate hypersurface V^{N-1} of rank n in the projective space P^N where $N = \frac{1}{2}(n+1)(n+2) - 1$. Consider an n-parameter family of hypercones in P^n whose vertices fill an n-dimensional domain. In the space P^N, to this family there corresponds a submanifold V^n lying on the hypersurface V^{N-1} and intersecting each generator E^L of V^{N-1}, where $L = \frac{1}{2}n(n+1) - 1$, at one point. At any point x of the submanifold V^n, we consider a hyperplane ξ tangent to the hypersurface V^{N-1}. Then the set of pairs (x, ξ) is a hyperband H^n, and the study of the family of hypercones in P^n indicated above is reduced to the study of the hyperband H^n.

2. A hyperband H^m is called *regular* if the tangent space E^m to the submanifold V^m and the generator E^{n-m-1} of the hypersurface V_m^{n-1} have no other common point except x. It is easy to see that a correlation $\mathcal{K}$ transfers a regular hyperband H^m into a regular hyperband ${}'H^m$.

We associate a family of moving frames with an element (x, ξ) of a regular hyperband H^m in such a way that the vertex A_0 of each frame coincides with the point x and the hyperplane $\alpha^n = A_0 \wedge A_1 \wedge \ldots \wedge A_{n-1}$ coincides with the hyperplane ξ:

$$A_0 = x, \quad \alpha^n = \xi.$$

In addition, we place the points A_i, $i = 1, \ldots, m$, of our moving frames into the tangent space $E^m = T_x(V^m)$. Since the hyperband H^m is regular, the plane generator $E^{n-m-1} = \tau_\xi(V_m^{n-1})$ of the hypersurface V_m^{n-1} has only one common point A_0 with the tangent space E^m. Hence the hyperplanes α^i of the family of tangential frames can be chosen in such a way that each of them contains the plane E^{n-m-1}.

By relations (1.65) and (1.71), the infinitesimal displacements of the point A_0 and the hyperplane α^n of moving frames of the family we have chosen have the form:

$$\begin{aligned} dA_0 &= \omega_0^0 A_0 + \omega_0^i A_i, \\ d\alpha^n &= -\omega_n^n \alpha^n - \omega_i^n \alpha^i. \end{aligned} \tag{7.301}$$

The forms of each of the systems, $\{\omega_0^i\}$ and $\{\omega_i^n\}$, are linearly independent because each of the objects, the point A_0 and the hyperplane α^n, depends on m parameters. Since they depend on the same parameters, we have:

$$\omega_i^n = a_{ij}\omega_0^j, \tag{7.302}$$

where a_{ij} is a nondegenerate tensor. This tensor is defined in a first order neighborhood of an element (x, ξ) of a hyperband H^m. It follows from equations (7.301) that a regular hyperband H^m is determined by the equations:

$$\omega_0^p = 0, \quad \omega_p^n = 0, \quad \omega_0^n = 0, \quad p = m+1, \ldots, n-1. \tag{7.303}$$

Exterior differentiation of the latter of these equations gives the exterior quadratic equation:

$$\omega_0^i \wedge \omega_i^n = 0, \tag{7.304}$$

which immediately leads to the symmetry of the tensor a_{ij}: $a_{ij} = a_{ji}$.

We now find the exterior differentials of the first two subsystems of system (7.303). They have the following form:

$$\omega_0^i \wedge \omega_i^p = 0, \quad \omega_p^i \wedge \omega_i^n = 0. \tag{7.305}$$

Since the forms of each of the systems $\{\omega_0^i\}$ and $\{\omega_i^n\}$, are linearly independent, applying Cartan's lemma, we find from equations (7.305) that

$$\begin{aligned} \omega_i^p &= b_{ij}^p \omega_0^j, \quad b_{ij}^p = b_{ji}^p, \\ \omega_p^i &= b_p^{ij} \omega_0^j, \quad b_p^{ij} = b_p^{ji}. \end{aligned} \tag{7.306}$$

The objects b_{ij}^p and b_p^{ij} are defined in a second order differential neighborhood of an element (x, ξ) of a hyperband H^m.

Let us find the second fundamental forms of the hyperband H^m. To this end, we differentiate equations (7.301) and projectivize the result by the subspaces E^m and E^{n-m-1}, respectively. As a result, we have:

$$\begin{aligned} d^2A_0/E^m &= b^p_{ij}\omega^i_0\omega^j_0 A_p + a_{ij}\omega^i_0\omega^j_0 A_n, \\ d^2\alpha^n/E^{n-m-1} &= b^{ij}_p\omega^n_i\omega^n_j\alpha^p + a^{ij}\omega^n_i\omega^n_j\alpha^0, \end{aligned} \tag{7.307}$$

where a^{ij} is the tensor inverse of the tensor a_{ij}. It follows from equations (7.307) that the forms

$$\Phi^p_{(2)} = b^p_{ij}\omega^i_0\omega^j_0, \quad \Phi^n_{(2)} = a_{ij}\omega^i_0\omega^j_0, \tag{7.308}$$

$$\Phi^{(2)}_p = b^{ij}_p\omega^n_i\omega^n_j, \quad \Phi^{(2)}_0 = a^{ij}\omega^i_n\omega^j_n \tag{7.309}$$

are the second fundamental forms of the hyperband H^m. Moreover, by relation (7.302), we have

$$\Phi^n_{(2)} = \Phi^{(2)}_0. \tag{7.310}$$

The further study of the differential geometry of a hyperband H^m is connected with the differential prolongation of equations (7.302) and (7.306). Exterior differentiation of these equations and application of the Cartan lemma to the exterior quadratic equations, obtained as a result of this differentiation, give:

$$\nabla a_{ij} = -a_{ij}(\omega^0_0 + \omega^n_n) + a_{ijk}\omega^k_0, \tag{7.311}$$

$$\nabla b^p_{ij} = -b^p_{ij}\omega^0_0 - a_{ij}\omega^p_n + b^p_{ijk}\omega^k_0, \tag{7.312}$$

$$\nabla b^{ij}_p = b^{ij}_p\omega^n_n + a^{ij}\omega^0_p - b^{ijk}_p\omega^n_k. \tag{7.313}$$

In formulas (7.311)–(7.313) the operator ∇ is constructed by means of the forms ω^j_i and ω^q_p (not the forms $\theta^j_i = \omega^j_i - \delta^j_i\omega^0_0$ and $\theta^q_p = \omega^q_p - \delta^q_p\omega^0_0$ as was in Sections **7.1**–**7.4** and preceding chapters 2). Thus, for example,

$$\nabla a_{ij} = da_{ij} - a_{ik}\omega^k_j - a_{kj}\omega^k_i.$$

Formulas (7.311) prove that the object a_{ij} is a tensor. Since this tensor is nondegenerate, it has the inverse tensor a^{ij} such that $a_{ik}a^{kj} = \delta^i_j$. It satisfies the following differential equations:

$$\nabla a^{ij} = a^{ij}(\omega^0_0 + \omega^n_n) - a^{ijk}\omega^n_k, \tag{7.314}$$

where

$$a^{ijk} = a^{il}a^{jg}a^{kf}a_{egf}; \tag{7.315}$$

The indices, l, g, f take the same values $1, \ldots, m$, as the indices i, j, k.

Moreover, equations (7.312) and (7.313) imply that the objects b_{ij}^p and b_p^{ij} are quasitensors. These quasitensors allow us to construct the quantities

$$b_p = \frac{1}{m} a_{ij} b_p^{ij}, \quad b^p = \frac{1}{m} a^{ij} b_{ij}^p, \tag{7.316}$$

which, by (7.314) and (7.315) satisfy the equations

$$\nabla_\delta b_p = -b_p \pi_0^0 + \pi_p^0 \tag{7.317}$$

and

$$\nabla_\delta b^p = b^p \pi_n^n - \pi_n^p. \tag{7.318}$$

Formulas (7.316)–(7.318) prove that the quantities b_p and b^p are geometric objects defined in a second order differential neighborhood of an element (x, ξ) of a hyperband H^m.

In a second order differential neighborhood we can also construct the following quantities:

$$B_{ij}^p = b_{ij}^p - b^p a_{ij}, \quad B_p^{ij} = b_p^{ij} - b_p a^{ij}, \tag{7.319}$$

which if $\omega^i = 0$, by (7.312), (7.313), (7.317) and (7.318), satisfy the equations

$$\nabla_\delta B_{ij}^p = -B_{ij}^p \pi_0^0, \quad \nabla_\delta B_p^{ij} = B_p^{ij} \pi_n^n, \tag{7.320}$$

and therefore, these quantities are relative tensors. Equations (7.319) and (7.316) imply that these tensors are connected with the tensors a^{ij} and a_{ij} by the following apolarity conditions:

$$B_{ij}^p a^{ij} = 0, \quad B_p^{ij} a_{ij} = 0. \tag{7.321}$$

Note that for $m = 1$ the tensors defined by formulas (7.170) are identical zeros. One-dimensional hyperbands in a three-dimensional Euclidean space were considered by Blaschke in the books [Bl 21] and [Bl 50]. In the space P^n, $n \geq 3$, one-dimensional hyperbands were studied by Vasilyan in [Va 77]. In this section we will assume that $m \geq 2$.

We will now establish a geometric meaning for the vanishing of the tensors B_{ij}^p and B_p^{ij} for $m \geq 2$. For this we consider two particular cases of hyperbands—the planar and conic hyperbands.

A hyperband H^m is called *planar* if its support submanifold V^m belongs to a subspace of dimension $m+1$. Since rank $(a_{ij}) = m \geq 2$, from relations (7.308) and the Segre theorem (see Theorem 2.2 in Section **2.5**) it follows that *a hyperband H^m is planar if and only if the following conditions hold*:

$$B_{ij}^p = 0. \tag{7.322}$$

Similarly, a hyperband H^m is called *conic* if the tangentially degenerate hypersurface V_m^{n-1} of rank m associated with H^m is a hypercone with $(n-m-2)$-dimensional vertex. The duality principle applied to the above condition for a hyperband H^m to be planar gives the following condition for a hyperband H^m to be conic: *a hyperband H^m is conic if and only if the following conditions hold*:

$$B_p^{ij} = 0. \tag{7.323}$$

3. The geometric objects b^p and b_p defined by formulas (7.316) allow us to construct invariant planes that are intrinsically associated with an element (x, ξ) of a hyperband H^m.

Consider the points

$$M_p = A_p - b_p A_0, \quad p = m+1, \ldots, n-1. \tag{7.324}$$

They define the plane $E^{n-m-2} = M_{m+1} \wedge M_{m+2} \wedge \ldots \wedge M_{n-1}$ belonging to the generator E^{n-m-1} of the tangentially degenerate hypersurface V_m^{n-1}. Let us prove that this $(n-m-2)$-plane is invariantly connected with an element (x, ξ) of a hyperband H^m.

In fact, by (7.303) and (7.306), we have

$$\delta A_p = \pi_p^0 A_0 + \pi_p^q A_q, \quad p, q = m+1, \ldots, n-1.$$

Using this and relations (7.317), we differentiate the points M_p with respect to the secondary parameters. As a result, we find that

$$\delta M_p = \pi_p^q M_q.$$

This means that the plane E^{n-m-2} defined by the points M_p remains constant when the secondary parameters vary, i.e. this plane is invariantly connected with an element (x, ξ) of a hyperband H^m.

In a similar manner we can prove that the hyperplanes

$$\mu^p = \alpha^p - b^p \alpha^n \tag{7.325}$$

define an $(m+1)$-plane $E^{m+1} = \mu^{m+1} \wedge \ldots \wedge \mu^{n-1}$ containing the tangent subspace $T_x(V^m)$ and invariantly connected with an element (x, ξ) of a hyperband H^m.

Thus we proved the following result.

Theorem 7.9 *A second order differential neighborhood of an element (x, ξ) of a hyperband H^m defines the normalizing planes $E^{n-m-2} \subset \tau_\xi(V_m^{n-1})$ and $E^{m+1} \supset T_x(V^m)$ that are intrinsically associated with an element (x, ξ) of a hyperband H^m and dual to one another.* ■

Let us find a geometric meaning for the normalizing planes which we constructed above. As we established in Chapter 4, the generator τ_ξ of the tangentially degenerate hypersurface V_m^{n-1} possesses the focus variety F_ξ. A point $X = x^0 A_0 + x^p A_p$ of the variety F_ξ satisfies the condition $dX \in \tau_\xi$. This condition implies the following equations:

$$x^0 \omega_0^i + x^p \omega_p^i = 0.$$

By relations (7.302) and (7.306), we can write the above equations in the form:

$$(x^0 \delta_j^i + x^p b_p^{ik} a_{kj}) \omega_0^k = 0. \tag{7.326}$$

Since system (7.326) must have a nontrivial solution with respect to ω_0^i, its determinant vanishes:

$$\det(x^0 \delta_j^i + x^p b_p^{ik} a_{kj}) = 0. \tag{7.327}$$

In the plane τ_ξ, the latter equation defines the algebraic variety F_ξ of dimension $n - m - 2$ and order m.

Using the same way which we used in Section **6.1**, we find the polar $(n - m - 2)$-dimensional plane of the point A_0 with respect to the focus variety F_ξ. The equation of the polar hyperplane is vanishing of the trace of the matrix of system (7.326):

$$m x^0 + a_{ij} b_p^{ij} x^p = 0.$$

But by (7.316), this equation is reduced to the form:

$$x^0 + b_p x^p = 0. \tag{7.328}$$

This equation must be combined with the equations

$$x^i = 0,$$

defining the generator $\tau_\xi(V_m^{n-1})$ of the hypersurface V_m^{n-1}. It can be immediately verified that the coordinates of the points M_p defined by formulas (7.324) satisfy equation (7.328), and thus *the normalizing plane* E^{n-m-2} *coincides with the polar* $(n - m - 2)$*-plane of the point* A_0 *with respect to the focus variety* F_ξ.

The dual image of the focus variety F_ξ is the focus cone Φ_x whose vertex is the tangent subspace $T_x(V^m)$ of the support submanifold V^m. Similarly to what was done above, one can prove that in tangential coordinates, the equation of the cone Φ_x has the form:

$$\det(\xi_n \delta_j^i + \xi_p b_{ik}^p a^{kj}) = 0. \tag{7.329}$$

The polar plane of the hyperplane $\xi = \alpha^n$ with respect to this cone is the $(m+1)$-dimensional plane passing through $T_x(V^m)$ and defined by the equation

$$m \xi_n + a^{ij} b_{ij}^p \xi_p = 0,$$

which by (7.316), can be written in the form:

$$\xi_n + b^p \xi_p = 0. \tag{7.330}$$

This equation must be combined with the equations

$$\xi_i = 0,$$

defining the tangent subspace $T_x(V^m)$ to the submanifold V^m.

One can easily prove that the tangential coordinates of the hyperplanes μ^p, defined by formulas (7.325), satisfy equation (7.330), and thus *the normalizing plane E^{m+1} coincides with the polar $(m+1)$-plane of the hyperplane α^n with respect to the focus cone Φ_x.*

It is easy to prove that for a planar hyperband H^m the normalizing plane E^{m+1} coincides with the $(m+1)$-dimensional subspace to which the support submanifold V^m belongs, and that for a conic hyperband H^m the normalizing plane E^{n-m-2} is the vertex of the cone into which the hypersurface V_m^{n-1} degenerates.

4. However, the normalizing planes E^{n-m-2} and E^{m+1} constructed above do not settle an invariant normalization of a hyperband. The further construction of an invariant normalization is connected with third and fourth order differential neighborhoods.

An invariant normalization of the support submanifold V^m of a hyperband H^m consists of first and second normals, n_x and l_x, which were defined in Chapter 6.

An invariant normalization of the tangentially degenerate hypersurface V_m^{n-1} is defined by the duality principle. The first normal ν_ξ of V_m^{n-1} is an $(m-1)$-dimensional plane belonging to the hyperplane $\xi = \alpha^n$ and not intersecting the generator τ_ξ of dimension $n-m-1$, and the second normal λ_ξ of the submanifold V_m^{n-1} is an $(n-m)$-dimensional plane not belonging to the hyperplane ξ and containing the generator τ_ξ.

A hyperband H^m is said to be *dual-normalized* if

$$\lambda_\xi = n_x, \quad \text{and} \quad \nu_\xi = l_x, \tag{7.331}$$

i.e if first and second normals of the submanifold V^m are second and first normals of the hypersurface V_m^{n-1}, respectively. Our aim is also to construct an invariant normalization of a hyperband H^m that is intrinsically connected with its geometry.

To this end, consider the tensor

$$J_q^p = B_{ij}^p B_q^{ij}. \tag{7.332}$$

By (7.320), this tensor satisfies the equations

$$\nabla_\delta J_q^p = J_q^p(\pi_n^n - \pi_0^0). \tag{7.333}$$

Note that for planar and conic hyperbands the tensor J^p_q is identically equal to zero. Thus, our further construction is not valid for these two classes of hyperbands.

Suppose that the rank ρ of the tensor J^p_q is positive. Then the relative invariant

$$J = J^{p_1}_{[q_1} J^{p_2}_{q_2} \dots J^{p_\rho}_{q_\rho]},$$

which is the sum of the diagonal minors of order ρ of the matrix (J^p_q), is different from zero. Differentiating the invariant J with respect to the secondary parameters by means of (7.333), we find that

$$\delta \ln |J|^{\frac{1}{\rho}} = \pi^n_n - \pi^0_0.$$

If the principal parameters are not fixed, then this equation takes the form

$$d \ln |J|^{\frac{1}{\rho}} = \omega^n_n - \omega^0_0 + J_k \omega^k. \tag{7.334}$$

or

$$d \ln |J|^{\frac{1}{\rho}} = \omega^n_n - \omega^0_0 + J^k \omega^n_k, \tag{7.335}$$

where

$$J^k = a^{kl} J_l.$$

If we prolong equations (7.334) and (7.335) and fix the principal parameters, we find that the objects J_k and J^k satisfy the equations:

$$\begin{aligned} \nabla_\delta J_k &= -J_k \pi^0_0 - \pi^0_k - a_{kl} \pi^l_n, \\ \nabla_\delta J^k &= J^k \pi^n_n - \pi^k_n - a^{kl} \pi^0_l. \end{aligned} \tag{7.336}$$

However, these objects do not yet allow us to construct normalizing objects containing the forms π^0_k or π^k_n alone.

So, we also consider the objects a_{ijk} and a^{ijk} defined by equations (7.311) and (7.314) and connected by relation (7.315). These objects as well as the objects J_k and J^k are defined in a third order differential neighborhood of an element (x, ξ) of a hyperband H^m. If we prolong equations (7.311) and (7.314) and fix the principal parameters, we find that the objects a_{ijk} and a^{ijk} satisfy the equations:

$$\begin{aligned} \nabla_\delta a_{ijk} &= -a_{ijk}(2\pi^0_0 + \pi^n_n) - 3a_{(ij}(\pi^0_{k)} - a_{k)l}\pi^l_n), \\ \nabla_\delta a^{ijk} &= a^{ijk}(\pi^0_0 + 2\pi^n_n) - 3a^{(ij}(\pi^{k)}_n - a^{k)l}\pi^0_l). \end{aligned} \tag{7.337}$$

Applying the same procedure, which we used for a hypersurface in Section **7.1** (cf. formula (7.24)), we construct the objects

$$a_i = \frac{1}{m+2} a^{jk} a_{ijk}, \quad a^i = \frac{1}{m+2} a_{jk} a^{ijk}. \tag{7.338}$$

By (7.315), they are connected by the relation:

$$a^i = a^{ij} a_j.$$

Equations (7.337) imply that these objects satisfy the equations:

$$\begin{aligned} \nabla_\delta a_k &= -a_k \pi_0^0 - \pi_k^0 + a_{kl} \pi_n^l, \\ \nabla_\delta a^k &= a^k \pi_n^n + \pi_n^k - a^{kl} \pi_l^0. \end{aligned} \tag{7.339}$$

It is clear now that the quantities a_k, a^k and J_k, J^k allow us to construct the following normalizing objects:

$$l_k = -\frac{1}{2}(a_k + J_k), \ \ l^k = -\frac{1}{2}(a^k + J^k),$$

which by (7.336) and (7.339), satisfy the equations

$$\nabla_\delta l_k = -l_k \pi_0^0 + \pi_k^0, \ \ \nabla_\delta l_k = -l_k \pi_0^0 + \pi_k^0. \tag{7.340}$$

The objects l_k and l^k, as well as the quantities a_k, a^k, J_k and J^k, are defined in a third order differential neighborhood of an element (x, ξ) of H^m.

Consider now the points

$$M_k = A_k - l_k A_0 \tag{7.341}$$

and the hyperplanes

$$\mu^k = \alpha^k - l^k \alpha^n. \tag{7.342}$$

One can easily check that they satisfy the equations:

$$\delta M_k = \pi_k^l M_l, \ \ \delta \mu^k = \pi_l^k \mu^l,$$

and therefore they define the invariant planes

$$E^{m-1} = M_1 \wedge \ldots \wedge M_m, \ \ E^{n-m} = \mu^1 \wedge \ldots \wedge \mu^m,$$

which are intrinsically connected with the geometry of a hyperband H^m. Since $E^{m-1} \subset T_x(V^m)$, $E^{m-1} \not\ni A_0 = x$ and $E^{n-m} \supset \tau_\xi(V_m^{n-1})$, $E^{n-m} \not\subset \alpha^n = \xi$, these planes are invariant normals of a hyperband H^m.

We proved the following result.

Theorem 7.10 *A third order differential neighborhood of an element (x, ξ) of a hyperband H^m defines a dual normalization of this hyperband by means of the normals*

$$E^{n-m} = n_x = \lambda_\xi, \ \ E^{m-1} = l_x = \nu_\xi,$$

defined by the points M_k and the hyperplanes μ^k, respectively, and intrinsically connected with the geometry of a hyperband H^m. ■

5. The points $A_0 = M_0, M_i, M_p$ and the hyperplanes $\alpha^n = \mu^n, \mu^i, \mu^p$ are the elements of invariant dual frames associated with an element (x, ξ) of a hyperband H^m. To complete our construction of invariant frames, we still need to find an invariant point M_n and an invariant hyperplane μ^0 connected by the incidence condition $(\mu^0, M_n) = 0$. These frame elements can be written in the form:

$$\begin{aligned} M_n &= A_n + b^p A_p + l^i A_i + l^0 A_0, \\ \mu^0 &= \alpha^0 + b_p \alpha^p + l_i \alpha^i + l_n \alpha^n. \end{aligned} \tag{7.343}$$

It follows from the incidence condition that the objects l^0 and l_n must be related by the following equation:

$$l^0 + l_n + b^p b_p + l^i l_i = 0. \tag{7.344}$$

The objects l^0 and l_n can be found in a fourth order differential neighborhood of an element (x, ξ) of a hyperband H^m. However, their explicit expressions in terms of the objects of order not exceeding four is rather complicated, and this is the reason that we will not write here these expressions.

As was the case for the original frames $\{A_u\}$ and $\{\alpha^u\}$, the elements of the frames $\{M_u\}$ and $\{\mu^u\}$ are connected by the incidence conditions $(\mu^v, M_u) = \delta_u^v$ which can be easily verified by means of formulas (7.324), (7.325), (7.341), (7.342), (7.343) and (7.344).

6. Applying the procedure which we used for a hypersurface in Section **7.1**, we can construct the Darboux tensor for a hyperband H^m. To this end, we will use the objects a_{ijk} and a_i defined by formulas (7.311) and (7.338) respectively. For $\omega^i = 0$, these two objects satisfy equations (7.337) and (7.339). These equations show that the quantities

$$A_{ijk} = a_{ijk} - 3a_{(ij}a_{k)},$$

constructed by means of the objects a_{ijk} and a_i, define a tensor associated with a second order differential neighborhood of an element (x, ξ) of a hyperband H^m. We will call this tensor the *Darboux tensor* of a hyperband H^m. It is easy to see that the Darboux tensor satisfies the apolarity condition

$$A_{ijk} a^{ij} = 0. \tag{7.345}$$

Similarly to the way we used in Section **7.2** for a hypersurface, we can prove that *the Darboux tensor A_{ijk} of a hyperband H^m vanishes if and only if this hyperband is quadratic.*

If a hyperband H^m is not quadratic, then the tensor A_{ijk} can be used for construction of normalizing objects of the hyperband. To this end, we construct the relative invariant

$$A = a^{il} a^{jg} a^{kf} A_{ijk} A_{lgf},$$

which by (7.337) and (7.316), satisfies the equation

$$\delta A = A(\pi_n^n - \pi_0^0).$$

If this invariant is different from zero, then

$$d \ln |A| = \omega_n^n - \omega_0^0 + A_k \omega^k.$$

The new object A_k, which appeared in the latter equation, has the same law of transformation as the object J_k had (see the first formula of (7.336)). This object A_k together with the object a_k allow us to construct normalizing objects satisfying the equations of type (7.340), and these normalizing objects define the invariant normals of a hyperband H^m.

Note that in general, for a nonquadratic hyperband the invariant A is different from zero. The proof of this fact is similar to the proof we gave in Section **7.3** for the invariant B_0.

The construction of normalizing objects given in this subsection is also valid if the rank ρ of the tensor J_q^p defined by formula (7.333) is equal to zero while the construction given in subsections **7.6.4** and **7.6.5** is not valid for $\rho = 0$. In particular, the construction given in this subsection is suitable for the planar and conic hyperbands.

NOTES

7.1. In Notes to Chapter 6 we indicated the developments in the theory of submanifolds V^m in different spaces (affine, projective, space with affine or projective connection etc.).

The theory of hypersurfaces was developed by tensor methods in the Fubini papers [Fu 18a] and [Fu 18b], as far back as 1918 (see also Chapter XII of [FČ 26]). Hlavatý in [H 33] and [H 39] and Bol in [Bo 50b] gave the further development of this theory. As we already mentioned in Notes to Section **6.1**, Schouten and Haantjes in [SH 36] considered the general normalized submanifolds V^m in spaces with a projective connection. As an application of their studies, they investigated a hypersurface in P^n. The theory of hypersurfaces was studied in the papers of Bortolotti (see [Bor 28] and [Bor 39]), Kanitani (see [Ka 25a], [Ka 25b], [Ka 26], [Ka 27], [Ka 28a], [Ka 28b], [Ka 28c], [Ka 28d], [Ka 29a], [Ka 29b], [Ka 29c], [Ka 29d] and [Ka 31]), Su Buchin (see [Su 43]) and Sasaki (see [Sas 88]).

The main results of the affine theory of surfaces in A^3 were extended to a hypersurface in the affine space A^n by Fernández in [Fer 55] and by Laugwitz in [Lau 57] and [Lau 59]. In the paper [Lap 58], Laptev constructed the differential geometry of a hypersurface in a space with a projective connection in an invariant form. His results (the Darboux tensor, curves and osculating hyperquadrics, the Fubini linear element, the pencil of projective normals etc.) are a direct generalization of the corresponding results from the theory of hypersurfaces in a projective space: the latter results are obtained from Laptev's results if the torsion-curvature tensor of the ambient space vanishes. Švec in [Šv 58], [Šv 61a], [Šv 61b] and [Šv 61c] and Cenkl in [Cen 62a] and [Cen 62b] also studied the geometry of a hypersurface in a space with a projective connection. However, their results are valid only if the ambient space has a nonvanishing the torsion-curvature tensor. For a hypersurface in a space with a projective connection, the Fubini projective linear element was generalized in [Šv 58] and [Ha 61] by Švec and Havelka, and the Darboux directions and curves for a hypersurface in a space with a projective connection were investigated by Švec in [Šv 59] and by Kanitani in [Ka 57].

7.2. The Darboux osculating hyperquadrics were considered by Kanitani in the book [Ka 31]. Hlavatý in [H 37] considered them in a space with an affine connection. The Darboux directions on a hypersurface and the Darboux cone associated with a hypersurface were the subject of studies of Kanitani and Švec (see [Ka 31], [Ka 67] and [Šv 59]). In our presentation we follow Laptev's paper [Lap 53] where the exterior differential forms and the moving frames were systematically used.

Švec in [61a] and Kanitani in [Ka 62] considered osculating hyperquadrics (in particular, Lie quadrics if $n = 3$) for a hypersurface in a space with a projective connection.

Note that in the metric space, Theorem 7.2 was first proved by Verbitsky in [Ve 49]. In the projective space this theorem was first proved by Laptev in the paper [Lap 53] mentioned above.

7.3. Laptev in [Lap 49] presented all known results for a surface in a three-dimensional projective space P^3 in an invariant form. This gave him an opportunity to extend the main facts of the theory of surfaces in P^3 to the theory of hypersurfaces in P^n (see [Lap 53]). In particular, in [Lap 53] Laptev constructed an invariant normalization of a hypersurface in a projective space P^n. Our exposition slightly modifies the construction of invariant normalization of a hypersurface given by Laptev in [Lap 53].

Izmailov constructed an invariant normalization of a hypersurface in a space with an affine connection (see his papers [Iz 56a], [Iz 56b], [Iz 57a] and [Iz 60]).

Kanitani (see [Ka 65]) generalized the canonical pencil for a hypersurface in a space with a projective connection. (Note that in [Ka 59] Kanitani generalized the canonical pencil for a hypersurface in an n-dimensional conformal space.)

For a more detailed description of the developments in the construction of an invariant normalization of hypersurfaces in different spaces see the survey papers [Lap 65] and [Lu 75] by Laptev and Lumiste.

7.4. É. Cartan in his book [Ca 37] gave the philosophy of studying uniqueness, existence and rigidity questions for submanifolds of a homogeneous space via the use of moving frames and the theory of Lie groups. In the paper [Gr 74] Griffiths gave un updated and clear exposition of this Cartan philosophy together with some applications to geometry.

We gave in the text a short description of developments on the rigidity problem (connected with preservation the Fubini linear element) for a hypersurface in a projective space. In addition, note that É. Cartan in [Ca 20c] indicated some omissions in the Fubini proof in [Fu 16] and presented his own proof based on his method of investigation of projective deformability of hypersurfaces in P^n first published in [Ca 16]). Fubini replied in [Fu 20] that his papers contain the right methods for the solution of the rigidity problem and indicated that his student Stipa in [Sti 20] determined all projectively deformable surfaces in P^3.

Note that our proof of Theorem 7.7 differs from the Laptev proof in [Lap 53].

7.5. In this section we again follow the Laptev paper [Lap 53] and by means of the formulas of Section **7.3** present the projective theory of a surface in the space P^3. In this presentation, with a surface in the space P^3, we associate the moving frame which was introduced and widely used by Finikov (see [Fi 37], p. 42).

Note that Su Buchin in his book [Su 83] gave a detailed investigation of surfaces $V^2 \subset A^3$ whose affine normal (see Bl 23]) coincide with one or another normal of the canonical pencil.

7.6. The bands in three-dimensional space were first studied by Blaschke (see [Bl 21] and [Bl 50]). Multidimensional hyperbands in the projective space P^n were first investigated by Wagner in [Wa 50] and after him by Chakmazyan (see [Cha 59]), Vasilyan (see [Va 70a], [Va 70b], [Va 71], [Va 73] and [Va 77]) Popov (see [Po 70a], [Po 70b], [Po 71], [Po 73] and [Po 83]), Stolyarov (see [Sto 75a], [Sto 75b], [Sto 75c], [Sto 76], [Sto 77], [Sto 78a], [Sto 78b], [Sto 82] and [Sto 92]), Terentjeva (see [Te 73]), and others. In [GC 72] Goldberg and Chakmazyan used hyperbands to construct a dual normalization of $V^2 \subset P^4$ in which the intersection of the osculating hyperplanes to the lines of the unique conjugate net of V^2 is taken as its first normal. The reader can find further references on the developments in the theory of multidimensional hyperbands, both regular and degenerate (we indicated here only papers dealing with regular hyperbands), in the projective space P^n in the survey paper [Sto 78] by Stolyarov and in the books [Po 83] and [Sto 92a] (see also the book [Sto 92b]).

See a geometric meaning of the apolarity conditions (7.311) and (7.345) in subsection **7.1.3**.

Chapter 8

Algebraization Problems in Projective Differential Geometry

The *algebraization problem* for a smooth manifold V of dimension r with a generating element α in a projective space P^n is the finding a criterion (in terms of the invariants and the tensors of V) for the manifold V to belong to an algebraic manifold of the same dimension and with the same generating element as the manifold V has. The conditions under which the manifold V becomes an algebraic manifold are called the *algebraizability conditions*.

Algebraization problems are important in differential geometry and, in particular, in projective differential geometry. First of all, problems of this kind were posed and solved in the theory of curves. It is well-known that if a plane curve is given in the Cartesian coordinates by the equation $y = f(x)$, then its rectifiability condition has the form $y'' = 0$, and the condition for the curve to belong to a second degree curve has the form:

$$9(y'')^2 y^{(\mathrm{v})} - 45 y'' y''' y^{(\mathrm{iv})} + 40 (y''')^3 = 0$$

(see, for example, [Kl 26]).

We will give a few more examples of algebraizability conditions which are known in projective differential geometry.

Let V^m be an m-dimensional point submanifold in P^n and b^α_{ij} are its second fundamental tensors. The submanifold V^m belongs to an m-dimensional subspace P^m if and only if $b^\alpha_{ij} = 0$ on V^m (see Section **2.2**).

A hypersurface V^{n-1} in P^n is a part of a hyperquadric Q if and only if the Darboux tensor b_{ijk} of V^{n-1} vanishes (see Section **7.2**).

The algebraizability conditions indicated above connect the values of derivatives or the tensors calculated at one point of a submanifold in question. This is the reason that these conditions are called *one-point conditions*. As we saw, even for second degree curves these conditions are very complex. As to higher degree curves, these conditions become extremely complicated.

However, there exist other, *multipoint conditions* that are necessary and sufficient for a system of d arcs of smooth curves in a projective plane P^2 to belong to a curve of degree d. These conditions can be formulated uniformly for any d, and they are the subject of the Abel theorem and the Reiss theorem.

The Abel theorem is: a system of d arcs C_α, $\alpha = 1, \ldots, d$, of smooth plane curves belongs to an algebraic curve $C_{(d)}$ of degree d if and only if one can chose a parametrization on the arcs C_α in such a way that for any straight line l meeting each of C_α at one point $z_\alpha = l \bigcap C_\alpha$ the following relation holds:

$$u_1(z_1) + u_2(z_2) + \ldots + u_d(z_d) = \text{const}, \tag{8.1}$$

where $u_\alpha(z_\alpha)$ is a value of the parameter u_α on the arc C_α at the point z_α.

In one direction this theorem was proved by Abel in 1829 (see [Ab 29] and [Ab 41]), and the converse of this theorem is contained in the papers of Lie and Scheffers (see [Lie 96] and [Sche 04]).

The Reiss theorem can be formulated as follows. Suppose again that C_α, $\alpha = 1, \ldots, d$, are arcs of smooth plane curves, l is an arbitrary straight line meeting each of the arcs C_α at one point $z_\alpha = l \bigcap C_\alpha$, and xOy is a Cartesian coordinate system whose axis Oy coincides with the straight line l (see Figure 8.1).

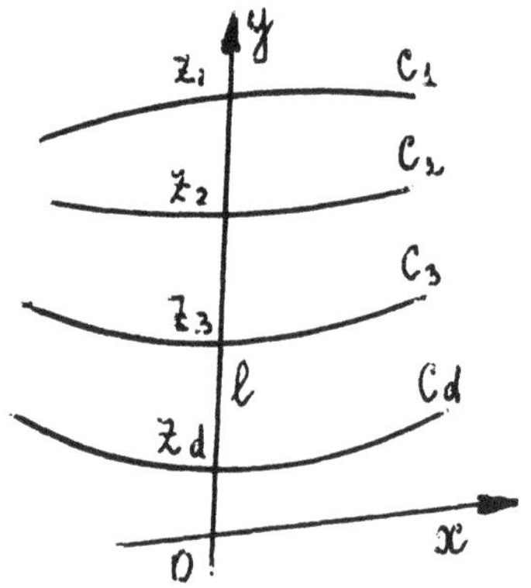

Figure 8.1

Then all arcs C_α belong to an algebraic curve $C_{(d)}$ of degree d if and only if for any straight line l the following condition holds:

$$y_1'' + y_2'' + \ldots + y_d'' = 0, \tag{8.2}$$

where $y_\alpha = y_\alpha(x)$ is the equation of the arc C_α in the coordinate system related with the straight line l, and y''_α is the second derivative of y_α at the point $C_\alpha \bigcap l$.

In one direction this theorem was proved by Reiss in 1837 and subsequently forgotten and rediscovered several times. Lie was the first who noted that the converse of the Reiss theorem is valid (see [Lie 82]), and this converse was proved by Engel in [En 39] and Teixidor in [Tei 53]. Modern proofs of this theorem may be found in [SegB 71] (§39) and [GH 78] (pp. 675–677).

In Sections **8.1** and **8.2** we consider the multidimensional generalizations of the Reiss theorem.

In addition, in the book [LS 96], Lie and Scheffers established the condition for a distribution of cones defined in a three-dimensional space by the Monge equation of the form $\Omega(x^i, dx^i) = 0,\ i = 1, 2, 3$, to belong to a complex of straight lines. In Section **8.3** we consider some algebraizability conditions connected with Monge's equation in an n-dimensional projective space.

8.1 The First Generalization of Reiss' Theorem

1. Consider some theorems generalizing Reiss' theorem on a system of curves in the plane P^2.

Let $\{V_1, V_2, \ldots, V_d\}$ be a system of hypersurfaces in a projective space P^{r+1}, $r \geq 2$, and l be a straight line intersecting each of the hypersurfaces V_α, $\alpha = 1, \ldots, d$, at a single point $x_\alpha = l \bigcap V_\alpha$ (see Figure 8.2).

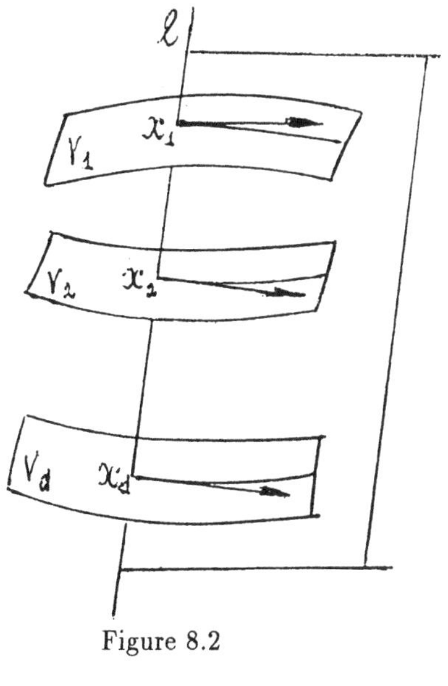

Figure 8.2

The set of straight lines with this property form a domain U on the Grassmannian $G(1, r+1)$ of straight lines in the P^{r+1}. Denote by ξ_α the tangent hyperplane to the hypersurface V_α at the point x_α, $\xi_\alpha = T_{x_\alpha}(V_\alpha)$, and let

$$b(x_\alpha) = < \xi_\alpha, d^2 x_\alpha > \tag{8.3}$$

be the second fundamental form of the hypersurface V_α at the point x_α.

Since two-dimensional planes passing through a straight line l establish a correspondence among directions on the hypersurfaces V_α emanating from the points $x_\alpha = l \bigcap V_\alpha$ (see Figure 8.2), the quadratic forms $b(x_\alpha)$ can be written in the form:

$$b(x_\alpha) = \underset{\alpha}{b_{ij}} \omega^i \omega^j, \quad i, j = 1, 2, \ldots, r, \tag{8.4}$$

where ω^i are 1-forms forming a common cobasis at the points x_α on these hypersurfaces. The tensors $\underset{\alpha}{b_{ij}}$ are the second fundamental tensors of the hypersurfaces V_α at the points x_α.

We will now prove the theorem, which can be called the *first generalized Reiss theorem*:

Theorem 8.1 *The hypersurfaces $V_1, V_2, \ldots, V_d$ of the space P^{r+1} belong to one and the same algebraic hypersurface of degree d if and only if for any straight line from some domain $U \subset G(1, r+1)$, the condition*

$$\sum_\alpha b(x_\alpha) = 0 \tag{8.5}$$

or the equivalent condition

$$\sum_\alpha \underset{\alpha}{b_{ij}} = 0 \tag{8.6}$$

holds.

Proof. We will prove this theorem only for $d = 2$. Let us associate with a pair of hypersurfaces V_1 and V_2 a family of moving frames in such a way that the vertices A_1 and A_2 of these frames coincide with the points of intersections of these hypersurfaces and the straight line l, and the points A_i, $i = 3, \ldots, r+2$, belong to the $(r-1)$-dimensional intersection of the tangent hyperplanes ξ_1 and

ξ_2 to the hypersurfaces V_1 and V_2 at the points A_1 and A_2 (Figure 8.3).

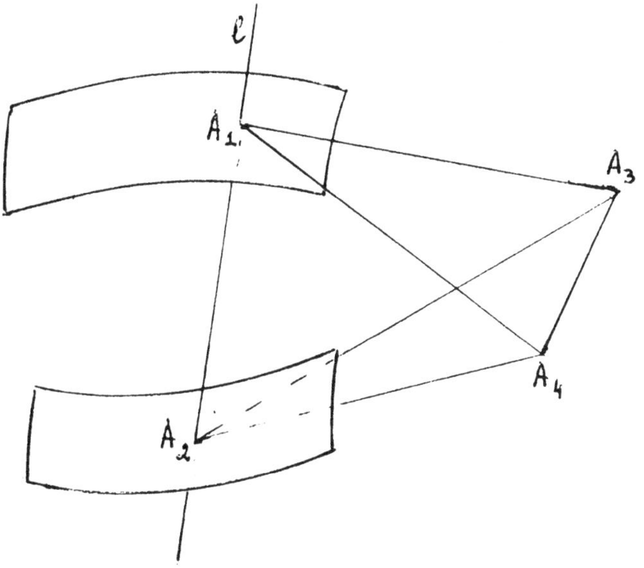

Figure 8.3

Then the differentials of the points A_1 and A_2 have the form:

$$dA_1 = \omega_1^1 A_1 + \omega_1^i A_i, \quad dA_2 = \omega_2^2 A_2 + \omega_2^i A_i, \tag{8.7}$$

and on the hypersurfaces V_1 and V_2 the equations

$$\omega_1^2 = 0, \quad \omega_2^1 = 0, \tag{8.8}$$

hold. The 1-forms ω_1^i are basis forms on V_1, and the 1-forms ω_2^i are basis forms on V_2. But for corresponding directions on V_1 and V_2 these forms coincide.

Exterior differentiation of equations (8.8) by means of structure equations (1.67) of P^n gives the following exterior quadratic equations:

$$\omega_1^i \wedge \omega_i^2 = 0, \quad \omega_2^i \wedge \omega_i^1 = 0.$$

Applying Cartan's lemma to these equations, we find that

$$\omega_i^2 = \underset{1}{b}_{ij} \omega_1^j, \quad \omega_i^1 = -\underset{2}{b}_{ij} \omega_2^j, \tag{8.9}$$

where $\underset{1}{b}_{ij} = \underset{1}{b}_{ji}$ and $\underset{2}{b}_{ij} = \underset{2}{b}_{ji}$ are the second fundamental tensors of hypersurfaces V_1 and V_2, and

$$b(A_1) = \underset{1}{b}_{ij}\omega_1^i\omega_1^j, \quad b(A_2) = \underset{2}{b}_{ij}\omega_2^i\omega_2^j$$

are their second fundamental forms. We can always assume that $\omega_1^i = \omega_2^i$.

We will now prove that *the hypersurfaces* V_1 *and* V_2 *belong to one and the same hyperquadric* Q *if and only if the tensors* $\underset{1}{b}_{ij}$ *and* $\underset{2}{b}_{ij}$ *satisfy the condition*

$$\underset{1}{b}_{ij} + \underset{2}{b}_{ij} = 0. \tag{8.10}$$

Necessity. Let the hypersurfaces V_1 and V_2 belong to a hyperquadric Q. We will write the equation of the hyperquadric Q relative to a moving frame $\{A_1, A_2, A_3, \ldots, A_{r+2}\}$ in the form:

$$a_{\alpha\beta}x^\alpha x^\beta = 0, \quad \alpha, \beta = 1, 2, \ldots, r+2. \tag{8.11}$$

The coefficients of this equation depend on the location of a frame $\{A_\alpha\}$ in the space P^{r+1}. Since the hyperquadric Q is invariant in P^{r+1}, it is easy to show (see Section **7.2**) that the coefficients of equation (8.11) satisfy the following differential equations:

$$da_{\alpha\beta} - a_{\alpha\gamma}\omega_\beta^\gamma - a_{\gamma\beta}\omega_\alpha^\gamma = \theta a_{\alpha\beta}. \tag{8.12}$$

Since the hyperquadric Q passes through the points A_1 and A_2 and is tangent at these points to the hyperplanes $A_1 \wedge A_3 \wedge \ldots \wedge A_{r+2}$ and $A_2 \wedge A_3 \wedge \ldots \wedge A_{r+2}$, the equation of this hyperquadric has the form:

$$2x_1x_2 + a_{ij}x^ix^j = 0, \quad i, j = 3, \ldots, r+2, \tag{8.13}$$

i.e. $a_{11} = a_{22} = 0$, $a_{1i} = a_{2i} = 0$ and $a_{12} = 1$. The latter conditions and equations (1.10) imply that

$$\omega_1^2 = 0, \quad \omega_2^1 = 0, \quad \theta = -\omega_1^1 - \omega_2^2, \tag{8.14}$$

$$\omega_i^2 = -a_{ij}\omega_1^j, \quad \omega_i^1 = -a_{ij}\omega_2^j, \tag{8.15}$$

$$da_{ij} - a_{ik}\omega_j^k - a_{kj}\omega_i^k + a_{ij}(\omega_1^1 + \omega_2^2) = 0. \tag{8.16}$$

Comparison of equations (8.15) and (8.9) yields

$$\underset{1}{b}_{ij} = -a_{ij}, \quad \underset{2}{b}_{ij} = a_{ij},$$

and condition (8.10) is satisfied.

Sufficiency. Suppose that condition (8.10) holds. By setting $\underset{1}{b}_{ij} = -a_{ij}$, $\underset{2}{b}_{ij} = a_{ij}$, we reduce equations (8.9) to the form (8.15). Exterior differentiation of equation (8.15) gives relation (8.16), and the latter relation together with equations (8.8) and (8.15) leads to the invariance of the hyperquadric

(8.13). Since the points A_1 and A_2 of the hypersurfaces V_1 and V_2 belong to hyperquadric (8.13), the sufficiency of the conditions of Theorem 8.1 is proved. ■

The method which was used in the proof of Theorem 8.1 for $d = 2$, can be also applied for any d. However, if $d > 2$, the calculations are much more complicated. Using this method, Botsu [Bot 74] proved Theorem 8.1 for $d = 3$, Goldberg [G 82b] proved it for $d = 4$, and finally Wood [Wo 82] proved it for any d.

If $d = 3$, Theorem 8.1 can be proved by another method based on the techniques of the web geometry (see [AS 92], p. 127).

Three hypersurfaces V_α, $\alpha = 1, 2, 3$, of the space P^{r+1} define a Grassmann three-web $GW(3, 2, r)$ of codimension r on the Grassmannian $G(1, r + 1)$ of dimension $2r$. The leaves of this web are the bundles of straight lines with their centers located on the hypersurfaces V_α. The curvature tensor of this web is determined by the formula:

$$b^i_{jkl} = \delta^i_j \underset{1}{b}_{kl} + \delta^i_k \underset{2}{b}_{lj} + \delta^i_l \underset{3}{b}_{jk}, \tag{8.17}$$

where $\underset{\alpha}{b}_{ij}$ is the second fundamental tensor of the hypersurface V_α. The hexagonality condition for an arbitrary three-web has the form:

$$b^i_{(jkl)} = 0.$$

By equation (8.17), for a Grassmann three-web the latter condition is reduced to the form:

$$\underset{1}{b}_{jk} + \underset{2}{b}_{jk} + \underset{3}{b}_{jk} = 0. \tag{8.18}$$

We will prove now that the Grassmann three-web $GW(3, 2, r)$ is hexagonal if and only if the hypersurfaces V_α generating this web belong to one and the same hypercubic. In fact, consider the plane P^2 in the space P^{r+1}. This plane intersects the hypersurfaces V_α along three curves γ_α, $\gamma_\alpha = V_\alpha \bigcap P^2$, $\alpha = 1, 2, 3$, and cut from the web $GW(3, 2, r)$ a three-web $GW(3, 2, 1)$ formed by three one-parameter families of pencils of straight lines, and the centers of these pencils are located on the curves γ_α (Figure 8.4). A correlative transformation κ sends the web $GW(3, 2, 1)$ to a rectilinear three-web formed by the tangents to the curves $\gamma^*_\alpha = \kappa(\gamma_\alpha)$ (Figure 8.5).

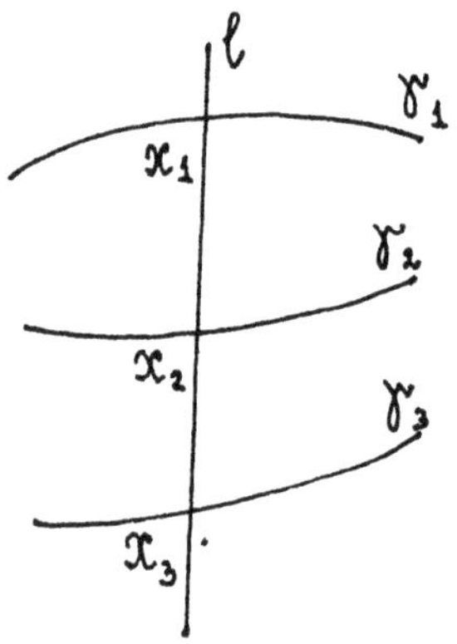

Figure 8.4

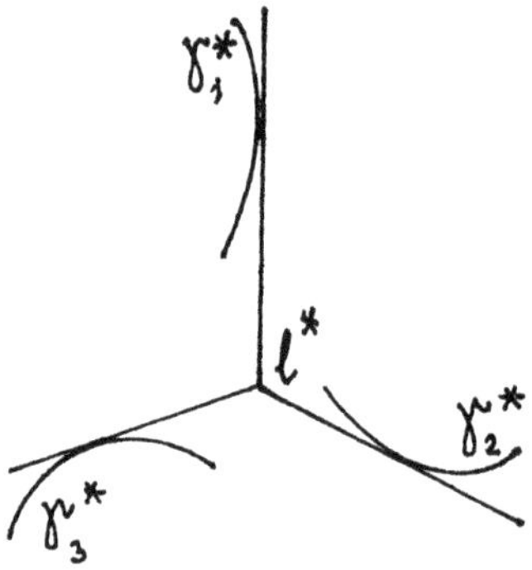

Figure 8.5

If the web $GW(3,2,r)$ is hexagonal, then the same is true for all two-dimensional three-webs $GW(3,2,1)$ which are cut from the web $GW(3,2,r)$ by 2-planes P^2. A correlation sends a hexagonal web $GW(3,2,1)$ to a hexagonal rectilinear three-web in the plane P^{2*}. However, according to the Graf–Sauer theorem, any hexagonal rectilinear three-web is formed by the straight lines belonging to a curve of third class [Bl 55]. Therefore, the curves γ_α generating the web $GW(3,2,1)$ belong to a curve of third degree. Thus, an arbitrary plane P^2 intersects the hypersurfaces V_α along three curves γ_α belonging to one and the same curve of third degree. This implies that the hypersurfaces V_α belong to one and the same hypercubic. The converse can be proved by the same arguments taken in the reverse order.

Wood in [Wo 84] gave another very simple proof of Theorem 1 for any d. His proof is based on application of a special affine coordinate system associated with the straight line l and the use of Viéta's theorem. This proof is close to Scheffers' proof given in [Sche 04] for $r = 1$ and $d = 4$.

Note also that Theorem 8.1 can be formulated in terms of multidimensional webs: *A Grassmann d-web $GW(d,2,r)$, generated in the space P^{r+1} by a system of d hypersurfaces V_α, $\alpha = 1,2,\ldots,d$, is algebraizable if and only if condition* (8.5) *holds.* In this form Theorem 8.1 was presented in the book [G 88].

8.2 The Second Generalization of Reiss' Theorem

1. Theorem 8.1 admits a generalization to a system of r-dimensional submanifolds $V_1, V_2, \ldots, V_d$ of projective space P^n where $r \leq n-1$.

Suppose that in P^n a system of smooth submanifolds V_α, $\alpha = 1, \ldots, d$, $\dim V_\alpha = r, 2 \leq r \leq n-1, d \geq n-r+1$, is given. Suppose also that there exists a subspace l of dimension $n-r$ which intersects each of the submanifolds V_α at a single point $x_\alpha = V_\alpha \bigcap L$, and that the points x_α are in general position in L. Denote by U a neighborhood of the subspace L in the Grassmannian $G(n-r, n)$ all subspaces of which also intersect submanifolds V_α at the points which are in general position. Let T_α be an r-dimensional subspace, which is tangent to the submanifold V_α at the point x_α: $T_\alpha = T_{x_\alpha}(V_\alpha)$.

Consider a subsystem of points x_{α_s}, $s = 1, \ldots, n-r-1$, of the system of points $x_\alpha = V_\alpha \bigcap L$. Since the number of points in this system is less than $n-r+1$, these points are linearly independent. Denote by Z the linear span of the points x_{α_s}, $s = 1, \ldots, n-r-1$, and by $\xi_{\alpha_u} = [Z, T_{\alpha_u}], u = n-r, \ldots, d$, the hyperplane passing through Z and the tangent subspace T_{α_u}. Then the quadratic form $b(x_{\alpha_u}) =< \xi_{\alpha_u}, d^2 x_{\alpha_u} >$ is the second fundamental form of the submanifold V_{α_u} at the point x_{α_u} with respect to the hyperplane ξ_{α_u}. Let further $\underset{\alpha_u}{b}_{ij}$ be the second fundamental tensor defined by this second fundamental form.

We will now prove the following theorem.

Theorem 8.2 *A system of r-dimensional submanifolds V_α of the space P^n is algebraizable if and only if for any subspace $L \in U$ and for distinct points $x_{\alpha_u} \in L$, $u = n-r, \ldots, d$, the condition*

$$\sum_{u=n-r}^{d} b(x_{\alpha_u}) = 0, \tag{8.19}$$

or the equivalent condition

$$\sum_{u=n-r}^{d} \underset{\alpha_u}{b}_{ij} = 0, \tag{8.20}$$

holds.

Proof. The proof of this theorem was first given by Akivis in his paper [A 83b]. Since the points x_α are in general position in L, the dimension of the linear span $[x_{\alpha_1}, \ldots, x_{\alpha_{n-r-1}}]$ is $n-r-2$. Consider a subspace $P^{r+1} \subset P^n$ which is complementary to Z and denote by

$$\pi : P^n \backslash Z \rightarrow P^{r+1}$$

the projection from the center Z. Under this projection, the projections of subspaces $L \subset U$ passing through Z are the lines $l \subset P^{r+1}, l = \pi(L)$. The projections of the submanifolds V_{α_u} are the hypersurfaces $\widetilde{V}_{\alpha_u} = \pi(V_{\alpha_u}) \subset P^{r+1}$, and the quadratic forms $b(x_{\alpha_u})$ are the second fundamental forms of these hypersurfaces at the points $\pi(x_{\alpha_u}) = \widetilde{V}_{\alpha_u} \bigcap l$. By Theorem 8.1, it follows from equations (8.18) and (8.19) that the hypersurfaces $\widetilde{V}_{\alpha_u}$ belong to an algebraic hypersurface of degree $d_1 = d - n + r + 1$ in the space P^{r+1}. Since the manifolds V_{α_u} were taken arbitrarily from the submanifolds V_α and the points x_{α_s} were taken arbitrarily on the rest of V_α, it follows that all of the submanifolds V_α belong to a projective algebraic variety V_d^r of dimension r and degree d of the space P^n, i.e. the system of submanifolds V_α is algebraizable. The sufficiency of the condition of Theorem 8.2 is proved. Its necessity follows easily from the necessity of conditions (8.5) and (8.6) for algebraization of a system of hypersurfaces. The scheme of the proof for the case $n = 4, r = 2, d = 4$ is presented in Figure 8.6. ■

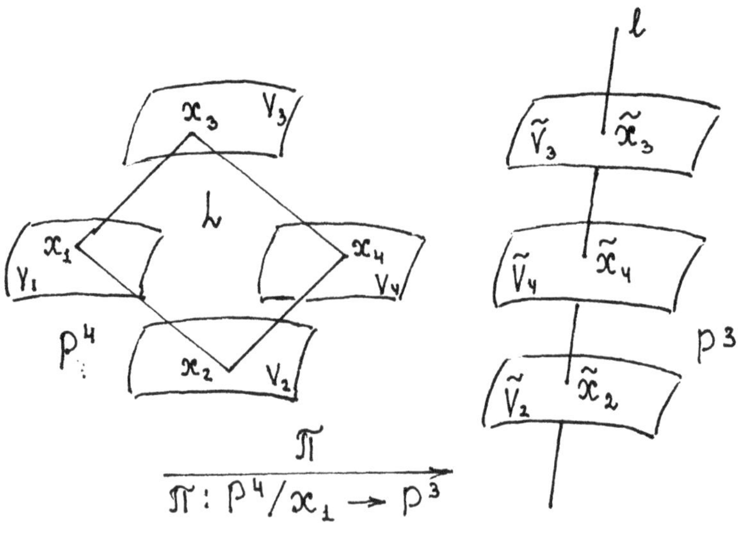

Figure 8.6

Note that the algebraization problem for a system of submanifolds V_α, $\alpha = 1, \ldots, d, \dim V_\alpha = r$, of projective space P^n is connected with the algebraization problem for a d-web $GW(d, n - r + 1, r)$ which is defined on the Grassmannian $G(n - r, n)$ of subspaces of dimension $n - r$ of P^n by means of a system of submanifolds V_α. A number of papers by Chern and Griffiths and by Goldberg are devoted to the latter algebraization problem (an extensive bibliography of this direction can be found in the book [G 88]). Unlike our presentation, in the papers indicated above the authors used the notion of the rank of a d-web and related the algebraization problem for webs with the problem of finding d-webs of maximum rank.

Note that in the paper [A 83b] as well as in the works [Wo 82] and [Wo 84] the theorems similar to Theorems 8.1 and 8.2 were called the Abel theorems. However, it is more accurate to call these theorems the Reiss theorems since the algebraizability conditions (8.8) and (8.20) in them are closer to the Reiss condition (8.2) than they are to the Abel condition (8.1).

8.3 Degenerate Monge's Varieties

1. Another series of algebraization problems is connected with Monge's equations. On an n-dimensional manifold M with coordinates x^i, $i = 1, \ldots, n$, these equations can be written in the form:

$$\Omega(x^i, dx^i) = 0. \tag{8.21}$$

The left-hand side of equation (8.21) is a homogeneous function, and at the tangent space to the manifold M at the point x, it determines a hypercone C_x with its vertex at the point x. As usually, we assume that the hypercone C_x is smooth and is defined by a differentiable function of the point x.

Suppose that the manifold M is a projective space P^n. Then equation (8.21) determines a hypercone C_x at each point x of a domain U of this space. Thus, in P^n a distribution of hypercones arises. This distribution is called the *Monge variety.* If the left-hand side of equation (8.21) is a homogeneous algebraic polynomial of degree p with respect to dx^i, then equation (8.21) defines a *distribution of algebraic hypercones* of order p in the domain U. In particular, if $p = 1$, then equation (8.21) has the form

$$P_i(x^k)dx^i = 0 \tag{8.22}$$

and is called the *Pfaffian equation.* This equation defines a *distribution of hyperplanar elements* in the domain U.

We will associate a family of moving frames with each generator l of hypercone (8.21) in such a way that the point A_0 of each of these frames coincides with the vertex x of the hypercone, the point A_1 lies on the generator l, and the points $A_2, \ldots, A_{n-1}$ lie in the tangent hyperplane to the hypercone along its generator $l = A_0A_1$ (see Figure 8.7).

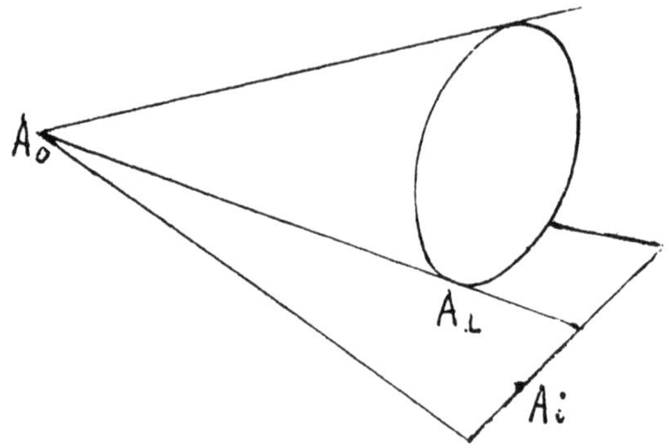

Figure 8.7

Then the differentials of the points A_0 and A_1 will have the form:

$$\begin{aligned} dA_0 &= \omega_0^0 A_0 + \omega_0^1 A_1 + \omega_0^i A_i + \omega_0^n A_n, \\ dA_1 &= \omega_1^0 A_0 + \omega_1^1 A_1 + \omega_1^i A_i + \omega_1^n A_n, \end{aligned} \tag{8.23}$$

where $i = 2, \ldots, n-1$. The 1-forms ω_0^1, ω_0^i and ω_0^n define a displacement of the vertex $x = A_0$ of the hypercone C_x. Since if the vertex x is fixed, the hyperplane $A_0 \wedge A_1 \wedge \ldots \wedge A_{n-1}$ is tangent to the hypercone C_x, we have

$$\omega_1^n \equiv 0 \pmod{\omega_0^1, \omega_0^i, \omega_0^n}.$$

Thus, the equations of the Monge variety relative to a moving frame of this family of moving frames has the form:

$$\omega_1^n = \lambda_1 \omega_0^1 + \lambda_i \omega_0^i + \lambda_n \omega_0^n. \tag{8.24}$$

Consider the manifold of straight lines A_0A_1— the generators of the hypercone C_x. A displacement of these straight lines is determined by the forms ω_0^i, ω_0^n, ω_1^i and ω_1^n. If these forms are linearly independent, then the straight lines A_0A_1 describe a $2(n-1)$-dimensional domain on the Grassmannian $G(1, n)$. This will be the case if the coefficient λ_1 in relation (8.24) is different from zero: $\lambda_1 \neq 0$. In this case a Monge variety is called *nondegenerate*. If on a Monge variety the condition

$$\lambda_1 = 0 \tag{8.25}$$

holds, then the generators A_0A_1 of the hypercone C_x depend on $2n-3$ parameters and describe a hypercomplex of straight lines in the space P^n. In this case a Monge variety is called *degenerate*.

An *integral curve* of the Monge equation is a line which, at each of its points, is tangent to the hypercone C_x. Along an integral curve, we have

$$A_0 = A_0(t), \quad dA_0 = \omega_0^0 A_0 + \omega_0^1 A_1, \tag{8.26}$$

and therefore,

$$\omega_0^i = 0, \quad \omega_0^n = 0 \tag{8.27}$$

along an integral curve.

Let us find the osculating plane of an integral curve of the Monge equation. By (8.26), we have

$$d^2 A_0 \equiv \omega_0^1(\omega_1^i A_i + \omega_1^n A_n) \pmod{A_0, A_1}.$$

Taking into account equations (8.24) and (8.27), we find from this that

$$d^2 A_0 \equiv \omega_0^1(\omega_1^i A_i + \lambda_1 \omega_0^1 A_n) \pmod{A_0, A_1}. \tag{8.28}$$

Thus, the osculating plane is defined by the points A_0, A_1 and $d^2 A_0$.

Definition 8.3 An integral curve of the Monge equation is called *asymptotic* if its osculating plane at any point A_0 belongs to the hyperplane $A_0 \wedge A_1 \wedge \ldots \wedge A_{n-1}$ which is tangent to the hypercone C_x along the straight line $A_0 A_1$.

Relation (8.28) implies the following theorem.

Theorem 8.4 *All integral curves of the Monge equation are asymptotic if and only if the Monge variety defined by this equation is degenerate, i.e. the manifold of straight lines defined by this equation in the space P^n is a hypercomplex.*

Proof. In fact, by (8.28), any integral curve of the Monge equation is asymptotic if and only if $\lambda_1 = 0$. But this condition is the condition of degeneracy of the Monge variety defined by this equation. ■

For $n = 3$, Theorem 8.4 has been proved in the book [LS 96]. However, the proof in [LS 96] is different from the above proof.

2. We now consider some particular cases. First, we suppose that the Monge equation (8.21) is reduced to the Pfaffian equation (8.22) (see [A 84a]). The hypercone C_x defined by this equation at the point x, becomes a hyperplanar element (x, ξ), and the Monge variety becomes a distribution Δ of hyperplanar elements.

We will associate a family of moving frames with each hyperplanar element (x, ξ) in such a way that the point A_0 of each of these frames coincides with the point x and the points A_i lie in the hyperplane ξ. Then the differentials of the points A_0 and A_i will have the form:

$$\begin{aligned} dA_0 &= \omega_0^0 A_0 + \omega_0^i A_i + \omega_0^n A_n, \\ dA_i &= \omega_i^0 A_0 + \omega_i^j A_j + \omega_i^n A_n. \end{aligned} \tag{8.29}$$

The forms ω_0^i and ω_0^n are linearly independent on the hyperdistribution Δ, and the forms ω_i^n are linear combinations of these independent forms:

$$\omega_i^n = b_{ij}\omega_0^j + b_i\omega_0^n. \tag{8.30}$$

Equations (8.29) and (8.30) imply that the integral curves of the hyperdistribution Δ are defined by the equations:

$$\omega_0^n = 0. \tag{8.31}$$

Since

$$d^2 A_0 \equiv \omega_0^i \omega_i^n A_n \pmod{A_0, A_i, \omega_0^n},$$

the equation of asymptotic lines of this hyperdistribution has the form:

$$\omega_0^i \omega_i^n = b_{ij}\omega_0^i \omega_0^j = 0. \tag{8.32}$$

This and Theorem 8.4 imply the following result.

Theorem 8.5 *On a hyperdistribution Δ in projective space P^n, the following three conditions are equivalent:*

a) $b_{(ij)} = 0$;

b) *all integral curves of the hyperdistribution Δ are asymptotic;*

c) *the hyperdistribution Δ is generated by a linear hypercomplex of straight lines in P^n.* ■

The part c) of Theorem 8.5 follows from the fact that in this case, the hypercones C_x belonging to the hypercomplex, which is defined by a degenerate Monge equation, are hyperplanes.

Thus, the condition $b_{(ij)} = 0$ is the algebraizability condition for a hyperdistribution Δ in P^n.

Note that the condition $b_{[ij]} = 0$ is the integrability condition for equation (8.31). This condition together with the algebraizability condition lead to the equation $b_{ij} = 0$. The latter condition means that the integral manifolds of equations (8.31) are hyperplanes of a pencil. In this case, the linear hypercomplex of straight lines, producing the hyperdistribution Δ, becomes a special linear hypercomplex which is formed by straight lines intersecting the $(n-2)$-dimensional axis of this pencil. In the paper [A 84a], Akivis also considered the case when the rank of the tensor b_{ij} is lower than maximal but the tensor is not zero.

3. Consider further the set of nondegenerate hypercones of second order of the space P^n (see the paper [Sa 70] by Safaryan).

We will associate a family of moving frames with each hypercone C_x of this manifold in such a way that the point A_0 of each of these frames coincides with the vertex x of the hypercone. The equations of infinitesimal displacement of this family of moving frames will have the form:

$$dA_0 = \omega_0^0 A_0 + \omega_0^i A_i, \quad dA_i = \omega_i^0 A_0 + \omega_i^j A_j, \quad i,j = 1, \dots, n. \tag{8.33}$$

The forms ω_0^i are basis forms in the manifold of hypercones. Suppose that the moving frames associated with the manifold of hypercones are normalized by the condition: $A_0 \wedge A_1 \wedge \dots \wedge A_n = 1$. This implies that the components of infinitesimal displacement of a moving frame are connected by the equation:

$$\omega_0^0 + \omega_i^i = 0. \tag{8.34}$$

The equation of a hypercone C_x relative to a moving frame of the chosen family has the form:

$$a_{ij} x^i x^j = 0, \quad i,j = 1, \dots, n, \tag{8.35}$$

where $\det(a_{ij}) \neq 0$. Equation (8.35) can be also normalized by the condition:

$$\det(a_{ij}) = 1. \tag{8.36}$$

Since a hypercone C_x is fixed if $\omega_0^i = 0$, the coefficients of equation (8.35) satisfy the differential equations:

$$\delta a_{ij} - a_{ij} \pi_j^k - a_{kj} \pi_i^k = \theta a_{ij}, \tag{8.37}$$

where δ is the symbol of differentiation for $\omega_0^i = 0$ (i.e. it is the symbol of differentiation with respect to the secondary parameters defining the location of a moving frame when the vertex A_0 is fixed), and $\pi_j^i = \omega_j^i(\delta)$. If we differentiate equation (8.36) and apply the formula for differentiation of determinant and relation (8.34), we arrive at the equation:

$$\theta = \frac{2}{n} \pi_0^0. \tag{8.38}$$

Equations (8.37) and (8.38) prove that the equations of a manifold of hypercones of second order in the space P^n have the form:

$$\nabla a_{ij} = da_{ij} - a_{ij} \omega_j^k - a_{kj} \omega_i^k = \frac{2}{n} \omega_0^0 a_{ij} + \lambda_{ijk} \omega_0^k. \tag{8.39}$$

The quantities λ_{ijk} are symmetric in the first two indices. In addition, normalization condition (8.36) implies

$$a^{ij} \lambda_{ijk} = 0,$$

where a^{ij} are entries of the inverse matrix (a^{ij}) of the matrix (a_{ij}).

The integral curves of a manifold of hypercones (8.35) are defined by the equation:

$$a_{ij} \omega_0^i \omega_0^j = 0. \tag{8.40}$$

Let us find the equation of asymptotic lines of this manifold. The equation of the tangent hyperplane of hypercone (8.35) along its generator, defined by the forms ω_0^i, has the form:

$$a_{ij}\omega_0^i x^j = 0. \tag{8.41}$$

For an asymptotic line, the points A_0, dA_0 and d^2A_0 belong to this hyperplane. Since relation (31) implies that

$$d^2A_0 \equiv (\nabla\omega_0^i + \omega_0^0\omega_i^0)A_i \pmod{A_0},$$

where

$$\nabla\omega_0^i = d\omega_0^i + \omega_0^j\omega_j^i,$$

we find from this that

$$a_{ij}\omega_0^i\nabla\omega_0^j = 0. \tag{8.42}$$

Differentiating equation (8.40) and applying (8.42), we obtain

$$\nabla a_{ij}\omega_0^i\omega_0^j = 0.$$

Substituting the expression of ∇a_{ij} from (8.39) into this equation, we find that

$$\lambda_{ijk}\omega_0^i\omega_0^j\omega_0^k = 0. \tag{8.43}$$

Thus, the asymptotic lines of a manifold of hypercones of second order are defined by the system of equations (8.40) and (8.43).

Any integral curve of the manifold of hypercones (8.35) will be an asymptotic line if and only if equation (8.43) is a consequence of equation (8.40). This is equivalent to the relation:

$$\lambda_{(ijk)} = \lambda_{(i}a_{jk)}, \tag{8.44}$$

connecting the quantities λ_{ijk} and a_{ij}. This and Theorem 8.4 imply the following result.

Theorem 8.6 *On a manifold of hypercones of second order in projective space P^n, the following three conditions are equivalent:*

a) *relation* (8.44),

b) *all integral curves of a manifold of hypercones are asymptotic,*

c) *a manifold of hypercones is generated by a quadratic hypercomplex of straight lines in P^n.* ■

The part c) of Theorem 8.6 follows from the fact that in this case, the hypercones C_x belonging to the hypercomplex, which is defined by a degenerate Monge equation, are hypercones of second order.

In addition, note that the quantities λ_{ijk} from equations (8.39) do not form a tensor since they depend on the location of the hyperplane $A_1 \wedge A_2 \wedge \ldots \wedge A_n$. However, these quantities allow us to construct the tensor by setting

$$b_{ijk} = \lambda_{ijk} - a_{ik}\lambda_j - a_{jk}\lambda_i + a_{ij}\mu_k$$

and finding the quantities λ_i and μ_i from the conditions

$$a^{ij} b_{ijk} = 0, \quad a^{jk} b_{ijk} = 0.$$

Then relation (8.44) is equivalent to the equation:

$$b_{(ijk)} = 0. \tag{8.45}$$

In addition to Theorem 8.6, we can also prove that the vanishing of the tensor b_{ijk} is equivalent to the fact that the manifold of hypercones (8.41) is conquadratic (see [Sa 70]).

8.4 Submanifolds with Degenerate Bisecant Varieties

1. Let us consider a smooth submanifold V^r in a projective space P^n where $n \geq 2r + 1$. Suppose that this submanifold does not belong to a space of dimension less than $2r + 1$. Let x and y be two points of the submanifold V^r, and let xy be the straight line joining these two points. The manifold, generated by the lines xy when the points x and y move along V^r, is called the *bisecant variety* of V^r and is denoted by $\sigma(V^r)$ (cf. Section **1.4**). It is easy to see that in general

$$\dim \sigma(V^r) = 2r + 1.$$

If $\dim \sigma(V^r) < 2r + 1$, we shall say that V^r has a *degenerate bisecant variety*.

As an example, we consider a two-dimensional Veronese variety $V(2) = S^2$ in the space P^5. This surface is defined by the parametric equations (1.106):

$$x^{ij} = t^i t^j, \quad i, j = 0, 1, 2, \tag{8.46}$$

(where $x^{ij} = x^{ji}$) and is the symmetric map of the projective plane P^2 into the space P^5 (see Section **1.4**). The bisecant variety $\sigma(S^2)$ is determined by the equations:

$$x^{ij} = t^i t^j + \lambda s^i s^j.$$

Since coordinates of points of this bisecant variety satisfy the equation (1.109):

$$\begin{vmatrix} x^{00} & x^{01} & x^{02} \\ x^{10} & x^{11} & x^{12} \\ x^{20} & x^{21} & x^{22} \end{vmatrix} = 0,$$

the variety $\sigma(S^2)$ is a cubic hypersurface in P^5 which we called in Section **1.4** the *cubic symmetroid*. Since $\dim \sigma(S^2) = 4 < 2\dim S^2 + 1 = 5$, the bisecant variety of the Veronese variety S^2 is degenerate.

In this section we will prove the following theorem.

Theorem 8.7 *In $P^n, n \geq 5$, any two-dimensional submanifold, with degenerate bisecant variety and with osculating spaces of maximal possible dimension five, belongs to the space P^5 and is a part of a two-dimensional Veronese variety.*

2. First, we will prove the following lemma.

Lemma 8.8 *Suppose that $n \geq 2r+1$ and $V^r \subset P^n$ but $V^r \not\subset P^{n_1}$ with $n_1 < 2r+1$. Then the bisecant variety of V^r is degenerate if and only if, for any two points $x, y \in V^r$, we have:*

$$T_x(V^r) \bigcap T_y(V^r) \neq \emptyset \tag{8.47}$$

Proof. With each point $x \in V^r$, we associate a family of moving frames $\{A_u\}$ in such a way that $A_0 = x$, $A_i \in T_x(V^r)$, and in a similar manner with each point $y \in V^r$, we associate a family of moving frames $\{B_u\}$ in such a way that $B_0 = y$, $B_i \in T_y(V^r)$. Then we have

$$dA_0 = \omega_0^0 A_0 + \omega_0^i A_i, \quad dB_0 = \theta_0^0 B_0 + \theta_0^i B_i,$$

where $i = 1, \ldots, r$. For an arbitrary point of the bisecant variety $\sigma(V^r)$, we have

$$d(A_0 + tB_0) = \omega_0^0(A_0 + tB_0) + (dt + t(\theta_0^0 - \omega_0^0))B_0 + \omega_0^i A_i + t\theta_0^i B_i,$$

and the 1-forms $dt + t(\theta_0^0 - \omega_0^0)$, ω_0^i and θ_0^i are linearly independent on $\sigma(V^r)$. If the points

$$A_0 + tB_0, \quad B_0, \quad A_i, \quad B_i \tag{8.48}$$

are linearly independent, then $\dim \sigma(V^r) = 2r+1$, and the $\sigma(V^r)$ is nondegenerate. Otherwise, the bisecant variety $\sigma(V^r)$ is degenerate. Thus, the bisecant variety $\sigma(V^r)$ of a submanifold V^r is degenerate if and only if points (8.48) are linearly dependent, and this is equivalent to condition (8.47). ■

Consider the Veronese surface S^2 defined by equation (8.46) in the space P^5. With each point of S^2, we associate a family of moving frames induced by the frames $\{B_i\}$, $i = 0, 1, 2$, in the plane P^2 whose embedding into P^5

generates S^2. For this, we set $B_{ij} = B_i \otimes B_j = B_{ji}$. The points B_{ii} of this frame belong to the surface S^2, and the points B_{ij}, $i \neq j$, lie on the cubic symmetroid F. Since the infinitesimal displacement of the moving frame $\{B_i\}$ in P^2 has the form

$$dB_i = \theta_i^j B_j,$$

for the frame $\{B_{ij}\}$ in P^5 we obtain:

$$dB_{ij} = (\delta_j^l \theta_i^k + \delta_i^l \theta_j^k) B_{kl}. \tag{8.49}$$

In particular, for the points B_{00} and B_{11}, we have

$$\begin{aligned} dB_{00} &= 2\theta_0^0 B_{00} + 2\theta_0^1 B_{01} + 2\theta_0^2 B_{02}, \\ dB_{11} &= 2\theta_1^1 B_{11} + 2\theta_1^0 B_{01} + 2\theta_1^2 B_{12}. \end{aligned}$$

Thus, the tangent planes to S^2 at arbitrary points B_{00} and B_{11} have the point B_{01} in common, and this point belongs to the cubic symmetroid F.

By equation (8.49), the matrix of components of infinitesimal displacement of the moving frame $\{B_{ij}\}$ associated with the surface S^2 has the form:

	B_{00}	B_{02}	B_{01}	B_{12}	B_{11}	B_{22}
dB_{00}	$2\theta_0^0$	$2\theta_0^2$	$2\theta_0^1$	0	0	0
dB_{02}	θ_2^0	$\theta_0^0 + \theta_2^2$	θ_2^1	θ_0^1	0	θ_0^2
dB_{01}	θ_1^0	θ_1^2	$\theta_0^0 + \theta_1^1$	θ_0^2	θ_0^1	0
dB_{12}	0	θ_1^0	θ_2^0	$\theta_1^1 + \theta_2^2$	θ_2^1	θ_1^2
dB_{11}	0	0	$2\theta_1^0$	$2\theta_1^2$	$2\theta_1^1$	0
dB_{22}	0	$2\theta_2^0$	0	$2\theta_2^1$	0	$2\theta_2^2$

(8.50)

Differentiating the expressions of dB_{00} and dB_{11}, we find that

$$\begin{aligned} d^2 B_{00} &\equiv 2(\theta_0^1)^2 B_{11} + 4\theta_0^1\theta_0^2 B_{12} + 2(\theta_0^2)^2 B_{22} \pmod{B_{00}, B_{01}, B_{02}}, \\ d^2 B_{11} &\equiv 2(\theta_1^0)^2 B_{00} + 4\theta_1^0\theta_1^2 B_{02} + 2(\theta_1^2)^2 B_{22} \pmod{B_{11}, B_{01}, B_{12}}. \end{aligned}$$

Thus, the second fundamental forms of the Veronese surface S^2 at the point B_{00} have the form:

$$\varphi^{11} = 2(\theta_0^1)^2, \;\; \varphi^{12} = 4\theta_0^1\theta_0^2, \;\; \varphi^{22} = 2(\theta_0^2)^2, \tag{8.51}$$

and at the point B_{11} they are:

$$\varphi^{00} = 2(\theta_1^0)^2, \;\; \varphi^{02} = 4\theta_1^0\theta_1^2, \;\; \varphi^{22} = 2(\theta_1^2)^2. \tag{8.52}$$

6. Proof of Theorem 8.7. We first recall that the equations of infinitesimal displacement of a moving frame $\{A_u\}$, $u = 0, 1, \ldots, n$, in projective space P^n have the form (1.65):

$$dA_u = \omega_u^v A_v, \tag{8.53}$$

and the 1-forms ω_u^v from (8.53) satisfy the structure equations (1.67) of P^n:

$$\omega_u^v = \omega_u^w \wedge \omega_w^v, \tag{8.54}$$

where $u, v, w = 0, 1, \ldots, n$.

Suppose that a submanifold $V^2 \subset P^n$ satisfies the conditions of Theorem 8.7. Let x and y be two arbitrary points of V^2. Then by Lemma 8.8, condition (8.47) holds, and it is possible to associate with the points x and y a family of moving frames $\{A_u\}$ in such a way that $A_0 = x, A_4 = y, T_x(V^2) = A_0 \wedge A_1 \wedge A_2$ and $T_y(V^2) = A_4 \wedge A_3 \wedge A_2$. This implies that the differentials of the points A_0 and A_1 have the form:

$$\begin{array}{ll} dA_0 = & \omega_0^0 A_0 + \omega_0^1 A_1 + \omega_0^2 A_2, \\ dA_4 = & \omega_4^2 A_2 + \omega_4^3 A_3 + \omega_4^4 A_4. \end{array} \tag{8.55}$$

So on the chosen family of frames the following equations hold:

$$\omega_0^4 = \omega_0^3 = \omega_0^\alpha = 0, \tag{8.56}$$

$$\omega_4^0 = \omega_4^1 = \omega_4^\alpha = 0, \tag{8.57}$$

where $\alpha = 5, \ldots, n$, and the forms ω_0^1, ω_0^2, ω_4^3 and ω_4^2 are linearly independent. Note that two equations (8.55) as well as equations (8.56) and (8.57) are transferred to each other under the substitution of indices:

$$0 \longleftrightarrow 4, \quad 1 \longleftrightarrow 3, \quad 2 \longleftrightarrow 2, \quad \alpha \longleftrightarrow \alpha. \tag{8.58}$$

All equations, which will be obtained below, are also invariant under substitution (8.58).

Exterior differentiation of equations (8.56) by means of structure equations (8.54) and application of Cartan's lemma to the resulting exterior quadratic equations gives:

$$\begin{array}{lll} \omega_1^4 = \lambda_{11}^4 \omega_0^1 + \lambda_{12}^4 \omega_0^2, & \omega_1^3 = \lambda_{11}^3 \omega_0^1 + \lambda_{12}^3 \omega_0^2, & \omega_1^\alpha = \lambda_{11}^\alpha \omega_0^1 + \lambda_{12}^\alpha \omega_0^2, \\ \omega_2^4 = \lambda_{21}^4 \omega_0^1 + \lambda_{22}^4 \omega_0^2, & \omega_2^3 = \lambda_{21}^3 \omega_0^1 + \lambda_{12}^3 \omega_0^2, & \omega_2^\alpha = \lambda_{21}^\alpha \omega_0^1 + \lambda_{22}^\alpha \omega_0^2, \end{array} \tag{8.59}$$

where

$$\lambda_{12}^4 = \lambda_{21}^4, \ \lambda_{12}^3 = \lambda_{21}^3, \ \lambda_{12}^\alpha = \lambda_{21}^\alpha.$$

In complete analogy with this, equations (8.57) imply:

$$\begin{array}{lll} \omega_3^0 = \mu_{33}^0 \omega_4^3 + \mu_{32}^0 \omega_4^2, & \omega_3^1 = \mu_{33}^1 \omega_4^3 + \mu_{32}^1 \omega_4^2, & \omega_3^\alpha = \mu_{33}^\alpha \omega_4^3 + \mu_{32}^\alpha \omega_4^2, \\ \omega_2^0 = \mu_{23}^0 \omega_4^3 + \mu_{22}^0 \omega_4^2, & \omega_2^1 = \mu_{23}^1 \omega_4^3 + \mu_{22}^1 \omega_4^2, & \omega_2^\alpha = \mu_{23}^\alpha \omega_4^3 + \mu_{22}^\alpha \omega_4^2, \end{array} \tag{8.60}$$

where

$$\mu^0_{32} = \mu^0_{23}, \ \mu^1_{32} = \mu^1_{23}, \ \mu^\alpha_{32} = \mu^\alpha_{23}.$$

Since in equations (8.59) and (8.60) the forms ω^α_2 appear twice, and they are linear combinations of different linearly independent forms, equations (8.59) and (8.60) imply that

$$\omega^\alpha_2 = 0, \tag{8.61}$$

and in addition that

$$\omega^\alpha_1 = \lambda^\alpha \omega^1_0, \ \omega^\alpha_3 = \mu^\alpha \omega^3_4, \tag{8.62}$$

where $\lambda^\alpha = \lambda^\alpha_{11}$ and $\mu^\alpha = \mu^\alpha_{33}$.

Let us find the osculating subspaces of the submanifold V^2 at the points A_0 and A_4 and the second fundamental forms of V^2 at these points. By (8.55), the second differentials of A_0 and A_4 can be written in the form:

$$\begin{aligned} d^2A_0 &\equiv \varphi^4 A_4 + \varphi^3 A_3 + \varphi^\alpha A_\alpha \quad (\text{mod } A_0, A_1, A_2), \\ d^2A_4 &\equiv \psi^0 A_0 + \psi^1 A_1 + \psi^\alpha A_\alpha \quad (\text{mod } A_4, A_3, A_2), \end{aligned}$$

where $\varphi^3, \varphi^4, \varphi^\alpha$ and $\psi^1, \psi^0, \psi^\alpha$ are the second fundamental forms of V^2 at the points A_0 and A_4. By (8.59), (8.60) and (8.62), these forms can be expressed as follows:

$$\begin{aligned} \varphi^4 &= \lambda^4_{11}(\omega^1_0)^2 + 2\lambda^4_{12}\omega^1_0\omega^2_0 + \lambda^4_{22}(\omega^2_0)^2, \\ \varphi^3 &= \lambda^3_{11}(\omega^1_0)^2 + 2\lambda^3_{12}\omega^1_0\omega^2_0 + \lambda^3_{22}(\omega^2_0)^2, \\ \varphi^\alpha &= \lambda^\alpha(\omega^1_0)^2. \end{aligned} \tag{8.63}$$

$$\begin{aligned} \psi^0 &= \mu^0_{33}(\omega^3_4)^2 + 2\mu^0_{32}\omega^3_4\omega^2_4 + \mu^0_{22}(\omega^2_4)^2, \\ \psi^1 &= \mu^1_{33}(\omega^3_4)^2 + 2\mu^1_{32}\omega^3_4\omega^2_4 + \mu^1_{22}(\omega^2_4)^2, \\ \psi^\alpha &= \mu^\alpha(\omega^3_4)^2. \end{aligned} \tag{8.64}$$

Since the dimensions of the osculating subspaces of the submanifold V^2 at the points $A_0 = x$ and $A_4 = y$ are equal to five, only three of forms (8.63) and three of forms (8.64) must be linearly independent. Thus the vectors λ^α and μ^α are different from zero, and the osculating subspaces are:

$$T_x(V^2) = A_0 \wedge A_1 \wedge A_2 \wedge A_3 \wedge A_4 \wedge \lambda^\alpha A_\alpha, \ T_y(V^2) = A_4 \wedge A_3 \wedge A_2 \wedge A_1 \wedge A_0 \wedge \mu^\alpha A_\alpha.$$

Exterior differentiation of equations (8.61) leads to the exterior quadratic equation:

$$\omega^1_2 \wedge \omega^\alpha_1 + \omega^3_2 \wedge \omega^\alpha_3 = 0.$$

Substituting the expressions of the forms ω^1_2 and ω^3_2 from (8.59) and (8.60) and the forms ω^α_1 and ω^α_3 from (8.62) into this equation, we find that

$$\mu^1_{22}\lambda^\alpha\omega^2_4 \wedge \omega^1_0 + (\mu^1_{32}\lambda^\alpha - \lambda^3_{12}\mu^\alpha)\omega^3_4 \wedge \omega^1_0 - \lambda^3_{22}\mu^\alpha\omega^3_4 \wedge \omega^2_0 = 0.$$

By the linear independence of the exterior products in this equation, we obtain:

$$\mu^1_{22}\lambda^\alpha = 0, \quad \mu^1_{32}\lambda^\alpha - \lambda^3_{12}\mu^\alpha = 0, \quad \lambda^3_{22}\mu^\alpha = 0. \tag{8.65}$$

Since the vectors λ^α and μ^α are different from zero, it follows from (8.65) that

$$\mu^1_{22} = 0, \quad \lambda^3_{22} = 0. \tag{8.66}$$

Since three of forms (8.63) and three of forms (8.64) must be linearly independent, the coefficients λ^3_{12} and μ^1_{32} must be different from zero. Thus by (8.65), the vectors λ^α and μ^α are collinear. This means that the *osculating subspaces* $T^{(2)}_x(V^2)$ *and* $T^{(2)}_y(V^2)$ *coincide.* Since x and y are two arbitrary points of the submanifold V^2, the *submanifold* V^2 *lies in the space* P^5. This proves the first part of Theorem 8.7.

Next we place the point A_5 of a moving frame into the space P^5 containing the submanifold V^2. This implies:

$$\lambda^5 = \lambda \neq 0, \quad \lambda^6 = \ldots = \lambda^n = 0, \quad \mu^5 = \mu \neq 0, \quad \mu^6 = \ldots = \mu^n = 0, \tag{8.67}$$

and in all preceding equations we can omit the terms with the index $\alpha > 5$. In particular, the second equation of (5.63) takes the form:

$$\mu^1_{23}\lambda - \lambda^3_{12}\mu = 0.$$

It follows that

$$\mu^1_{23} = k\mu, \quad \lambda^3_{12} = k\lambda, \quad k \neq 0, \tag{8.68}$$

and by (8.68) and (8.66), equations (8.59) and (8.60) give

$$\omega^3_2 = k\lambda\omega^1_0, \quad \omega^1_2 = k\mu\omega^3_4. \tag{8.69}$$

Note that a frame $\{A_u\}$, $u = 0, 1, \ldots, 5$ is associated with the submanifold $V^2 \subset P^5$ in the same way as a frame $\{B_{ij}\}$, $i, j = 0, 1, 2$ is associated with the Veronese surface S^2. The correspondence between the vertices of these two frames is established by the following table:

A_0	A_1	A_2	A_3	A_4	A_5
B_{00}	B_{02}	B_{01}	B_{12}	B_{11}	B_{22}

(8.70)

We will now study a surface described by the point A_2 of intersection of the tangent subspaces $T_x(V^2)$ and $T_y(V^2)$. It follows from equation (8.61) that

$$dA_2 = \omega^0_2 A_0 + \omega^1_2 A_1 + \omega^2_2 A_2 + \omega^3_2 A_3 + \omega^4_2 A_4.$$

By relations (8.59), (8.60) and (8.66), we have

$$\omega_2^0 \wedge \omega_2^1 \wedge \omega_2^3 \wedge \omega_2^4 = \mu_{22}^0 \mu_{23}^1 \lambda_{21}^3 \lambda_{22}^4 \omega_4^2 \wedge \omega_4^3 \wedge \omega_0^1 \wedge \omega_0^2.$$

We already proved that the coefficients μ_{23}^1 and λ_{22}^3 are different from zero. Since, for $\alpha = 5$, the second fundamental forms (8.63) and (8.64) are independent, the coefficients μ_{22}^0 and λ_{22}^4 must also be different from zero. So the forms ω_2^0, ω_2^1, ω_2^3 and ω_2^4 are linearly independent, and the *point* A_2 *describes a four-dimensional submanifold* V^4, *i.e. a hypersurface in* P^5.

Let us find the second fundamental form of V^4. We have

$$d^2 A_2 \equiv (\omega_2^1 \omega_1^5 + \omega_2^3 \omega_3^5) A_5 \pmod{T_{A_2}(V_4)}.$$

By (8.62) and (8.69), the quadratic form in this relation takes the form:

$$\omega_2^1 \omega_1^5 + \omega_2^3 \omega_3^5 = 2k\lambda\mu\omega_0^1 \omega_4^3.$$

Since this form is expressed in terms of only two of the four basis forms of V^4, *the hypersurface* V^4 *is tangentially degenerate of rank two* (see Section **4.1**). It is easy to verify that the system of equations

$$\omega_0^1 = 0, \;\; \omega_4^3 = 0$$

is completely integrable on V^4 and defines a fibration of V^4 into two-dimensional planes $A_0 \wedge A_2 \wedge A_4$ along which the tangent subspace $A_0 \wedge A_1 \wedge A_2 \wedge A_3 \wedge A_4$ to V^4 is fixed.

By (8.66), (8.67) and (8.68), the second differentials $d^2 A_0$ and $d^2 A_4$ can be written in the form:

$$\begin{aligned} d^2 A_0 \equiv \;& (\lambda_{11}^4 (\omega_0^1)^2 + 2\lambda_{12}^4 \omega_0^1 \omega_0^2 + \lambda_{22}^4 (\omega_0^2)^2) A_4 \\ & + (\lambda_{11}^3 (\omega_0^1)^2 + 2k\lambda\omega_0^1 \omega_0^2) A_3 + \lambda(\omega_0^1)^2 A_5 \pmod{A_0, A_1, A_2}, \end{aligned} \tag{8.71}$$

$$\begin{aligned} d^2 A_4 \equiv \;& (\mu_{33}^0 (\omega_4^3)^2 + 2\mu_{32}^0 \omega_4^3 \omega_4^2 + \mu_{22}^0 (\omega_4^2)^2) A_0 \\ & + (\mu_{33}^1 (\omega_4^3)^2 + 2k\mu\omega_4^3 \omega_4^2) A_1 + \mu(\omega_4^3)^2 A_5 \pmod{A_4, A_3, A_2}. \end{aligned} \tag{8.72}$$

We will make a further specialization of a moving frame $\{A_u\}$, $u = 0, 1, \ldots, 5$, associated with V^2, choosing the frame $\{B_{ij}\}$ associated with the Veronese surface S^2 as a sample and taking into account the fact that there is a factor 2 in formulas (8.50) which does not appear in equations (8.55).

First, taking into account that $\lambda \neq 0$ and $\mu \neq 0$, we can normalize the points A_0 and A_1 by setting:

$$\widetilde{A}_0 = \frac{1}{2\lambda} A_0, \;\; \widetilde{A}_4 = \frac{1}{2\mu} A_4.$$

Then, by relations (8.71), (8.72) and (8.68), we find that

$$
\begin{aligned}
d^2\widetilde{A}_0 \equiv\ & \tfrac{1}{2}(A_5 + \tfrac{\lambda^4_{11}}{\lambda}A_4 + \tfrac{\lambda^3_{11}}{\lambda}A_3)(\omega^1_0)^2 + (kA_3 + \tfrac{\lambda^4_{12}}{\lambda}A_4)\omega^1_0\omega^2_0 \\
& + \tfrac{1}{2}\tfrac{\lambda^4_{22}}{\lambda}A_4(\omega^2_0)^2 \pmod{A_0, A_1, A_2}, \\
d^2\widetilde{A}_4 \equiv\ & \tfrac{1}{2}(A_5 + \tfrac{\mu^1_{33}}{\mu}A_1 + \tfrac{\mu^0_{33}}{\mu}A_0)(\omega^3_4)^2 + (kA_1 + \tfrac{\mu^0_{23}}{\mu}A_0)\omega^2_4\omega^3_4 \\
& + \tfrac{1}{2}\tfrac{\mu^0_{22}}{\mu}A_0(\omega^2_4)^2 \pmod{A_4, A_3, A_2}.
\end{aligned} \tag{8.73}
$$

Define:

$$
\begin{aligned}
&\widetilde{A}_5 = A_5 + \tfrac{\lambda^4_{11}}{\lambda}A_4 + \tfrac{\lambda^3_{11}}{\lambda}A_3 + \tfrac{\mu^1_{33}}{\mu}A_1 + \tfrac{\mu^0_{33}}{\mu}A_0, \\
&\widetilde{A}_3 = kA_3 + \tfrac{\lambda^4_{12}}{\lambda}A_4, \\
&\widetilde{A}_1 = kA_1 + \tfrac{\mu^0_{23}}{\mu}A_0.
\end{aligned}
$$

Since $k \neq 0$ and the point $\widetilde{A}_3$ remains a basis point in the tangent subspace $T_{A_4}(V^2)$ as well as the point $\widetilde{A}_1$ remains a basis point in the tangent subspace $T_{A_0}(V^2)$, this specialization of a moving frame can be made. For brevity, we also set:

$$
\frac{\lambda^4_{22}}{\lambda} = \widetilde{\lambda}^4_{22}, \quad \frac{\mu^0_{22}}{\mu} = \widetilde{\mu}^0_{22}.
$$

These coefficients are also different from zero.

As a result of this specialization, relations (64) take the form:

$$
\begin{aligned}
d^2A_0 &\equiv \tfrac{1}{2}(\omega^1_0)^2A_5 + \omega^1_0\omega^2_0A_3 + \tfrac{1}{2}\lambda^4_{22}(\omega^2_0)^2A_4 \pmod{A_0, A_1, A_2}, \\
d^2A_4 &\equiv \tfrac{1}{2}(\omega^3_4)^2A_5 + \omega^2_4\omega^3_4A_1 + \tfrac{1}{2}\mu^0_{22}(\omega^2_4)^2A_0 \pmod{A_4, A_3, A_2},
\end{aligned}
$$

where we omitted the symbol $\sim$ over the points and the coefficients. Moreover, the second fundamental forms of V^2 at the points A_0 and A_4 take the form:

$$
\begin{aligned}
&\varphi^5 = \tfrac{1}{2}(\omega^1_0)^2, \quad \varphi^3 = \omega^1_0\omega^2_0, \quad \varphi^4 = \tfrac{1}{2}\lambda^4_{22}(\omega^2_0)^2, \\
&\psi^5 = \tfrac{1}{2}(\omega^3_4)^2, \quad \psi^1 = \omega^3_4\omega^2_4, \quad \psi^0 = \tfrac{1}{2}\mu^0_{22}(\omega^2_4)^2.
\end{aligned}
$$

This gives the following expressions for the 1-forms in the left-hand side of equations (8.59) and (8.60):

$$
\omega^5_1 = \frac{1}{2}\omega^1_0, \quad \omega^5_3 = \frac{1}{2}\omega^3_4, \tag{8.74}
$$

$$
\omega^5_2 = 0, \tag{8.75}
$$

$$
\omega^3_1 = \frac{1}{2}\omega^2_0, \quad \omega^1_3 = \frac{1}{2}\omega^2_4, \tag{8.76}
$$

$$
\omega^3_2 = \frac{1}{2}\omega^1_0, \quad \omega^1_2 = \frac{1}{2}\omega^3_4, \tag{8.77}
$$

$$\omega_1^4 = 0, \quad \omega_3^0 = 0, \tag{8.78}$$

$$\omega_2^4 = \frac{1}{2}\lambda_{22}^4 \omega_0^2, \quad \omega_2^0 = \frac{1}{2}\mu_{22}^0 \omega_4^2. \tag{8.79}$$

Equations (8.74)–(8.79) absorbed equations (8.61), (8.62), (8.66), and (8.69) which we had earlier.

Next, we will find the differential prolongations of equations (8.74)–(8.79). Taking exterior derivatives of (8.74), we find that

$$(\omega_0^0 - 2\omega_1^1 + \omega_5^5) \wedge \omega_0^1 = 0, \quad (\omega_4^4 - 2\omega_3^3 + \omega_5^5) \wedge \omega_4^3 = 0. \tag{8.80}$$

Exterior differentiation of equation (8.75) leads to the identity, and exterior differentiation of equation (8.76) gives

$$\begin{array}{l} (\omega_0^0 - \omega_1^1 - \omega_2^2 + \omega_3^3) \wedge \omega_0^1 + (\omega_5^3 - 2\omega_1^2) \wedge \omega_0^1 = 0, \\ (\omega_4^4 - \omega_3^3 - \omega_2^2 + \omega_1^1) \wedge \omega_4^2 + (\omega_5^1 - 2\omega_3^2) \wedge \omega_4^3 = 0. \end{array} \tag{8.81}$$

Differentiating equations (8.77), we obtain:

$$\begin{array}{l} (\omega_0^0 - \omega_1^1 - \omega_2^2 + \omega_3^3) \wedge \omega_0^1 + (1 - \lambda_{22}^4)\omega_0^2 \wedge \omega_4^3 = 0, \\ (\omega_4^4 - \omega_3^3 - \omega_2^2 + \omega_1^1) \wedge \omega_4^3 + (1 - \mu_{22}^0)\omega_4^2 \wedge \omega_0^1 = 0. \end{array}$$

It follows that

$$\lambda_{22}^4 = 1, \quad \mu_{22}^0 = 1 \tag{8.82}$$

and

$$(\omega_0^0 - \omega_1^1 - \omega_2^2 + \omega_3^3) \wedge \omega_0^1 = 0, \quad (\omega_4^4 - \omega_3^3 - \omega_2^2 + \omega_1^1) \wedge \omega_4^3 = 0. \tag{8.83}$$

By (8.82), equations (8.79) take the form:

$$\omega_2^4 = \frac{1}{2}\omega_0^2, \quad \omega_2^0 = \frac{1}{2}\omega_4^2. \tag{8.84}$$

Finally, differentiating equations (8.78) and (8.84), we arrive at the equations:

$$\begin{array}{l} (\omega_3^4 - \omega_1^2) \wedge \omega_0^2 + \omega_5^4 \wedge \omega_0^1 = 0, \\ (\omega_1^0 - \omega_3^2) \wedge \omega_4^2 + \omega_5^0 \wedge \omega_4^3 = 0, \end{array} \tag{8.85}$$

and

$$\begin{array}{l} (\omega_0^0 - 2\omega_2^2 + \omega_4^4) \wedge \omega_0^2 + (\omega_3^4 - \omega_1^2) \wedge \omega_0^1 = 0, \\ (\omega_4^4 - 2\omega_2^2 + \omega_0^0) \wedge \omega_4^2 + (\omega_1^0 - \omega_3^2) \wedge \omega_4^3 = 0. \end{array} \tag{8.86}$$

Applying Cartan's lemma to equations (8.80), (8.81), (8.83), (8.85) and (8.86), we obtain:

$$\omega_3^4 - \omega_1^2 = a_0\omega_0^1, \quad \omega_1^0 - \omega_3^2 = a_4\omega_4^3, \tag{8.87}$$

$$\omega_5^4 = a_0\omega_0^2 + b_0\omega_0^1, \quad \omega_5^0 = a_4\omega_4^2 + b_4\omega_4^3, \tag{8.88}$$

$$\omega_5^3 - 2\omega_1^2 = c_0\omega_0^1, \quad \omega_5^1 - 2\omega_3^2 = c_4\omega_4^3, \tag{8.89}$$

$$\omega_0^0 - \omega_1^1 - \omega_2^2 + \omega_3^3 = 0, \quad \omega_4^4 - \omega_3^3 - \omega_2^2 + \omega_1^1 = 0, \tag{8.90}$$

$$\omega_1^1 - \omega_2^2 + \omega_3^3 - \omega_5^5 = 0. \tag{8.91}$$

Next, differentiating equations (8.90) and (8.91), we find that

$$a_0 = a_4 = a, \quad c_0 = c_4 = c. \tag{8.92}$$

Taking exterior derivatives (8.87) and (8.89), substituting the values (8.92) into equations the equations obtained and applying Cartan's lemma, we find the equations:

$$\begin{aligned} da + a(\omega_2^2 - \omega_5^5) - \tfrac{1}{2}\omega_5^2 &= \tfrac{1}{2}(b_0\omega_4^3 + b_4\omega_0^1), \\ dc + c(\omega_2^2 - \omega_5^5) - \tfrac{3}{2}\omega_5^2 &= b_0\omega_4^3 + b_4\omega_0^1, \end{aligned} \tag{8.93}$$

and the relation:

$$c = 3a. \tag{8.94}$$

From relations (8.92) and (8.93) it follows that

$$b_0 = b_4 = 0, \tag{8.95}$$

and

$$da + a(\omega_2^2 - \omega_5^5) - \frac{1}{2}\omega_5^2 = 0. \tag{8.96}$$

It is easy to verify that the latter equation is completely integrable by means of preceding equations.

The form ω_5^2 contains a secondary parameter defining the displacement of the point A_5 towards to the point A_2. Equation (8.94) shows that using this parameter, we can reduce the coefficient a to zero. As a result, the form ω_5^2 also becomes zero.

As this specialization of frames takes place, by relations (8.92), (8.94) and (8.95), equations (8.87)–(8.89) and (8.96) take the form:

$$\begin{array}{ll} \omega_3^4 = \omega_1^2, & \omega_1^0 = \omega_3^2, \\ \omega_5^4 = 0, & \omega_5^0 = 0, \\ \omega_5^3 = 2\omega_1^2, & \omega_5^1 = 2\omega_3^2, \\ \multicolumn{2}{c}{\omega_5^2 = 0.} \end{array} \tag{8.97}$$

In addition, equations (8.90) and (8.91) imply that

$$\omega_1^1 = \frac{1}{2}(\omega_0^0 + \omega_5^5), \quad \omega_3^3 = \frac{1}{2}(\omega_4^4 + \omega_5^5), \quad \omega_2^2 = \frac{1}{2}(\omega_0^0 + \omega_4^4). \tag{8.98}$$

Our considerations prove that the system of equations (8.56), (8.57), (8.74)–(8.78), (8.84), (8.97) and (8.98) is completely integrable. This system contains 27 linearly independent equations. Thus, in the space P^5, this system defines a family of moving frames (and a submanifold V^m along with this family) which depends on 27 constants.

By equations (8.56), (8.57), (8.74)-(8.78), (8.84), (8.97) and (8.98), the matrix (ω_v^u), $u, v = 0, 1, \ldots, 5$, of components of infinitesimal displacement of a frame $\{A_u\}$, associated with the surface V^2 satisfying the conditions of Theorem 8.7, takes the form:

$$(\omega_v^u) = \begin{pmatrix} \omega_0^0 & \omega_0^1 & \omega_0^2 & 0 & 0 & 0 \\ \omega_3^2 & \frac{1}{2}(\omega_0^0 + \omega_5^5) & \omega_1^2 & \frac{1}{2}\omega_0^2 & 0 & \frac{1}{2}\omega_0^1 \\ \frac{1}{2}\omega_4^2 & \frac{1}{2}\omega_4^3 & \frac{1}{2}(\omega_0^0 + \omega_4^4) & \frac{1}{2}\omega_0^1 & \frac{1}{2}\omega_0^2 & 0 \\ 0 & \frac{1}{2}\omega_4^2 & \omega_3^2 & \frac{1}{2}(\omega_4^4 + \omega_5^5) & \omega_1^2 & \frac{1}{2}\omega_4^3 \\ 0 & 0 & \omega_4^2 & \omega_4^3 & \omega_4^4 & 0 \\ 0 & 2\omega_3^2 & 0 & 2\omega_1^2 & 0 & \omega_5^5 \end{pmatrix}. \tag{8.99}$$

Comparing this matrix with matrix (8.50) of components of infinitesimal displacement of a frame, associated with the Veronese surface S^2, we readily find that these two matrices coincide if we set:

$$\begin{array}{llll} \omega_0^1 = 2\theta_0^2, & \omega_0^2 = 2\theta_0^1, & \omega_4^2 = 2\theta_1^0, & \omega_4^3 = 2\theta_1^2, \\ \omega_1^2 = \theta_2^1, & \omega_3^2 = \theta_2^0, & & \\ \omega_0^0 = 2\theta_0^0, & \omega_4^4 = 2\theta_1^1, & \omega_5^5 = 2\theta_2^2. & \end{array} \tag{8.100}$$

Thus, *the submanifold V^2 satisfying the conditions of Theorem* 8.7 *is a Veronese surface S^2 in P^5*. This concludes the proof of Theorem 8.7. ■

NOTES

8.1–8.2. J.B. Little communicated us some interesting history behind the Abel papers [Ab 29] and [Ab 41]. The proof of Abel's theorem appears in [Ab 41]. That paper was apparently written in 1826 and submitted to the Paris Academy at that time. Fourier who was the secretary of the Academy referred the paper to Legendre and Cauchy, who in essence laid the paper aside and forgot about it. It was only rediscovered after Abel's death and published in 1841. As to the paper [Ab 29], this paper was really a summary of [Ab 41].

Different generalizations of the Abel theorem and other results on it can be found in [Gr 76], [Gr 77], [CG 78a], [CG 78b], [Lit 83], [Lit 84], [Lit 85], [Lit 87] and [Lit 89].

The "Reiss-type" relations were studied by Bäcklund in [Bä 69], Holst in [Ho 82a], [Ho 82b] and [Ho 82c], Bompiani in [Bom 39], [Bom 40] and [Bom 46], Engel in [En 39], Kubota in [Ku 39], B. Segre in [SegB 47] (see also his book [SegB 71]), Griffiths and Harris in [GH 78] and by J.B. Little in [Lit 84], [Lit 85] and [Lit 87]. Bompiani in [Bom 46] gave an interesting history of the Reiss theorem. The relationship between the Reiss relation (8.2) and the Graf-Sauer theorem was established by Liebmann in [Lieb 27] (see also [Bl 55]).

In this chapter we follow the Akivis paper [A 92b].

On the algebraization problem for multidimensional webs and its connection with the algebraization problem for submanifolds in a projective space see the papers [A 73], [A 80], [A 81a], [A 82a], [A 82d], [A 83a], [A 83b] by Akivis, [AG 74] by Akivis and Goldberg, [Bot 74] by Botsu, [C 82] by Chern, [CG 77], [CG 78a], [CG 78b], [CG 81] by Chern and Griffiths, [G 75], [G 82a], [G 82b], [G 90] by Goldberg, [Wo 82], [Wo 84] by Wood, and the books [G 88] by Goldberg and [AS 92] by Akivis and Shelekhov. The algebraizability problem for webs is connected with the fact the G-structure associated with a web is closed (see [A 75] and [A 83a]).

Some other problems of algebraizability were investigated by Akivis in [A 85] and [A 87a] and by Baimuratov in [B 74].

8.3. For $n = 3$ Theorem 8.4 has been proved in the book [LS 96] written by Lie and Scheffers.

The case when the Monge variety becomes a distribution Δ of hyperplanar elements was considered by Akivis in [A 84a]). In the same paper Akivis also considered the case when the rank of the tensor b_{ij} is lower than maximal but the tensor is not zero.

Safaryan in [Sa 70] proved Theorem 8.6 for the set of nondegenerate hypercones of second order of the space P^n.

8.4. Theorem 8.7 in the form presented in this section was proved by Akivis in [A 92b]. This theorem is somewhat stronger than the similar theorem in [Se 01] and [GH 79] where the assumption $V^2 \subset P^5$ has been made from the beginning. Lemma 8.8 is due to Griffiths and Harris (see [GH 79], p. 427). Other sufficient conditions for the Veronese embedding were considered in [LP 71] by J.A. Little and Pohl and in [Sas 91] by Sasaki. These conditions are different from the conditions given in [Se 01], [GH 78], [GH 79] and [A 92b]. Nomizu and Yano in [NY 74], Nomizu in [Nom 76] and Chern, do Carmo and Kobayashi in [CDK 70] characterized Veronese varieties from the metric point of view.

Bibliography

[Ab 41] Abel, N: *Mémoire sur une propriété générale d'une classe très-étendue de fonctions transcendantes*, Mémoires des savants etrangers, **7** (1841), 176–204; see also *Œuvres Complétes de Niels Henrik Abel*, Nouv. ed., vol. 1, Grondal et Son, Christiania, pp. 145–211. (F.[1] **13**, p. 20.)

[Ab 29] Abel, N: *Démonstration d'une propriété générale d'une certaine classe de fonctions transcendantes*, J. Reine Angew. Math. **4** (1829), 212–215; see also *Œuvres Complétes de Niels Henrik Abel*, Nouv. ed., vol. 1, Grondal et Son, Christiania, pp. 515–517. (F. **13**, p. 20.)

[A 49] Akivis, M.A: *A focal family of rays as the image of a pair of T-complexes*, (Russian) Dokl. Akad. Nauk SSSR **65** (1949), no. 4, 429–432. (MR **11**, p. 134; Zbl. **41**, p. 292.)

[A 50] Akivis, M.A: *Pairs of T-complexes.* (Russian) Mat. Sb. (N.S.) **27** (69) (1950), no. 3, 365–378. (MR **13**, p. 152; Zbl. **38**, pp. 340–341.)

[A 57] Akivis, M.A: *Focal images of a surface of rank r*, (Russian) Izv. Vyssh. Uchebn. Zaved. Mat. **1957**, no. 1, 9–19. (MR **25** #498; Zbl. **94**, p. 186.)

[A 61a] Akivis, M.A: *On multidimensional surfaces carrying a net of conjugate lines*, (Russian) Dokl. Akad. Nauk SSSR **139** (1961), no. 6, 1279–1282; English transl: Soviet Math. Dokl. **2** (1961), no. 4, 1065–1068. (MR **24** #A2908; Zbl. **134**, p. 169.)

[A 61b] Akivis, M.A: *Transformations of surfaces by a focal family of rays*, (Russian) Uspekhi Mat. Nauk **16** (1961), no. 1, 193–195.

[A 62a] Akivis, M.A: *Transformations of surfaces by a focal family of rays*, (Russian) Izv. Vyssh. Uchebn. Zaved. Mat. **1962**, no. 6 (31), 3–13. (MR **27** #1898; Zbl. **126**, p. 375.)

[A 62b] Akivis, M.A: *On a class of tangentially degenerate surfaces*, (Russian) Dokl. Akad. Nauk SSSR **146** (1962), no. 3, 515–518; English transl: Soviet Math. Dokl. **3** (1962), no. 5, 1328–1331. (MR **25** #5467; Zbl. **136**, p. 173.)

[1] In the bibliography we will use the following abbreviations for the review journals: F. for *Jahrbuch für die Fortschritte der Mathematik*, MR for *Mathematical Reviews*, and Zbl. for *Zentralblatt für Mathematik und ihren Grenzgebiete.*

[A 62c] Akivis, M.A: *On the Voss normals of a surface carrying a net of conjugate lines*, (Russian) Mat. Sb. (N.S.) **58** (100) (1962), no. 2, 695-706. (MR **26** #688; Zbl. **117**, p. 384.)

[A 63] Akivis, M.A: *On the structure of surfaces carrying a net of conjugate lines*, (Russian) Moskov. Gos. Ped. Inst. Uchen. Zap. No. **208** (1963), 31–47.

[A 66] Akivis, M.A: *On the structure of two-component conjugate systems*, (Russian) Trudy Geometr. Sem **1** (1966), 7–31. (MR **34** #8295; Zbl. **178**, p. 246.)

[A 70a] Akivis, M.A: *The structure of conjugate systems on multidimensional surfaces*, (Russian) Izv. Vyssh. Uchebn. Zaved. Mat. **1970**, no. 10 (101), 3–11. (MR **44** #2147; Zbl. **209**, p. 260.)

[A 70b] Akivis, M.A: *The invariant framing of a surface carrying a net of conjugate lines*, (Russian) Moskov. Gos. Ped. Inst. Uchen. Zap. **374** (1970), no. 1 Voprosy Differentsial. Geom., 18–27. (MR **58** #30833.)

[A 73] Akivis, M.A: *The local differentiable quasigroups and three-webs that are determined by a triple of hypersurfaces*, (Russian) Sibirsk. Mat. Zh. **14** (1973), no. 3, 467-474; English transl: Siberian Math. J. **14** (1973), no. 3, 319-324. (MR **48** #2911; Zbl. 267.53005 & 281.53002.)

[A 75] Akivis, M.A: *Closed G-structures on a differentiable manifold*, (Russian) Problems in Geometry, Vol. **7**, 69–79. Akad. Nauk SSSR Vsesoyuz. Inst. Nauchn. i Tekhn. Informatsii, Moscow, 1975. (MR **57** #17549; Zbl. 549.53032.)

[A 77] Akivis, M.A: *Multidimensional differential geometry*, (Russian) Kalinin. Gos. Univ., Kalinin, 1977, 99 pp. (MR 82k:53001; Zbl. 459.53001.)

[A 80] Akivis, M.A: *Webs and almost-Grassmann structures*, (Russian) Dokl. Akad. Nauk SSSR **252** (1980), no. 2, 267–270; English transl: Soviet Math. Dokl. **21** (1980), no. 3, 707–709. (MR 82a:53016; Zbl. 479.53015.)

[A 81a] Akivis, M.A: *A class of three-webs that are determined by a triple of hypersurfaces*, (Russian) Sibirsk. Mat. Zh. **22** (1981), no. 1, 3–7; English transl: Siberian Math. J. **22** (1981), no. 1, 1–4. (MR 82c:53014; Zbl. 456.53006 & 472.53028.)

[A 81b] Akivis, M.A: *On a multidimensional generalization of Koenigs' net*, (Russian) Izv. Vyssh. Uchebn. Zaved. Mat. **1981**, no. 9 (232), 3–4, English transl: Soviet Math. (Iz. VUZ) **25** (1981), no. 9, 1–3. (MR 83f:53004; Zbl. 476.53007 & 498.53006.)

[A 82a] Akivis, M.A: *Webs and almost-Grassmann structures*, (Russian) Sibirsk. Mat. Zh. **23** (1982), no. 6, 6–15. English transl: Siberian Math. J. **23** (1982), no. 6, 763–770. (MR 84b:53018; Zbl. 505.53004 & 516.53013.)

[A 82b] Akivis, M.A: *Differential geometry of the Grassmannian*, (Russian) "Trudy 3 Omsk. Obl. Mat. Konf. Omsk, April 9-11, 1981", Omsk, 1982, 2-7. Bibl. 3 titles. Dep. in VINITI on 2/24/83 under no. 1018-83.

[A 82c] Akivis, M.A: *On the differential geometry of the Grassmann manifold*, (Russian) Tensor **38** (1982), 273-282. (MR 87e:53021; Zbl. 504.53010.)

[A 83a] Akivis, M.A: *Differential geometry of webs*, (Russian) Problems in Geometry, Vol. 15, 187–213, Itogi Nauki i Tekhniki, Akad. Nauk SSSR, Vsesoyuz. Inst. Nauchn. i Tekhn. Informatsii, Moscow, 1983; English transl: J. Soviet Math. **29** (1985), no. 5, 1631–1647. (MR 85i:53019; Zbl. 567.53014.)

[A 83b] Akivis, M.A: *The local algebraizability condition for a system of submanifolds of a real projective space*, (Russian) Dokl. Akad. Nauk SSSR **272** (1983), no. 6, 1289–1291. English transl: Soviet Math. Dokl. **28** (1983), no. 2, 507–509. (MR 85c:53018; Zbl. 547.53006.)

[A 84a] Akivis, M.A: *Plane hyperdistributions in P^n*, (Russian) Mat. Zametki **36** (1984), no. 2, 213–222; English transl: Math. Notes **36** (1984), no. 1-2, 599–604. (MR 86b:53009; Zbl. 563.53012.)

[A 84b] Akivis, M.A: *Projective generalization of D.F. Egorov's transforms*, (Russian) Izv. Vyssh. Uchebn. Zaved. Mat. **1984**, no. 7 (266), 3–10. English transl: Soviet Math. (Iz. VUZ) **28** (1984), no. 7, 1–10. (MR 86b:53008; Zbl. 554.53006 & 572.53006.)

[A 85] Akivis, M.A: *A simplest condition of algebraizability of an n-dimensional manifold of null-pairs*, (Russian). Problems in the Theory of Webs and Quasigroups, 3–7, Kalinin. Gos. Univ., Kalinin, 1985. (MR 87i:53026; Zbl. 577.53010.)

[A 87a] Akivis, M.A: *The conditions of algebraizability of a triple of curves in a three-dimensional projective space*, (Russian) Webs and Quasigroups, Kalinin. Gos. Univ., Kalinin, 1987, 129-136. (MR 88i:53019; Zbl. 617.53012.)

[A 87b] Akivis, M.A: *On multidimensional strongly parabolic surfaces*, (Russian) Izv. Vyssh. Uchebn. Zaved. Mat. **1987**, no. 5 (300), 3–10. English transl: Soviet Math. (Iz. VUZ) **31** (300) (1987), no. 5, 1–11. (MR 89g:53016; Zbl. 632.53012.)

[A 88] Akivis, M.A: *A projective differential geometry of submanifolds*, (Russian) Tezisy Dokl., 9th All-Union Geom. Conference, Kishinev, 1988, 6–7.

[A 92a] Akivis, M.A: *Secant degenerate submanifolds*, (Russian) Webs and Quasigroups, Tver Gos. Univ., Tver, 1992, 15–27.

[A 92b] Akivis, M.A: *On some algebraizability problems in projective differential geometry*, (Russian) Izv. Vyssh. Uchebn. Zaved. Mat. **1992**, no. 6 (361), 3–14. English transl: Soviet Math. (Iz. VUZ) **31** (1992), no. 6.

[AB 75] Akivis, M.A. and Baimuratov, H.A: *A class of complexes of oriented spheres*, (Russian) Izv. Vyssh. Uchebn. Zaved. Mat. **1975**, no. 8 (159), 3–9. English transl: Soviet Math. (Iz. VUZ) **19** (1975), no. 8. (MR **54** #5992; Zbl. 324.50029.)

[AC 75] Akivis, M.A. and Chakmazyan, A.V: *On normalized submanifolds of an affine space admitting a parallel normal vector field*, Akad. Nauk Armyan. SSR Dokl. **60** (1975), no. 3, 137–143. (MR **53** # 6451; Zbl. 325.53007.)

[AG 74] Akivis, M.A. and Goldberg, V.V: *The four-web and the local differentiable ternary quasigroup that are determined by a quadruple of surfaces of codimension two*, (Russian) Izv. Vyssh. Uchebn. Zaved. Mat. **1974**, no. 5 (144), 12–24. English transl: Soviet Math. (Iz. VUZ) **18** (1974), no. 5, 9–19. (MR **50** #8321; Zbl. 297.53037.)

[AK 93] Akivis, M.A. and Konnov, V.V: *Some local aspects of the theory of conformal structures*, (Russian) Uspekhi Mat. Nauk **48** (1993), no. 1, 3–40.

[AR 64] Akivis, M.A. and Ryzhkov, V.V: *Multidimensional surfaces of special projective types*, (Russian) Proc. Fourth All-Union Math. Congr. (Leningrad, 1961), Vol. II, pp. 159–164, Izdat. "Nauka", Leningrad, 1964. (MR **36** #3251; Zbl. **192**, p. 277.)

[AS 92] Akivis, M.A. and Shelekhov, A.M: *Geometry and algebra of multidimensional three-webs*, Kluwer Academic Publishers, Dordrecht-Boston-London, 1992, xii+358 pp.

[At 52] Atanasyan, L.S: *Normalized manifolds of a special type in a multidimensional affine space*, (Russian) Trudy Sem. Vektor. Tenzor. Anal. **9** (1952), 351–410. (MR **14** #796; Zbl. **48**, p. 401.)

[At 58a] Atanasyan, L.S: *Construction of inner geometry of surfaces in a multidimensional affine space*, (Russian) Moskov. Gos. Ped. Inst. Uchen. Zap. **108** (1958), 177–217. (MR **22** #22474; Zbl. **96**, p. 158.)

[At 58b] Atanasyan, L.S: *On the theory of normalized surfaces in a multidimensional projective space*, (Russian) Moskov. Gos. Ped. Inst. Uchen. Zap. **108** (1958), 3–44. (MR **22** #22472; Zbl. **96**, p. 158.)

[AV 63] Atanasyan, L.S. and Vorontsova, N.S: *Special normalizations of degenerate hypersurfaces of an $(n+1)$-dimensional projective space*, (Russian) Volzh. Mat. Sb. Vyp. 1 (1963), 5–9. (MR **34** #706; Zbl. **173**, p. 495.)

[AV 65] Atanasyan, L.S. and Vorontsova, N.S: *A construction of an invariant normalization of an r-degenerate hypersurface of a multidimensional projective space*, (Russian) Moskov. Gos. Ped. Inst. Uchen. Zap. No. 243 (1965), 5–28. (MR **34** #738; Zbl. **177**, p. 435.)

[Bä 69] Bäcklund, A.V: *Einige Sätze über die Normalen algebraischer Kurven*, Lund. Universitets Ørsskrift **5** (1869).

[B 74] Baimuratov, Kh.A: *The differential-geometric characterization of three-dimensional biquadratic surfaces*, (Russian) Collection of articles on differential geometry, 21–35. Kalinin. Gos. Univ., Kalinin, 1974. (MR **53** #11505.)

[B 75a] Baimuratov, Kh.A: *The geometry of a three-dimensional surface V_3 which carries four families of rectilinear generators in the projective space P_5*, (Russian) Izv. Vyssh. Uchebn. Zaved. Mat. **1975**, no. 10 (161), 3–14; English transl: Soviet Math. (Iz. VUZ) **19** (1975), no. 10, 1–10. (Zbl. 321.53009.)

[B 75b] Baimuratov, Kh.A: *The geometry of a three-dimensional surface of general form in a five-dimensional space*, (Russian) Collection of articles on differential geometry, no. 2, 3–14. Kalinin. Gos. Univ., Kalinin, 1975. (MR **58** #30796.)

[Bal 50] Baldassarri, M: *Le varietà pluririgate a tre dimensioni*, Rend. Sem. Mat. Univ. Padova **19** (1950), 172–200. (MR **12**, p. 438; Zbl. **38**, p. 318.)

[BW 69] Barner, M. and Walter, R: *Zur projective Differentialgeometrie der Regelflächen im vierdimensionalen Raum*, Math. Nachr. **41** (1969), no. 4-6, 191–211. (MR **41** #6066; Zbl. **182**, p. 548.)

[Ba 53] Bazylev, V.T: *Quasi-Laplacian transformations of p-surfaces of a space P_n*, (Russian) Dokl. Akad. Nauk SSSR **92** (1953), 453–455. (MR **16**, p. 70; Zbl. **53**, p. 296.)

[Ba 55a] Bazylev, V.T: *Quasi-Laplacian transformations of multidimensional surfaces*, (Russian) Uspekhi Mat. Nauk **10** (1955), no. 2 (64), 214–215. (Zbl. **64**, p. 161.)

[Ba 55b] Bazylev, V.T: *Quasi-Laplacian transformations of p-dimensional surfaces of an n-dimensional projective space*, (Russian) Moskov. Gorod. Ped. Inst. Uchen. Zap., vol. **35** (1955), no. 4, 261–322.

[Ba 61] Bazylev, V.T: *On a class of higher-dimensional surfaces*, (Russian) Izv. Vyssh. Uchebn. Zaved. Mat. **1961**, no. 1 (20), 27–35. (MR **25** #514; Zbl. **134**, p. 390.)

[Ba 64] Bazylev, V.T: *On planar nets associated with a Cartan surface*, (Russian) Sibirsk. Mat. Zh. **5** (1964), 729–737. (MR **31** #2670; Zbl. **173**, p. 495.)

[Ba 65a] Bazylev, V.T: *Multidimensional nets and their transformations*, (Russian), pp. 138–164. Akad. Nauk SSSR Inst. Nauchn. Informatsii, Moscow, 1965. (MR **33** #3526.)

[Ba 65b] Bazylev, V.T: *The geometry of planar multidimensional nets*, (Russian), Moskov. Gos. Ped. Inst. Uchen. Zap. No. 243 (1965), 29–37. (MR **33** #6527.)

[Ba 66] Bazylev, V.T: *Nets on multidimensional surfaces of a projective space*, (Russian), Izv. Vyssh. Uchebn. Zaved. Mat. **1966**, no. 2 (51), 9–19. (MR **34** #707; Zbl. **147**, p. 210.)

[Ba 67] Bazylev, V.T: *Fields of conjugate directions on multidimensional surfaces of full rank*, (Russian) Moskov. Gos. Ped. Inst. Uchen. Zap. No. 271 (1967), 7–33. (MR **40** #6384.)

[Bel 54] Bell, P.O: *A theorem on generalized conjugate nets in projective n-space*, Duke Math. J., **21** (1954), no. 2, 323–327. (MR **16**, p. 168; Zbl. **55**, p. 401.)

[Ber 64a] Berezina, L.Ya: *An invariant normalization of m-dimensional surfaces in an n dimensional projective space where* $n < \frac{1}{2}m(m+3)$, (Russian) II All-Union Geom. Confer. (Kharkov, 1964) Tezisy Dokl., Kharkov. Univ., Kharkov, 1964, pp. 17–18.

[Ber 64b] Berezina, L.Ya: *An invariant normalization of an m-dimensional surfaces in an n dimensional projective space where* $n < \frac{1}{2}m(m+3)$, (Russian) III Sibirsk. Konfer. Mat. & Mekh. Dokl., Tomsk. Univ., Tomsk, 1964, pp. 180–181.

[Bert 24] Bertini, E: *Einführung in die projective Geometrie mehrdimensionalen Räume*, Seidel u. Sohn, Wien, 1924, xxii+480 pp. (F. **49**, p. 484; **50**, p. 591.)

[Bla 71] Blank, Ya.P: *On a generalization of S. Lie problem on translation surfaces*, (Russian) Trudy Geometr. Sem. **3** (1971), 5–27. (MR **47** #4159.)

[BG 70] Blank, Ya.P. and Gormasheva, N.M: *Peterson hypersurfaces in* P^4, (Russian) Ukrain. Geom. Sb. Vyp. 9 (1970), 20–29. (MR **45** #2590; Zbl. 222.53008.)

[Bl 21] Blaschke, W: *Vorlesungen über Differentialgeometrie und geometrische Grundlagen von Einsteins Relativitätstheorie*, Band I, Springer-Verlag, Berlin, 1921, xii+230 pp. (F. **48**, pp. 1305–1306); 2nd ed., 1924, xii+242 pp. (F. **50**, pp. 452–453); 3rd ed., 1930, x+311 pp. (F. **56**, p. 588); republished by Dover Publ., New York, 1945, xiv + 322 pp. (MR **7**, p. 391; Zbl. **63/I**, p. A85.); 4th ed., 1945, x+311 pp.

[Bl 23] Blaschke, W: *Vorlesungen über Differentialgeometrie und geometrische Grundlagen von Einsteins Relativitätstheorie*, Band 2: Affine Differentialgeometrie, Springer-Verlag, Berlin, 1st and 2nd ed., 1923, x+259 pp. (F. **49**, pp. 499–501.)

[Bl 50] Blaschke, W: *Einführung in die Differentialgeometrie*, Springer-Verlag, Berlin/Göttingen/Heidelberg, 1950, vii+146 pp. (MR **13**, p. 274; Zbl. **41**, p. 288); 2nd ed., 1960, viii+173 pp; Russian translation, GITTL, Moskva, 1957, 223 pp. (MR **13**, p. 274; Zbl. **91**, p. 340.)

[Bl 55] Blaschke, W: *Einführung in die Geometrie der Waben*, Birkhäuser-Verlag, Basel-Stuttgart, 1955, 108 pp. (MR **17**, p. 780; Zbl. **68**, p. 365.); Russian translation, GITTL, Moskva, 1959, 144 pp. (MR **22** #2942; Zbl. **68**, p. 365.)

[BB 38] Blaschke, W. and Bol, G: *Geometrie der Gewebe*, Springer-Verlag, Berlin, 1938, viii+340 pp. (Zbl. **20**, p. 67.)

[Bo 50a] Bol, G: *Projektive Differentialgeometrie*, Vandenhoeck & Ruprecht, Göttingen, vol. 1, 1950, vii+365 pp. (MR **11**, p. 539; Zbl. **35**, p. 234.); vol. 2, 1954, v+372 pp. (MR **16**, p. 1150; Zbl. **59**, p. 155.); vol. 3, 1967, viii+527 pp. (MR **37** #840; Zbl. **173**, p. 233.)

[Bo 50b] Bol, G: *Zur tensoriellen Behandlung der projectiven Flächentheorie*, Math. Ann. **122** (1950), 279–295. (MR **12**, p. 441; Zbl. **31**, p. 274.)

[Bom 21] Bompiani, E: *Proprietà differenziale caratteristiche di enti algebriche*, Mem. Roy. Accad. Lincei (5) **13** (1921), facs. 8, 26 pp. (F. **48**, p. 851.)

[Bom 39] Bompiani, E: *Über zwei Kalotten einer Hyperquadric*, Jahresber. Deutsch. Math.-Verein. **49** (1939), 143–145. (MR **1**, p. 83; Zbl. **21**, p. 352.)

[Bom 40] Bompiani, E: *Callotte a centri allineati di superficie algebriche*, Atti Accad. Italia Rend. Cl. Sci. Fis. Mat. Nat. (7) **1** (1940), 93–101. (MR **1**, p. 167; Zbl. **24**, p. 279.)

[Bom 46] Bompiani, E: *Elementi differenziali regolari e non regolari nel piano e loro applicazioni alle curve algebrische piano*, Univ. Roma Ist. Naz. Alta Mat. Rend. Mat. e Appl. (5) **5** (1946), 1–46. (MR **9**, p. 60.)

[Bori 82] Borisenko, A.A: *Multidimensional parabolic surfaces in Euclidean space*, (Russian) Ukrain. Geom. Sb. No. 25 (1982), 3–5. (MR 84j:53006; Zbl. 511.53003.)

[Bori 85] Borisenko, A.A: *Complete parabolic surfaces in Euclidean space*, (Russian) Ukrain. Geom. Sb. No. 28 (1985), 8–19. (MR 86k:53079; Zbl 577.53006.)

[Bori 92] Borisenko, A.A: *Unique determination of multidimensional submanifolds in a Euclidean space from the Grassmann image*, (Russian) Mat. Zametki **51** (1992), no. 1, 8–15; English transl: Math. Notes **51** (1992), no. 1.

[Bor 28] Bortolotti, E: *Alcuni risultati di geometria proiettivo-differenziale*, Boll. Un. Mat. Ital. **7** (1928), 178–184. (F. **54**, p. 778.)

[Bor 39] Bortolotti, E: *Sulla geometria proiettiva differenziale delle transformazioni dualistiche*, Atti Accad. Naz. Lincei Rend. Cl. Sci. Fis. Mat. Natur. (6) **28** (1939), 224–230. (Zbl. **20**, p. 397.)

[Bot 74] Botsu, V.P: *A direct proof of the generalized theorem of Graf–Sauer*, (Russian) Collection of Articles on Differential Geometry, Kalinin. Gos. Univ., Kalinin, 1974, 36–51. (MR **58** #30792.)

[Br 38] Brauner, K: *Über Mannigfaltigkeiten, deren Tangentialmannigfaltigkeiten ausgeart sind*, Monatsh. Math. Phys. **46** (1938), 335–365. (Zbl. **19**, p. 232.)

[BCGGG 91] Bryant, R.L., Chern, S.S., Gardner, R.B., Goldsmith, H.L. and Griffiths, P.A: *Exterior differential systems*, Springer-Verlag, New York, 1991, vii+475 pp. (MR 92h:58007; Zbl. 726.58002.)

[Bu 91] Bubyakin, I.V: *Geometry of five-dimensional complexes of two-dimensional planes in projective space*, (Russian), Funktsional. Anal. i Prilozhen. **25** (1991), no. 3, 73–76; English transl: Functional Analysis and Its Appl. **25** (1991), no. 3, 223–224. (MR 92i:53014; Zbl. 736.53015.)

[BN 48] Bushmanova, G.V. and Norden A.P: *Projective invariants of a normalized surface*, (Russian) Dokl. Akad. Nauk SSSR **60** (1948), no. 8, 1309–1312. (MR **10**, p. 478; Zbl. **395**, p. 173.)

[Ca 16] Cartan, É: *La deformation des hypersurfaces dans l'espace euclidien reel à n dimensions*, Bull. Soc. Math. France **44** (1916), 65–99. (F. **46**, p. 1129.)

[Ca 18] Cartan, É: *Sur les variétés à* 3 *dimensions*, C.R. Acad. Sc. **167** (1918), 357–360. (F. **46**, p. 1128.)

[Ca 19] Cartan, É: *Sur les variétés de courbure constante d'un espace euclidien ou non-euclidien*, Bull. Soc. Math. France **47** (1919), 125–160; **48** (1920), 132–208. (F. **47**, pp. 692–693.)

[Ca 20a] Cartan, É: *Sur la déformation projective des surfaces*, C.R. Acad. Sc. **170** (1920), 1439–1441. (F. **47**, p. 656.)

[Ca 20b] Cartan, É: *Sur l'applicabilité projective des surfaces*, C.R. Acad. Sc. **171** (1920), 27–29. (F. **47**, p. 656.)

[Ca 20c] Cartan, É: *Sur la déformation projective des surfaces*, Ann. Sci. École Norm. Sup. **37** (1920), 259–356. (F. **47**, pp. 656–657.)

[Ca 20d] Cartan, É: *Sur le problèm général de la déformation*, C.R. Congrès Strasbourg, 1920, 397–406. (F. **48**, p. 817.)

[Ca 24] Cartan, É: *Sur la connexion projective des surfaces*, C.R. Acad. Sc. **178** (1924), 750–752. (F. **47**, p. 656.)

[Ca 27] Cartan, É: *Sur un problème du calcul des variations en géométrie projective plane*, Recueil Soc. Math. Moscou **34** (1927), 349–364. (F. **53**, p. 486.)

[Ca 31] Cartan, É: *Sur les développantes d'une surface réglée*, Bull. Acad. Roumaine **14** (1931), 167–174. (Zbl. **3**, p. 130.)

[Ca 37] Cartan, É: *La théorie de groupes finis et continus et la géométrie différentielle traitées par la méthode di repère mobile*, Gauthier-Villars, Paris, 1937, vi+269 pp. (Zbl. **18**, p. 298.)

[Ca 44] Cartan, É: *Sur une classe de surfaces apparentées aux surfaces R et aux surfaces de Jonas*, Bull. Sc. Math. **68** (1944), 41–50. (MR **7**, p. 78; Zbl. **63/I**, p. A145.)

[Ca 45a] Cartan, É: *Les systèmes différentiels extérieurs et leurs applications géometriques*, Hermann, Paris, 1945, 214 pp. (MR **7**, p. 520); 2nd ed., 1971, 210 pp. (Zbl. **211**, p. 127.)

[Ca 45b] Cartan, É: *Sur un problème de géométrie differentielle projective*, Ann. Ec. Norm. **62** (1945), 205-231. (MR **8**, p. 92; Zbl. **63/I**, p. A145.)

[Ca 60] Cartan, É: *Riemannian geometry in orthogonal frame*, (Russian) Izdat. Moscow. State Univ., Moscow, 1960, 307 pp. (MR **23** #A1316; Zbl. **99**, p. 372.)

[CaH 67] Cartan, H: *Formes différentielles. Application élémentaires au calcul des variations at à la théorie des courbes et surfaces*, Hermann, Paris, 1967, 186 pp. (MR **37** (1969) #6358; Zbl. **184**, p. 127.) English transl: *Differential forms*, H. Mufflin Co., Boston, 1970, 166 pp. (MR **42** #2379; Zbl. **213**, p. 370.)

[Cas 50] Casanova, G: *La notion de pôle harmonique*, Rev. math. spec. **65** (1950), no. 6, 437–440.

[Če 26] Čech, E: *Projektivni diferenciálni geometrie*, (Czech) Mákladem. Jednoty Çeskoslovensnských Matematiku a fysiku, Praha, 1926, 406 pp. (F. **52**, p. 753.)

[Cen 62a] Cenkl, B: *Homographies conservant l'élément du troième ordre d'une surface dans un espace à connexion projective*, Czechoslovak. Math. J. **12** (1962), no. 2, 288-293. (MR **26** #2964; Zbl. **139**, p. 396.)

[Cen 62b] Cenkl, B: *La normale d'une surface dans l'espace à connexion projective*, Czechoslovak. Math. J. **12** (1962), no. 4, 582–606. (MR **26** #6900; Zbl. **129**, p. 143.)

[Cha 59] Chakmazyan, A.V: *Dual normalization*, (Russian) Akad. Nauk Armyan. SSR Dokl. **28** (1959), no. 4, 151–157. (MR **22** #11329a; Zbl. **86**, p. 365.)

[Cha 76] Chakmazyan, A.V: *Submanifolds with a parallel p-dimensional subbundle of the normal bundle*, (Russian) Izv. Vyssh. Uchebn. Zaved. Mat. **1976**, no. 8 (171), 107–110. English translation: Soviet Math. (Iz. VUZ) **20** (1976), no. 8, 84–86. (MR **57** #17547; Zbl. 343.53009.)

[Cha 77a] Chakmazyan, A.V: *Submanifolds of a space of constant curvature with a parallel subbundle of the normal bundle*, (Russian) Ukrain. Geom. Sb. **20** (1977), 132–140. (MR **58** #18287; Zbl. 431.53046.)

[Cha 77b] Chakmazyan, A.V: *A class of submanifolds in* V_c^n *with a parallel p-dimensional subbundle of the normal bundle*, (Russian) Mat. Zametki **22** (1977), no. 4, 477–483. English translation: Math. Notes **22** (1977), no. 4, 757–761. (MR **57** #13570; Zbl. 379.53006.)

[Cha 77c] Chakmazyan, A.V: *Normalized submanifolds of an affine space with flat normal affine connection*, (Russian) Differential Geometry, Kalinin, 1977, 120–129. (MR 82f:53027.)

[Cha 77d] Chakmazyan, A.V: *A submanifold V^m in P_n normalized in the sense of Norden with a parallel normal subbundle*, (Russian) Mat. Zametki **22** (1977), no. 5, 649–662. English translation: Math. Notes **22** (1977), no. 5, 853–860. (MR **57** #7420; Zbl. 363.53003.)

[Cha 77e] Chakmazyan, A.V: *Norden-normalized submanifold with a parallel field of normal directions in P^n*, (Russian) Dokl. Akad. Nauk SSSR **236** (1977), no. 4, 816–819. English translation: Soviet Math. Dokl. **18** (1977), no. 5, 1289–1293. (MR **57** #1295; Zbl. 399.53003.)

[Cha 78a] Chakmazyan, A.V: *On submanifolds with parallel field of normal p-directions in a space of constant curvature*, (Russian) Tartu Riikl. Ül. Toimitised Vih. 464, Trudy Mat. i Meh., no 22 (1978), 137–145. (MR 81j:53016; Zbl. 407.53015.)

[Cha 78b] Chakmazyan, A.V: *A connection in the normal bundles of a normalized submanifold V_m in P_n*, (Russian) Problems in Geometry, Vol. **10**, 55–74. Akad. Nauk SSSR Vsesoyuz. Inst. Nauchn. i Tekhn. Informatsii, Moscow, 1978; English transl: J. Soviet Math. **14** (1980), no. 3, 1205–1216. (MR 80i:53010; Zbl. 405.53007 & 443.53006.)

[Cha 80] Chakmazyan, A.V: *A normalized (in the Norden sense) submanifold with a parallel field of normal directions in P_n*, (Russian) Izv. Vyssh. Uchebn. Zaved. Mat. **1980**, no. 1 (212), 57–63; English transl: Soviet Math. (Iz. VUZ) **24** (1980), no. 1, 69–77. (MR 81g:53006; Zbl. 464.53015.)

[Cha 83] Chakmazyan, A.V: *Normalized submanifolds with flat normal connection in a projective space*, (Russian) Mat. Zametki **33** (1983), no. 2, 281–287. English transl: Math. Notes **33** (1983), no. 2, 140–144. (MR 85g:53018; Zbl. 517.53009 & 526.53007.)

[Cha 84a] Chakmazyan, A.V: *A normal connection of a normalized manifold of planes in a projective space*, Izv. Vyssh. Uchebn. Zaved. Mat. **1984**, no. 7 (266), 74–79. English transl: Soviet Math. (Iz. VUZ) **28** (1984), no. 7, 97–104. (MR 86b:53014; Zbl. 548.53016.)

[Cha 84b] Chakmazyan, A.V: *The affine geometry of a normalized submanifold with parallel field of normal p-directions*, (Russian) Tartu Riikl. Ül. Toimitised Vih. 665 (1984), 81–89. (MR 86a:53017; Zbl. 586.53018.)

[Cha 87] Chakmazyan, A.V: *On framings with flat normal connection of submanifolds of an affine space*, Izv. Vyssh. Uchebn. Zaved. Mat. **1987**, no. 1 (296), 48–53. English translation: Soviet Math. (Iz. VUZ) **31** (1987), no. 1, 64–70. (MR 88d:53012; Zbl. 615.53002.)

[Cha 89] Chakmazyan, A.V: *Normal connection in the geometry of normalized submanifolds of an affine space*, (Russian) Problems in Geometry, Vol. **21**, 93–107. Akad. Nauk SSSR Vsesoyuz. Inst. Nauchn. i Tekhn. Informatsii, Moscow, 1989; English transl: J. Soviet Math. **55** (1991), no. 6, 2131–2140. (MR 90k:53043; Zbl. 711.53010 & 729.53016.)

[Cha 90] Chakmazyan, A.V: *Normal connections in the geometry of submanifolds*, (Russian) Armyanskij Gosudarstvennyj Pedagogicheskij Institut Im. Kh. Abovyana, Erevan, 1990, 116 p. (MR 92g:53047; Zbl. 704.53016.)

[Ch 73] Chen, B.Y: *Geometry of submanifolds*, Marcel Dekker, Inc., New York, 1973, vii+298 pp. (MR **50** #5697; Zbl. 262.53036.)

[Ch 81] Chen, B.Y: *Geometry of submanifolds and its applications*, Science University of Tokyo, Tokyo, 1981, iii+ 96 pp. (MR 83m:53051; Zbl. 474.53050.)

[CV 80] Chen, B.Y. and Vanhecke, L: *Geodesic spheres and locally symmetric spaces*, C.R. Math. Rep. Acad. Sci. Canada **2** (1980), no. 2, 63–66. (MR 81e:53026; Zbl. 431.53047.)

[CV 81] Chen, B.Y. and Vanhecke, L: *Differential geometry of geodesic spheres*, J. Reine Angew. Math. **325** (1981), 28–67. (MR 82m:53038; Zbl. 503.53013.)

[C 44] Chern, S.S: *Laplace transforms of a class of higher dimensional varieties in a projective space of n dimensions*, Proc. Nat. Acad. Sci. USA **30** (1944), 95–97. (MR **5**, p. 217; Zbl. **63/I**, p. A165.)

[C 47] Chern, S.S: *Sur une classe remarquable de variétés dans l'espace projectif à n dimensions*, Science Reports Tsing Hua Univ. **4** (1947), 328–336. (MR **10**, p. 65.)

[C 78] Chern, S.S: *A summary of my scientific life and works*, In *Selected Papers*, xxi–xxxi, Springer-Verlag, New York-Heidelberg-Berlin, 1978, xxxi+476 pp. (MR 82h:01074; Zbl. 403.07012.)

[C 82] Chern, S.S: *Web geometry*, Bull. Amer. Math. Soc. (N.S.) **6** (1982), no. 1, 1–8. (MR 84g:53024; Zbl. 483.53012.)

[CDK 70] Chern, S.S., do Carmo, M. and Kobayashi, S: *Minimal submanifolds on the sphere with second fundamental form of constant length*, in *Functional Analysis and Related Fields* (Proc. Conf. for M. Stone, Univ. Chicago, Chicago, Ill, 1968), pp. 59–75, Springer-Verlag, New York, 1970. (MR **43** #8424; Zbl. **216**, p. 440.)

[CG 77] Chern, S.S. and Griffiths, P.A: *Linearization of webs of codimension one and maximum rank*, Proc. Intern. Symp. Algebraic Geometry 1977, Kyoto, Japan, 85–91. (MR 81k:53010; Zbl. 406.14003.)

[CG 78a] Chern, S.S. and Griffiths, P.A: *Abel's theorem and webs*, Jahresber. Deutsch. Math.-Verein. **80** (1978), no. 1–2, 13–110. (MR 80b:53008; Zbl. 386.14002.)

[CG 78b] Chern, S.S. and Griffiths, P.A: *An inequality for the rank of a web and webs of maximum rank*, Ann. Scuola Norm. Sup. Pisa Cl. Sci. (4) **5** (1978), no. 3, 539–557. (MR 80b:53009; Zbl. 402.57001.)

[CG 81] Chern, S.S.and Griffiths, P.A: *Corrections and addenda to our paper "Abel's theorem and webs"*, Jahresber. Deutsch. Math.-Verein. **83** (1981), 78–83. (MR 82k:53030; Zbl. 474.14003.)

[CK 52] Chern, S.S. and Kuiper, N.H: *Some theorems on isometric imbeddings of compact Riemannian manifolds in Euclidean space*, Ann. of Math. (2) **56** (1952), 422–430. (MR **14**, p. 408; Zbl. **52**, p. 276.)

[Da 72] Darboux, G: *Leçons sur la théorie générale des surfaces*, 4 vols., 3rd ed., Chelsea Publ. Co., Bronx, NY, 1972. (MR **53** # 79, 80, 81, 82; Zbl. 257.53001.)

[De 67] Degen, W: *Konjugierte Systeme und Kegelshattengrenzen auf Hyperflächen im projektiven n-dimensionalen Raum*, Math. Z. **97** (1967), no. 2, 105–122. (MR **35** #895; Zbl. **143**, p. 445.)

[De 68] Degen, W: *Zur projektiven Kinematik der Hüllflächen gewisser Hyperquadriksharen*, J. Reine Angew. Math. **229** (1968), 209–220. (MR **37** #2106; Zbl. **173**, p. 236.)

[D 64] Dieudonné, J: *Algèbre linéaire et géométrie élémentaire*, Hermann, Paris, 1964, 223 pp. (MR **30** #2015; Zbl. **185**, p. 488); English transl: *Linear algebra and geometry*, Houghton Mufflin Co., Boston, MA, 207 pp. (MR **42** #6004; Zbl. **185**, p. 488)

[D 71] Dieudonné, J: *Elements d'analyse*, vol. 3, Gauthier-Villars, Paris, 1971 (MR **42** #5266); English transl: *Treatise on analysis*, vol. 3, Academic Press, New York/London, 1972, xviii+388 pp. (MR **50** #3261; Zbl. 268.58001.)

[EH 87] Eisenbud, D. and Harris, J: *On varieties of minimal degree (a centennial account)*, Algebraic Geometry, Bowdoin, 1985 (Brunswick, Maine, 1985), 3–13, Proc. Sympos. Pure Math. **46**, Part 1, Amer. Math. Soc., Providence, RI, 1987. (MR 89f:14042; Zbl. 646.14036.)

[En 39] Engel, F: *Die Umkehrung des Reissschen Satzes über Kurven n-ten Ordnung*, Deutsche Math. **4** (1939), 340–347. (Zbl. **21**, p. 256.)

[E 65] Ermakov, Yu. I: *On the invariant framing of some surfaces of a special type in projective space*, (Russian) Dokl. Akad. Nauk SSSR **162** (1965), no. 6, 1234–1237; English transl: Soviet Math. Dokl. **6** (1965), no. 3, 840–843. (MR **31** #6171; Zbl. **133**, p. 148.)

[E 66] Ermakov, Yu. I: *Surfaces of a special form in the affine and projective spaces*, (Russian) Izv. Vyssh. Uchebn. Zaved. Mat. **1966**, no. 3 (52), 60–67. (MR **34** #3455; Zbl. **173**, p. 495.)

[Fer 55] Fernández, G: *Geometria diferencial afin hipersperficies*, Rev. Un. Mat. Argentina **17** (1955), 29–38. (MR **18**, p. 670; Zbl. **72**, p. 166.)

[Fi 37] Finikov, S.P: *Projective differential geometry*, (Russian) ONTI, Moscow/
Leningrad, 1937, 263 pp. (Zbl. **17**, p. 421.)

[Fi 48] Finikov, S.P: *Cartan's method of exterior forms in differential geometry. The theory of compatibility of systems of total and partial differential equations*, (Russian) OGIZ, Moscow/Leningrad, 1948, 432 pp. (MR **11**, p. 597; Zbl. **33**, pp. 60–61.)

[Fi 50] Finikov, S.P: *Theory of congruences*, (Russian) Gostekhizdat, Moscow/Leningrad, 1950, 528 pp. (MR **12**, p. 744.) German transl. by G. Bol, Akademie Verlag, 1959, xv+491 pp. (Zbl. **85**, p. 367.)

[Fi 56] Finikov, S.P: *Theory of pairs of congruences*, Gostekhizdat, Moscow/ Leningrad, 1956, 443 pp. (MR **19**, p. 676; Zbl. **72**, p. 168.) French transl. by M. Decuyper, U.E.R. Mathematiques Pures et Appliquees, No. 68, Université des Sciences et Techniques de Lille I, Villeneuve d'Ascq, Vol. 1 & 2, 1976, xxix+616 pp. (MR **55** #4023a & #4023b; Zbl. 342.53010.)

[Fu 16] Fubini, G: *Applicabilitá proiettiva di due superficie*, Rend. Circ. Mat Palermo **41** (1916), 135–162. (F. **46**, pp. 1098–1099.)

[Fu 18a] Fubini, G: *Studî relativi all' elemento lineare proiettivo di una ipersuperficie*, Atti Accad. Naz. Lincei Rend. Cl. Sci. Fis. Mat. Natur. (5) **27** (1918), 99–106. (F. **46**, p. 1095.)

[Fu 18b] Fubini, G: *Il problema della deformazione proiettivo delle ipersuperficie. Le varietà a un qualsiasi numero di dimensioni*, Atti Accad. Naz. Lincei Rend. Cl. Sci. Fis. Mat. Natur. (5) **27** (1918), 147–155. (F. **46**, pp. 1095–1097.)

[Fu 20] Fubini, G: *Sur les surfaces projectivement applicables*, C.R. Acad. Sc. **171** (1920), 27–29. (F. **47**, p. 656.)

[FČ 26] Fubini, G. and Čech, E: *Geometria proiettiva differenziale*, Zanichelli, Bologna, vol. 1, 1926, 394 pp., vol. 2, 1927, 400 pp. (F. **52**, pp. 751–752.)

[FČ 31] Fubini, G. and Čech, E: *Introduction à la géométrie projective différentielle des surfaces*, Gauthier-Villars & Cie, Paris, 1931, vi+291 pp. (Zbl. **2**, p. 351.)

[Ge 57] Geidelman, R.M: *Multidimensional systems R*, (Russian) Uspekhi Mat. Nauk **12** (1957), no. 3, 285–290. (MR **19**, p. 879; Zbl. **80**, p. 370.)

[Gl 70] Gluzdov, V.A: *Some classes of families of* $(2-2)$*-planes in* P_5, (Russian) Uchen. Zap. Gor'kovsk. Gos. Ped. Inst. Vyp. 123 (1970), 169–172.

[Go 28] Godeaux, L: *Sur la surfaces ayant mêmes quadriques de Lie, I, II, III*, Acad. Roy. Belg. Bull. Cl. Sci. (5) **14**, 158–173, 174–187, 345–348. (F. **54**, p. 735.)

[Go 64] Godeaux, L: *La géométrie différentielle des surfaces considérées dans l'espace réglé*, Acad. Roy. Belg. Cl. Sci. Mém. (2) **34**, no. 6, Brussel, 1964, 84 pp.(MR **29** #5169; Zbl. **123**, p. 149.)

[G 66] Goldberg, V.V: *On a normalization of* p*-conjugate systems of an* n*-dimensional projective space*, (Russian) Trudy Geom. Sem. **1** (1966), 89–109. (MR **35** #900; Zbl. **168**, p. 425.)

[G 69] Goldberg, V.V: *Pairs of* p*-conjugate systems with a common conjugately harmonic normalization*, (Russian) Trudy Geometr. Sem. **2** (1969), 95–118. (MR **41** #936; Zbl. 246.53012.)

[G 70] Goldberg, V.V: *The invariant rigging of a Cartan variety* V_p *in a projective space* P_{2p}, (Russian) Izv. Vyssh. Uchebn. Zaved. Mat. **1970**, no. 12 (103), 11–21. (MR **43** #8013; Zbl. **219**, p. 344.)

[G 75] Goldberg, V.V: *The* $(n+1)$*-web determined by* $n+1$ *surfaces of codimension* $n-1$, (Russian) Problems in Geometry, Vol. 7, 173–195, Akad. Nauk SSSR Vsesoyuz. Inst. Nauchn. i Tekhn. Inform., Moscow, 1975. (MR **57** #17537; Zbl. 548.53013.)

[G 82a] Goldberg, V.V: *The solutions of the Grassmannization and algebraization problems for* $(n+1)$*-webs of multidimensional surfaces*, Tensor (N.S.) **36** (1982), no. 1, 9–21. (MR 87a:53027; Zbl. 479.53014.)

[G 82b] Goldberg, V.V: *Grassmann and algebraic four-webs in a projective space*, Tensor (N.S.) **38** (1982), 179–197. (MR 87e:53024; Zbl. 513.53009.)

[G 88] Goldberg, V.V: *Theory of multicodimensional* $(n+1)$*-webs*, Kluwer Academic Publishers, Dordrecht-Boston-London, 1988, xxii+466 pp. (MR 90h:53021; Zbl. 668.53001.)

[G 90] Goldberg, V.V: *Local differentiable quasigroups and webs*, Chapter X in the book *Quasigroups and Loops: Theory and Applications*, 263–311, Heldermann Verlag Berlin, 1990. (Zbl. 737.53015.)

[GC 72] Goldberg, V.V. and Chakmazyan, A.V: *Certain classes of doubly normalized surfaces of a four-dimensional projective space*, (Russian) Comment. Math. Univ. Carolin. **13** (1972), no. 2, 325–332. (MR **46** #4389; Zbl. 239.53006.)

[Gr 74] Griffiths, P.A: *On Cartan's method of Lie groups and moving frames as applied to uniqueness and existence question in differential geometry*, Duke Math. J. **41** (1974), no. 4, 775–814. (MR **53** #14355; Zbl. 294.53034.)

[Gr 76] Griffiths, P.A: *Variations on a theorem of Abel*, Invent. Math. **35** (1976), 321–390. (MR **55** #8036; Zbl. 339.14003.)

[Gr 77] Griffiths, P.A: *On Abel's differential equations*, Algebraic Geometry, J.J. Sylvester Sympos., Johns Hopkins Univ., Baltimore, Md., 1976, 26–51. Johns Hopkins Univ. Press, Baltimore, Md, 1977. (MR **58** #655; Zbl. 422.14016.)

[Gr 83] Griffiths, P.A: *Exterior differential systems and the calculus of variations*, Birkhäuser, Boston/Basel/Stuttgart, 1983, viii+335 pp. (MR 84h:58007; Zbl. 512.49003.)

[GH 78] Griffiths, P.A. and Harris, J: *Principles of algebraic geometry*, Wiley-Interscience (John Wiley & Sons), New York, 1978, xii+813 pp. (MR 80b:14001; Zbl. 408.14001.) .

[GH 79] Griffiths, P.A. and Harris, J: *Algebraic geometry and local differential geometry*, Ann. Sci. École Norm. Sup. (4) **12** (1979), 355–452. (MR 81k:53004; Zbl. 426.14019.)

[GJ 87] Griffiths, P.A. and Jensen, G.R: *Differential systems and isometric embeddings*, Princeton University Press, Princeton, N. J., 1987, 225 pp. (MR 88k:53041; Zbl. 637.53001.)

[Gro 39] Grove, V.G: *A tensor analysis for a V_k in a projective space S_n*, Bull. Amer. Math. Soc. **45** (1939), 385–398. (Zbl. **21**, p. 259.)

[Ha 61] Havelka, J: *Élément projectif linéaire d'une hypersurface plongée dans un espace à connexion projective*, Czechoslovak. Math. J. **11 (86)** (1961), no. 2, 249–257. (MR **23** #A4082; Zbl. **103**, p. 153.)

[H 33] Hlavatý, V: *Invariants projectifs d'une hypersurface*, Rend. Circ. Mat. Palermo **57** (1933), 402–430. (Zbl. **8**, p. 224.)

[H 37] Hlavatý, V: *Faisceaux de Darboux et questions connexes dans l'espace affine courbe*, Časopis Pěst. Mat. **66** (1937), 229–260. (Zbl **16**, p. 326.)

[H 39] Hlavatý, V: *Hypersurfaces in a projective curved space*, Ann. of Math. (2) **39** (1939), 725–761. (Zbl. **20**, p. 071.)

[H 45] Hlavatý, V: *Differentielle Liniengeometrie*, P. Noordhoff Ltd., Groningen, 1945, xxii+566 pp. (MR **8**, p. 346; Zbl. **63/I**, p. A407.) English transl: *Differential line geometry*, P. Noordhoff,Ltd., Groningen, 1953, x+495 pp. (MR **15**, p. 252; Zbl. **51**, p. 391.)

[H 49] Hlavatý, V: *Affine embedding theory. I. Affine normal spaces*, Nederl. Akad. Wetensh. Proc. (Amsterdam) **52** (1949), 505–517 = Indagationes Math. **11** (1949), 165–177. (MR **11**, p. 54; Zbl. **33**, p. 397.)

[HP 47] Hodge, W.V.D. and Pedoe, D: *Methods of algebraic geometry*, vol. 1, Cambridge Univ. Press, Cambridge & Macmillan Comp., 1947, viii+440 pp. (MR **10**, p. 396; Zbl. **157**, p. 275.)

[Ho 82a] Holst, E: *Ein Beitrag zur methodischen Behandlung der metrischen Eigenschaften algebraischer Kutven*, Archiv for Math. og Naturvid. **7** (1882), 109–114. (F. **14**, p. 600.)

[Ho 82b] Holst, E: *Analytischer Beweis eines geometrischen Satzes*, Archiv for Math. og Naturvid. **7** (1882), 117–178. (F. **14**, pp. 600–601.)

[Ho 82c] Holst, E: *Ein Paar syntetischer Methoden in der metrischen Geometrie mit Anwendungen*, Archiv for Math. og Naturvid. **7** (1882), 240–362. (F. **14**, pp. 599–600.)

[IK 67] Ivanova-Karatopraklieva, I: *Variétés de droites à trois dimensions dans P_4*, Annuaire Univ. Sofia Fac. Math. 1965–1966 **60** (1967), 257–267. (MR **38** #3783; Zbl. **165**, p. 555.)

[IK 68] Ivanova-Karatopraklieva, I: *Quelques researches sur la géométrie des systèmes à deux paramèters de droites dans P_4*, Annuaire Univ. Sofia Fac. Math. **61** (1966/1967), 257–267 (1968). (MR **38** #3783; Zbl. **165**, pp. 244–245.)

[Iv 90] Ivlev, E.T: *On some fields of projectively invariant geometric images on a normalized multidimensional surface*, (Russian) in *Phys. i Matem. Modelirovanie teplovykh i gidrodinam. processov*, pp. 171–176, Tomsk. Polytechn. Inst., Tomsk, 1990.

[IL 67] Ivlev, E.T. and Luchinin, A.A: *On frames of a surface S_n in $P_{n+2} (n \geq 2)$*, (Russian) Izv. Vyssh. Uchebn. Zaved. Mat. **1967**, no. 9 (64), 48–59. (MR **36** #2085; Zbl. **158**, p. 400.)

[Iz 54] Izmailov, V.D:*On a system of pseudotensors on two-dimensional surfaces of affine spaces*, (Russian) Dokl. Akad. Nauk SSSR **94** (1954), no. 1, 9–12. (MR **15**, p. 899; Zbl. **58**, p. 156.)

[Iz 56a] Izmailov, V.D:*The problem of the interior normalization of a hypersurface in a space of affine connection*, (Russian) Dokl. Akad. Nauk SSSR **109** (1956), 906–909. (MR **18**, p. 932; Zbl. **75**, p. 315.)

[Iz 56b] Izmailov, V.D:*On the theory of a hypersurface in a space with an affine connection*, (Russian) Proc. III All-Union Math. Congr. (Moskva, 1956), Vol. I, p. 153, Akad. Nauk SSSR, Moskva, 1956.

[Iz 57a] Izmailov, V.D:*On the geometry of a hypersurface in a space with an affine connection*, (Russian) Sverdlovsk. Gos. Ped. Inst. Uchen. Zap. **13** (1957), 3–22.

[Iz 57b] Izmailov, V.D:*On the affine theory of two-dimensional surfaces*, (Russian) Moskov. Gos. Ped. Inst. Uchen. Zap. **108** (1957), 219–259. (MR **22** #12475; Zbl. **108**, p. 343.)

[Iz 60] Izmailov, V.D:*On the theory of a hypersurface in a space with an affine connection*, (Russian) Uspekhi Mat. Nauk **15** (1960), no. 5, 171–178. (MR **23** #A3528; Zbl. **105**, p. 159.)

[Iz 61] Izmailov, V.D: *Two-dimensional surfaces in spaces with affine connection*, (Russian) Mat. Sb. **54** (1961), no. 3, 311–330. (MR **25** #2546; Zbl. **114**, p. 378.)

[JM 92] Jensen, G.R. and Musso, E: *Rigidity of hypersurfaces in complex projective space*, Preprint, 1992, 21 pp.

[J 89] Jijtchenko, A.B: *On embedded tangential and secant varieties in projective algebraic varieties*, Theta functions-Bowdoin 1987, Part 1 (Brunswick, ME, 1987), 709–718, Proc. Sympos. Pure Math. **49**, Part 1, Amer. Math. Soc., Providence, RI, 1989 (MR 90i:14055; Zbl. 698.14007.)

[Ju 63a] Juza, M: *Le système monoparamétrique des plans dans l'espace* S_6, (Czech) Mat.-Fyz. Časopis Sloven. Akad. Vied. **13** (1963), no. 2, 125–136. (MR **28** #3377; Zbl. **171**, pp. 419-420.)

[Ju 63b] Juza, M: *Le système monoparamétrique des plans dans l'espace* S_6, Comment. Math. Univ. Carolin. **4** (1963), no. 3, 99–101. (MR **29** #5176; Zbl. **129**, p. 139.)

[Ka 25a] Kanitani, J: *On the projective line element upon a hypersurface*, Mem. College Sci. Kyoto Imperial Univ. **8** (1925), 357–381. (F. **51**, pp. 580–581.)

[Ka 25b] Kanitani, J: *On Darboux curves on a hypersurface in the n-dimensional space*, Mem. College Sci. Kyoto Imperial Univ. **9** (1925), 125–151. (F. **51**, p. 580.)

[Ka 26] Kanitani, J: *Absolute differential calculus and its application to projective differential geometry of hypersurfaces in four-dimensional space*, Mem. College Sci. Kyoto Imperial Univ. **9** (1926), 253–283. (F. **52**, p. 762.)

[Ka 27] Kanitani, J: *Sur le rang de la forme de Darboux de l'hypersurface*, Bull. Soc. Math. France **55** (1927), 206–217. (F. **53**, p. 644.)

[Ka 28a] Kanitani, J: *Sur l'hypersurface de trois dimensions dont la forme de Darboux se décompose en trois facteurs linéaires*, Mem. of the Ryojun College of Engineering **1** (1928), 295–306. (F. **57**, p. 944.)

[Ka 28b] Kanitani, J: *Sur la transformation des hypersurfaces conservant les courbes asymptotiques ainsi que les que les courbes de Darboux*, Mem. of the Ryojun College of Engineering **1** (1928), 307-317. (F. **57**, p. 944.)

[Ka 28c] Kanitani, J: *Hypersurface regardeées comme envelopes de leurs hyperquadriques osculatrices*, Mem. of the Ryojun College of Engineering **1** (1928), 319–350. (F. **57**, pp. 944–945.)

[Ka 28d] Kanitani, J: *Une interprétation géométrique de l'élément linéaire projectif de l'hypersurface*, Atti Accad. Naz. Lincei Rend. Cl. Sci. Fis. Mat. Natur. (6) **8** (1928), 208–210. (F. **54**, p. 781.)

[Ka 29a] Kanitani, J:*Sur quelques repères mobiles*, Mem. of the Ryojun College of Engineering **2** (1929), 227–247. (F. **57**, p. 945.)

[Ka 29b] Kanitani, J:*Sur un réseau conjugué*, Mem. of the Ryojun College of Engineering **2** (1929), 249–257. (F. **57**, p. 945.)

[Ka 29c] Kanitani, J:*Sur une forme quadratique intrinsèque par rapport a l'hypersurface dans l'espace projectif à plusieurs dimensions*, Atti Accad. Naz. Lincei Rend. Cl. Sci. Fis. Mat. Natur. (6) **9** (1929), 30–32. (F. **55**, p. 1061.)

[Ka 29d] Kanitani, J:*Sur l'hypersurface dont la forme de Darboux est un cube parfait*, Ann. Fac. Sci. Toulouse (3) (1929), 27–42. (F. **55**, p. 416.)

[Ka 31] Kanitani, J:*Géométrie différentielle projectif des hypersurfaces*, Ryojun College of Engineering, Port Arthur, 1931, 140 pp. (Zbl. **5**, p. 261.)

[Ka 57] Kanitani, J: *Sur la forme de Darboux relative à une variété différentiable*, Mem. Coll. Sci. Univ. Kyoto Ser. A, Math. **30** (1956-1957), no. 3, 203–216. (MR **19**, p. 1193; Zbl. **96**, p. 158.)

[Ka 59] Kanitani, J: *Sur le faisceau canonique d'une variété différentiable*, Mem. Coll. Sci. Univ. Kyoto Ser. A **32** (1959), 153–170. (MR **21** #7529; Zbl. **95**, p. 364.)

[Ka 62] Kanitani, J: *Sur la structure projective d'espace fibré*, Ann. Mat. Pura Appl. (4) **57** (1962), 151–172. (MR **25** #3467; Zbl. **105**, p. 349.)

[Ka 65] Kanitani, J: *Sur la arêtes principales à un space fibr'e*, J. Math. Kyoto Univ. **4** (1965), 487–492. (MR **34** #5020; Zbl. **134**, p. 402.)

[Ka 67] Kanitani, J: *Sur la hypersurfaces à $2r-1$ dimensions admettant des espaces générateurs à $r-1$ dimensions*, J. Math. Kyoto Univ. **7** (1967), 295–313. (MR **37** #2108; Zbl. **183**, p. 499.)

[Kh 74] Khasin, G.B: *Three-parameter families of two-dimensional planes in a four-dimensional projective space*, (Russian) Izv. Vyssh. Uchebn. Zaved. Mat. **1974**, no. 6 (145), 80–90, English transl: Soviet Math. (Iz. VUZ) **18** (1974), no. 9, . (MR **52** #11750; Zbl. 294.53011.)

[Kl 26] Klein, F: *Vorlesungen über höhere Geometrie*, Springer, Berlin, 1926, 405 pp. (F. **30**, p. 624), reprinted by Springer, Berlin/Heidelberg, viii+405 pp. 1968 (MR **37** #2065; Zbl. **159**, p. 222) and by Chelsea Pub. Co., New York, 1949. (Zbl. **41**, p. 081.)

[Kli 52] Klingenberg, W: *Über das Einspannungsproblem in der projektiven und affinen Differentialgeometrie*, Math. Z. **55** (1952), no. 1, 321–345. (MR **14**, p. 206; Zbl **46**, p. 396.)

[KN 63] Kobayashi, S. and Nomizu, K: *Foundations of differential geometry*, 2 vols., Wiley–Interscience, New York/London/Sydney, 1963, xi+329 pp., (MR **27** #2945; Zbl. **119**, p. 375.); Vol. 2, 1969, xv+470 pp. (MR **38** #6501; Zbl. **175**, p. 465.)

[Kor 55a] Korovin, V.I: *Closed Laplace sequences*, (Russian) Dokl. Akad. Nauk SSSR **101** (1955), no. 4, 605–606. (MR **17**, p. 187; Zbl. **67**, p. 142.)

[Kor 55b] Korovin, V.I: *The systems R in a four-dimensional projective space*, (Russian) Dokl. Akad. Nauk SSSR **101** (1955), no. 5, 797–799. (MR **17**, p. 187; Zbl. **64**, p. 161.)

[Ko 63] Kovantsov, N.I: *Theory of complexes*, (Russian) Izdat. Kiev. Univ., Kiev, 1963, 292 pp. (MR **33** #4816; Zbl. **118**, p. 380.)

[Koz 47] Koz′mina, T.L: *The Laplace transformation of three-conjugate systems*, (Russian) Dokl. Akad. Nauk SSSR **55** (1947), no. 3, 183–185. (MR **8**, p. 531; Zbl. **35**, p. 378.)

[Kr 67a] Kruglyakov, L.Z: *A canonical frame of a nonfocal two-parameter family of straight lines in a five-dimensional projective space*, (Russian) Trudy Tomsk. Gos. Univ. Ser. Mat.-Meh. **191** (1967), 36–47. (MR **36** #4453; Zbl. **159**, p. 232.)

[Kr 67b] Kruglyakov, L.Z: *Semi-focal 2-families of straight lines in P_5*, (Russian) Izv. Vyssh. Uchebn. Zaved. Mat. **1967**, no. 11 (66), 35–42. (MR **37** #4727; Zbl. 162, p. 248.)

[Kr 68] Kruglyakov, L.Z: *Pseudofocal 2-families of straight lines in P_5*, Trudy Tomsk. Gos. Univ. Ser. Mat.-Meh. **196** (1968), 70–78. (MR **43** #1061.)

[Kr 69] Kruglyakov, L.Z: *Pseudofocal families of d-planes in P_n*, (Russian) Izv. Vyssh. Uchebn. Zaved. Mat. **1969**, no. 10 (89), 70–77. (MR **41** #938; Zbl. 191, p. 198.)

[Kr 71a] Kruglyakov, L.Z: *On the projective differential geometry of families of multidimensional planes*, (Russian) Dokl. Akad. Nauk SSSR **196** (1971), no. 2, 282–284; English transl: Soviet Math. Dokl. **12** (1971), no. 1, 103–106. (MR **43** #3941; Zbl. 231.53009.)

[Kr 71b] Kruglyakov, L.Z: *Generalized pseudofocal families of d-planes in P_N*, (Russian) Izv. Vyssh. Uchebn. Zaved. Mat. **1971**, no. 11 (114), 78–84. (MR **45** #9255; Zbl. 232.53007.)

[Kr 73] Kruglyakov, L.Z: *Certain classes of nonfocal 2-families of straight lines in a five-dimensional space*, (Russian) Geometry Collection, No. 11, Trudy Tomsk. Gos. Univ. **235** (1973), 36–49. (MR **51** #4069; Zbl. 321.53011.)

[Kr 80] Kruglyakov, L.Z: *Foundations of projective differential geometry of families of multidimensional planes*, (Russian) Izdat. Tomsk. Univ., Tomsk, 1980, 111 pp. (MR 83h:53003; Zbl. 499.53005.)

[KR 73] Kruglyakov, L.Z. and Rybchenko, L.V: *Pseudofocal a-families of straight lines in an $(a+2)$-dimensional projective space*, (Russian) Geometry Collection, No. 13, Trudy Tomsk. Gos. Univ. **246** (1973), 65–86. (MR **51** #4070; Zbl. 481.53008.)

[Ku 37] Kubota, T: *Einige Kennzeichnende Eigenschaft der ebenen algebraischen Kurven und der algebraischen Flächen*, Sci. Rep. Tôhoku Emp. Univ. (1) **26** (1937), 243–247. (Zbl. **17**, p. 278.)

[La 32] Lane, E.P: *Projective differential geometry of curves and surfaces*, University of Chicago Press, Chicago, 1932, 321 pp. (Zbl. **5**, p. 025.)

[La 42] Lane, E.P: *A treatise on projective differential geometry*, University of Chicago Press, Chicago, 1942, ix+466 pp. (MR **4**, p. 114; Zbl. **63/I**, p. A525.)

[Lap 49] Laptev, G.F: *An invariant construction of the projective differential geometry of a hypersurface*, (Russian) Dokl. Akad. Nauk SSSR **65** (1949), no. 2, 121–124. (MR **11**, p. 53; Zbl. **41**, p. 59.)

[Lap 51] Laptev, G.F: *On fields of geometric objects on imbedded manifolds*, (Russian) Dokl. Akad. Nauk SSSR **78** (1951), no. 2, 197–200. (MR **13**, p. 280; Zbl. **44**, p. 358.)

[Lap 53] Laptev, G.F: *Differential geometry of imbedded manifolds. Group-theoretic method of differential geometry investigations*, (Russian) Trudy Moskov. Mat. Obshch. **2** (1953), 275–382. (MR **15**, p. 254; Zbl. **53**, p. 428.)

[Lap 58] Laptev, G.F: *A hypersurface in a space with an projective connection*, (Russian) Dokl. Akad. Nauk SSSR **121** (1958), no. 1, 41–44. (MR **20** #6137; Zbl. **85**, p. 164.)

[Lap 59] Laptev, G.F: *On invariant normalization of a surface with an affine connection*, (Russian) Dokl. Akad. Nauk SSSR **126** (1959), no. 3, 490–493. (MR **22** #954; Zbl. **87**, p. 141.)

[Lap 64] Laptev, G.F: *Manifolds imbedded in generalized spaces*, (Russian) Proc. Fourth All-Union Math. Congr. (Leningrad, 1961), Vol. II, pp. 226–233, Izdat. "Nauka", Leningrad, 1964. (MR **36** #4473; Zbl. 178, pp. 247–248.)

[Lap 65] Laptev, G.F: *Differential geometry of multidimensional surfaces*, (Russian) Itogi Nauki, Geometriya, 5–64, Akad. Nauk SSSR Inst. Nauchn. Informatsii, Moscow, 1965. (MR **33** #4817.)

[Lap 66] Laptev, G.F: *Fundamental differential structures of higher orders on a smooth manifold*, (Russian) Trudy Geom. Sem. Inst. Nauchn. Informatsii, Akad. Nauk SSSR **1** (1966), 139–190. (MR **34** #6681; Zbl. **171**, p. 423.)

[Lap 69] Laptev, G.F: *The structure equations of the principal fibre bundle*, (Russian) Trudy Geom. Sem. Inst. Nauchn. Informatsii, Akad. Nauk SSSR **2** (1969), 161–178. (MR **40** #8074; Zbl. 253.53032.)

[LO 71] Laptev, G.F. and Ostianu, N.M: *Distributions of m-dimensional linear elements in a space with projective connection I*, (Russian) Trudy Geometr. Sem. **3** (1971), 49–94. (MR **46** #9884.)

[LZ 63] Laptev, G.F. and Zaits, A: *On a normalization of a surface in a space with an affine connection*, (Russian) Litovsk. Mat. Sb. **3** (1963), no. 2, 212.

[Lau 57] Laugwitz, D: *Zur Differentialgeometrie der Hyperflächen in Vectorräumen und zur affingeometrischen Deutung der Theorie der Finsler Räume*, Math. Z. **67** (1957), no. 1, 63–74. (MR **18**, p. 927; Zbl. **77**, p. 160.)

[Lau 59] Laugwitz, D: *Beiträge zur affinen Flächentheorie mit Anwendungen auf die allgemeinmetrische Differentialgeometrie*, Bayer. Akad. Wiss. Math.-Natur. Kl. Abh. (N. F.) **93** (1959), 59 pp. (MR **21** #5977; Zbl. **100**, pp. 174–175.)

[Lib 52] Liber, A.E: *On the theory of surfaces in a centroaffine (vector) space*, (Russian) Dokl. Akad. Nauk SSSR **85** (1952), no. 1, 37–40. (MR **14**, p. 319; Zbl. **49**, p. 332.)

[Lib 53] Liber, A.E: *On the theory of surfaces in a projective space*, (Russian) Dokl. Akad. Nauk SSSR **90** (1953), no. 2, 137–140. (MR **15**, p. 253; Zbl. **51**, p. 389.)

[Lib 55] Liber, A.E: *On the geometry of surfaces in affine spaces*, (Russian) Scientific Yearbook, Saratov. Gos. Univ., Saratov, 1955, 669–671.

[Lib 56] Liber, A.E: *On the geometry of m-surfaces in affine and projective spaces*, (Russian) Proc. III All-Union Math. Congr. (Moskva, 1956), Vol. I, pp. 157–158, Akad. Nauk SSSR, Moskva, 1956.

[Lib 64] Liber, A.E: *On the theory of two-dimensional surfaces in affine and projective n-spaces*, (Russian) III Sibirsk. Konfer. Mat. & Mekh. Dokl., Tomsk. Univ., Tomsk, 1964, pp. 194–195.

[Lib 66] Liber, A.E: *On the theory of surfaces in affine and centroaffine n-spaces*, (Russian) Trudy Sem. Vektor. Tenzor. Anal. **13** (1966), 407–446. (MR **34** &6664; Zbl. **162**, p. 430.)

[Lich 55] Lichnerowicz, A: *Théorie globale des connexions et des groupes d'holonomie*, Edizioni Cremonese, Roma, 1955. (MR **19**, p. 453; Zbl. **116**, p. 391.) English translation: *Global theory of connections and holonomy groups*, Noordhof Intern. Pub., 1976. (Zbl. 337.53031.)

[Lie 82] Lie, S: *Bestimmung aller Flächen, die in mehrfacher Weise durch Translationsbewegung einer Kurve erzeugt werden*, Archiv for Math., **7** (1882), 155–176 (F. **14**, pp. 642–643); see also *Gesammelte Abhandlungen*, **I**, part 1, Abt. I, B.G. Teubner, Leipzig & Ascheoug & Co., Oslo, 1937, 450–467. (Zbl. **9**, p. 318.)

[Lie 96] Lie, S: *Die Theorie der Translationsflächen und das Abel'sche Theorem*, Leipziger Berichte, **48** (1896), Hefte II, III, 141–198 (F. **27**, pp. 326–327); see also *Gesammelte Abhandlungen*, **II**, part 2, Abt. XII, B.G. Teubner, Leipzig & Ascheoug & Co., Oslo, 1937, 526–579. (Zbl. **17**, p. 180.)

[LS 96] Lie, S. and Scheffers, G: *Geometrie der Berührungstransformationen, I*, B. G. Teubner, Leipzig, 1896, vol. 1, xi+694 pp. (F. **27**, pp. 547–556); 2nd corrected ed., Chelsea Pub. Co., Bronx, N.Y., 1977, xii+694 pp. (MR **57**, #45; Zbl. 406.01015.)

[Lieb 27] Liebmann, H: *Bestimmung der geradlinigen Dreiecksnetze aus der Krümmungselementen der Hüllkurven*, Münch. Ber. **1927**, 73–87. (F. **53**, p. 654.)

[LP 71] Little, J.A. and Pohl, W.F: *On tight immersions of maximal codimension*, Invent. Math. **11** (1971), 179–204. (MR **45** #2722; Zbl. **217**, p. 191.)

[Lit 83] Little, J.B: *Translation manifolds and the converse of Abel's theorem*, Compositio Math. **49** (1983), no. 2, 147–171. (MR 85d:14041; Zbl. 521.14017.)

[Lit 84] Little, J.B: *On the converse of Abel's theorem in characteristic p*, Manuscripta Math. **46** (1984), no. 1–3, 27–63. (MR 85g:14056; Zbl. 541.14037.)

[Lit 85] Little, J.B: *On analogs of the Reiss relation for curves on rational ruled surfaces*, Duke Math. J. **52** (1985), no. 4, 909–922. (MR 87e:14034; Zbl. 618.53001.)

[Lit 87] Little, J.B: *On Lie's approach to the study of translation manifolds*, J. Differential Geom. **26** (1987), no. 2, 253–272. (MR 90m:14024; Zbl. 593.14038 & 617.14028.)

[Lit 89] Little, J.B: *Translation manifolds and the Shottke problem*, Theta functions—Bowdoin, 1987, Part 1 (Brunswick, ME, 1987), 517–529, Proc. Sympos. Pure Math. **49** (1989), Part I, Amer. Math. Soc., Providence, RI. (MR 90m:14024; Zbl. 737.14010.)

[Lo 50a] Lopshits, A.M: *On the theory of a hypersurface in $(n+1)$-dimensional equiaffine space*, (Russian) Trudy Sem. Vektor. Tenzor. Anal. **8** (1940), 273–285. (MR **12**, pp. 636–637; Zbl. **41**, p. 506.)

[Lo 50b] Lopshits, A.M: *On the theory of a surface of n dimensions in an equicentroaffine space of $(n+2)$ dimensions*, (Russian) Trudy Sem. Vektor. Tenzor. Anal. **8** (1940), 286–295. (MR **12**, p. 637; Zbl. **41**, p. 506.)

[Lu 59] Lumiste, Ü.G: *n-dimensional surfaces with asymptotic fields of p-directions*, (Russian) Izv. Vyssh. Uchebn. Zaved. Mat. **1959**, no. 1 (8), 105–113. (MR **25** #3451; Zbl. **136**, p. 172.)

[Lu 65] Lumiste, Ü.G: *Induced connections in imbedded affine and projective bundles*, (Russian) Tartu Riikl. Ül. Toimitised Vih. 177, Trudy Mat. i Meh. (1965), 6–42. (MR **35** #2238; Zbl. **145**, p. 424.)

[Lu 67] Lumiste, Ü.G: *Stratifiable families of 1-pairs of a four-dimensional projective space*, (Russian) Tartu Riikl. Ül. Toimitised Vih. 206, Trudy Mat. i Meh. (1967), 10–21. (MR **41** #4402; Zbl. **165**, p. 244.)

[Lu 67] Lumiste, Ü.G: *Stratifiable families of 1-pairs of a four-dimensional projective space*, (Russian) Tartu Riikl. Ül. Toimitised Vih. 206, Trudy Mat. i Meh. (1967), 10–21. (MR **41** #4402; Zbl. **165**, p. 244.)

[Lu 75] Lumiste, Ü.G: *Differential geometry of submanifolds*, (Russian) Itogi Nauki i Tekhniki, Algebra, Topologiya, Geometriya, vol. 13, 273–340, Akad. Nauk SSSR Inst. Nauchn. Informatsii, Moscow, 1975; English. transl: J. Soviet Math. **7** (1977), no. 4, 654–677. (MR **55** #4008; Zbl. 421.53036.)

[LC 74] Lumiste, Ü.G. and Chakmazyan, A.V: *Submanifolds with a parallel normal vector field*, (Russian) Izv. Vyssh. Uchebn. Zaved. Mat. **1974**, no. 5 (144), 148–157. English translation: Soviet Math. (Iz. VUZ) **18** (1974), no. 5, 123–131. (MR **51** #13944; Zbl. 291.53028.)

[LC 81] Lumiste, Ü.G. and Chakmazyan, A.V: *The normal connection and submanifolds with parallel normal vector fields in a space of constant curvature*, (Russian) Problems in Geometry, Vol. **12**, 3–30. Akad. Nauk SSSR Vsesoyuz. Inst. Nauchn. i Tekhn. Informatsii, Moscow, 1981; English transl: J. Soviet Math. **21** (1983), no. 2, 107–127. (MR 83a:53052; Zbl. 479.53042 & 502.53041.)

[Mi 58] Mihăilescu, T: *Geometrie differenţiela proiectivă*, (Roumanian) Biblioteca Matematică, Vol. II, Editura Acad. R.P.R., Bucharest, vol. 1, 1958, 494 pp. (MR **30** #579; Zbl. **82**, p. 370.); vol. 2, 1963, 232 pp. (MR **22** #2937; Zbl. **119**, p. 167.)

[M 51a] Muracchini, I: *Le varietà V_5 i cui spazi tangenti reciprocono una varietà W di dimensione inferiore alla ordinaria*, Boll. Un. Mat. Ital. (3) **6** (1951), 97–103. (MR **13**, p. 272; Zbl. **43**, p. 370.)

[M 51b] Muracchini, I: *Ricerche sulle varietà quasi-asintotiche. I. Quasi asintotiche $\sigma_{1,2}$*, Boll. Un. Mat. Ital. (3) **6** (1951), 198–205. (MR **14**, p. 205; Zbl. **45**, p. 250.)

[M 51c] Muracchini, I: *Ricerche sulle varietà quasi-asintotiche. II. Quasi asintotiche $\sigma_{r,s}$ di specie maxima*, Boll. Un. Mat. Ital. (3) **6** (1951), 299–304. (MR **14**, p. 205; Zbl. **45**, p. 251.)

[M 51d] Muracchini, I: *La varietà V_5 i cui spaci tangenti ricoprono una varietà W de dimensione inferiore alla ordinaria. I*, Rivista Mat. Univ. Parma **2** (1951), 435–462. (MR **14**, p. 79; Zbl. **54**, p. 68.)

[M 52] Muracchini, I: *La varietà V_5 i cui spaci tangenti ricoprono una varietà W de dimensione inferiore alla ordinaria. II*, Rivista Mat. Univ. Parma **3** (1952), 75–89. (MR **14**, p. 316; Zbl. **48**, p. 396.)

[M 53a] Muracchini, I: *Sulle varietà V_3 analitiche pluririgate*, Boll. Un. Mat. Ital. (3) **8** (1953), no. 2, 138–144. (MR **15**, p. 154; Zbl. **51**, p. 390.)

[M 53b] Muracchini, I: *Sulle varietà del Veronese*, Atti del Quatro Congresso dell' Unione Matematica Italiana, Taormina, 1951, vol. II, Casa Editrice Perrella, Roma, 1953. (MR **15**, p. 59; Zbl. **56**, p. 155.)

[M 58] Muracchini, I: *Sur les varietà più volte striate, ed alcune caratterizzazione delle varietà di Segre*, Rend. Mat. et Appl. (5) **17** (1958), 15–34. (MR **20** #4556; Zbl. **94**, p. 167.)

[Nom 76] Nomizu, K: *A characterization of Veronese varieties*, Nagoya Math. J. **60** (1976), 181–188. (MR **52** #15325; Zbl. 305.53046 & 312.53044.)

[NY 74] Nomizu, K. and Yano, K: *On circles and spheres in Riemannian geometry*, Math. Ann. **210** (1974), 163–170. (MR **50** # 1171; Zbl. 273.53039 & 282.53037.)

[N 35] Norden, A.P: *Die relative Geometrie der Flächen im projektiven Raume*, Trudy Sem. Vektor. Tenzor. Anal. **2/3** (1935), 229–268. (Zbl. **12**, p. 225.)

[N 37] Norden, A.P: *Über Paare konjugierter Parallelübertragungen*, Trudy Sem. Vektor. Tenzor. Anal. **4** (1937), 205–252. (Zbl. **17**, p. 227.)

[N 45a] Norden, A.P: *The affine connection on surfaces of a projective and conformal space*, (Russian) Dokl. Akad. Nauk SSSR **48** (1945), 539–541. (MR **8**, p. 193; Zbl. **60**, p. 392.)

[N 45b] Norden, A.P: *On pairs of conjugate parallel translations in n-dimensional spaces*, (Russian) Dokl. Akad. Nauk SSSR **49** (1945), no. 9, 625–628. (MR **8**, p. 93; Zbl. **60**, p. 392.)

[N 47] Norden, A.P: *La connexion affine sur les surfaces de l'espace projectif*, Mat. Sb. **20** (60) (1947), 263–281. (MR **9**, p. 67; Zbl. **41**, p. 306.)

[N 48] Norden, A.P: *Riemannische Metrik auf Flächen des projektiven Raumes*, (Russian) Dokl. Akad. Nauk SSSR **60** (1948), 345–347. (MR **10**, p. 478; Zbl **35**, p. 377.)

[N 76] Norden, A.P: *Spaces with affine connection*, (Russian) Second edition, Izdat. "Nauka", Moscow, 1976, 432 pp. (MR **57** #7421.)

[NC 74] Norden, A.P. and Chebysheva, B.P: *Inner geometry of a hypersurface in a space with a degenerate metric*, (Russian) Izv. Vyssh. Uchebn. Zaved. Mat. **1974**, no. 10 (149), 57–60. English translation: Soviet Math. (Iz. VUZ) **18** (1974), no. 10, 47–49. (MR **51** # 8970; Zbl. 296.53008.)

[Os 64] Ostianu, N.M: *On an invariant normalization of a surface in a projective space*, (Russian) II All-Union Geom. Confer. (Kharkov, 1964) Tezisy Dokl., Kharkov. Univ, Kharkov, 1964, p. 203.

[Os 66] Ostianu, N.M: *Geometry of a multidimensional surface in a projective space*, (Russian) Trudy Geometr. Sem. **1** (1966), 239–263. (MR **34** #6668; Zbl. **171**, p. 420.)

[Os 69] Ostianu, N.M: *An invariant normalization of a family of multidimensional planes in a projective space*, (Russian) Trudy Geometr. Sem. **2** (1969), 247–262. (MR **40** #6410; Zbl. 248.53006.)

[Os 70] Ostianu, N.M: *An invariant normalization of a surface carrying a net*, (Russian) Izv. Vyssh. Uchebn. Zaved. Mat. **1970**, no. 7 (98), 72–82. (MR **44** #933; Zbl. **201**, p. 4.)

[Os 71] Ostianu, N.M: *Distributions of m-dimensional linear elements in a space with projective connection II*, (Russian) Trudy Geometr. Sem. **3** (1971), 95–114. (MR **46** #9885.)

[Os 73] Ostianu, N.M: *Distribution of hyperplanar elements in a projective space*, (Russian) Trudy Geometr. Sem. **4** (1973), 71–120. (MR **54** #11191; Zbl. 308.53002.)

[OB 78] Ostianu, N.M. and Balazyuk, T.N: *Manifolds imbedded in a space with a projective connection*, (Russian) Trudy Geometr. Sem. **10** (1978), 75–115; English transl: J. Soviet Math. **14** (1980), no. 3, 1217–1241. (MR 80g:53014; Zbl. 402.53018.)

[P 88] Polovtseva, M.A: *Projective differential geometry of three-dimensional manifolds*, (Russian) Ph. D. Thesis, Moskov. Gos. Ped. Inst., 1988.

[P 90] Polovtseva, M.A: *Three-dimensional manifolds of a space P^N having a six-dimensional first osculating subspace*, (Russian) in *Qualitative theory of geometry and analysis* , Khabarovsk. Gos. Ped. Inst., Khabarovsk, 1990, 69–78.

[P 91a] Polovtseva, M.A: *Three-dimensional manifolds of a projective space having a six-dimensional osculating subspace and a family of two-dimensional asymptotic directions*, (Russian) Khabarovsk. Gos. Ped. Inst., Khabarovsk, 1991, 12 pp., bibl. 6 titles. Dep. in VINITI on 10/17/91 under no. 4005-B91.

[P 91b] Polovtseva, M.A: *Three-dimensional manifolds of a projective space having a six-dimensional osculating subspace and two families of two-dimensional asymptotic directions*, (Russian) Khabarovsk. Gos. Ped. Inst., Khabarovsk, 1991, 13 pp., bibl. 6 titles. Dep. in VINITI on 10/17/91 under no. 4006-B91.

[Po 70a] Popov, Yu. I: *On the theory of a framed hyperband in a multidimensional projective space*, (Russian) Moskov. Gos. Ped. Inst. Uchen. Zap. **374** (1970), no. 1 Voprosy Differentsial. Geom., 102–117. (MR **58** #30812.)

[Po 70b] Popov, Yu.I: *Hyperbands with common normalization in a multidimensional projective space*, (Russian) Moskov. Gos. Ped. Inst. Uchen. Zap. **374** (1970), no. 1 Voprosy Differentsial. Geom., 118–129. (MR **58** #30813.)

[Po 71] Popov, Yu.I: *Introduction of an invariant normalization on a regular hyperband in a multidimensional projective space*, (Russian) Moskov. Gos. Ped. Inst. by Correspondence, Uchen. Zap. **30** (1971), 286–296.

[Po 73] Popov, Yu.I: *The theory of normalized regular hyperbands with an associated connection in a multidimensional projective space*, (Russian) Kaliningrad. Gos. Univ. Differentsial'naya Geom. Mnogoobraz. Figur, Vyp. 3 (1973), 81–96. (MR **57** #13730.)

[Po 83] Popov, Yu.I: *General theory of regular hyperbands*, (Russian) Kaliningrad. Gos. Univ., Kaliningrad, 1983, 83 pp. (MR 89c:53009; Zbl. 653.53008.)

[Pya 53a] Pyasetsky S.A: *On the differential geometry of a hypersurface in an affine complex space*, (Russian) Moskov. Gos. Ped. Inst. Uchen. Zap. **71** (1953), no. 1, 99–126. (MR **18**, p. 598.)

[Pya 53b] Pyasetsky S.A: *On the foundations of the differential geometry of a hypersurface in a centroaffine space*, (Russian) Moskov. Gos. Ped. Inst. Uchen. Zap. **71** (1953), no. 1, 127–153. (MR **18**, p. 598.)

[Pya 60] Pyasetsky S.A: *An invariant normalization of an m-surface in a complex centroaffine space* E_{m+2}, (Russian) Mordovsk. Univ. Uchen. Zap. **8** (1960), 79–86.

[Pya 61] Pyasetsky S.A: *An analytic construction of the second centroaffine normal of an m-surface in* E_{m+2}, (Russian) Mordovsk. Univ. Uchen. Zap. **18** (1961), 171–174.

[Re 37] Reiss, M: *Mémoire sur les proprétés générales des courbes argebriques*, Correspondance Mathématique et Physique de Quetelet **9** (1837), 249–308.

[Ro 66] Rosenfeld, B.A: *Multidimensional spaces*, (Russian) Nauka, Moskva, 1966, 647 pp. (MR **34** #6594; Zbl. **143**, p. 438.)

[Ru 59] Ruscior, Ş: *Variétés doublement réglés dans* S_4, Bul. Inst. Politehn. Iaşi Sect. 1 **5 (9)** (1959), no. 1–2, 13–18. (Zbl. **132**, p. 154.)

[Ru 63] Ruscior, Ş: *On the characteristic hypersurface of a ruled hypersurface in* S_4, Bul. Inst. Poltehn. Iaşi Sect. 1 **9 (13)** (1963), no. 3–4, 59–66. (MR **31** #2664; Zbl. **134**, p. 389.)

[Ry 56] Ryzhkov, V.V: *Conjugate systems on multidimensional surfaces*, Uspekhi Mat. Nauk **11** (1956), no. 4 (70), 180–181. (Zbl. **71**, p. 146)

[Ry 58] Ryzhkov, V.V: *Conjugate nets on multidimensional surfaces*, (Russian) Trudy Moskov. Mat. Obshch. **7** (1958), 179–226. (MR **21** #5978; Zbl. **192**, p. 224.)

[Ry 60] Ryzhkov, V.V: *Tangential degenerate surfaces*, (Russian) Dokl. Akad. Nauk SSSR **135** (1960), no. 1, 20–22. English transl: Soviet Math. Dokl. **1** (1960), no. 1, 1233–1236. (MR **23** #A2143; Zbl. **196**, p. 538.)

[Sac 60] Sacksteder, R: *On hypersurfaces with no negative sectional curvature*, Amer. J. Math. **82** (1960), no. 3, 609–630. (MR **22** #7087; Zbl. **194**, p. 227.)

[Sa 70] Safaryan, L.P: *Certain classes of manifolds of cones of the second order in P_n*, (Russian) Akad. Nauk Armyan. SSR Dokl. **50** (1970), no. 2, 83–89. (MR **44** #3221; Zbl. **219**, p. 333.)

[Sas 88] Sasaki, T: *On the projective geometry of hypersurfaces*, Équations différentielles dans le champ complexe, Vol. III (Strasbourg, 1985), 115–161, Publ. Inst. Rech. Math. Av., Univ. Louis Pasteur, Strasbourg, 1988. (MR 93d:530212.)

[Sas 91] Sasaki, T: *On the Veronese embedding and related system of differential equation*, Global Differential Geometry and Global Analysis (Berlin, 1990), 210–297. Lecture Notes in Math. **1481**, Springer-Verlag, Berlin et al, 1991. (Zbl. 741.53039.)

[Sav 57] Savelyev, S.I: *Surfaces with plane generators along which the tangent plane is constant*, (Russian) Dokl. Akad. Nauk SSSR **115** (1957), no. 4, 663–665. (MR **20** #2004; Zbl. **85**, p. 368.)

[Sav 60] Savelyev, S.I: *On surfaces with plane generators along which the tangent plane is constant*, (Russian) Izv. Vyssh. Uchebn. Zaved. Mat. **1960**, no. 2 (15), 154–167. (MR **25** #4436; Zbl. **99**, p. 170.)

[Sche 04] Scheffers, G: *Das Abel'sche Theorem und das Lie'sche Theorem über Translationsflächen*, Acta Math. **28** (1904), 65–92. (F. **35**, p. 426.)

[S 54] Schouten, J.A: *Ricci calculus. An introduction to tensor analysis and its geometrical applications*, 2nd ed., Springer-Verlag, Berlin-Gottingen-Heidelberg, 1954, xx+516 pp. (MR **16**, p. 521; Zbl. **57**, p. 378)

[SH 36] Schouten, J.A. and Haantejes, J: *Zur allgemeinen projektiven Differentialgeometrie*, Compositio Math. **3** (1936), 1–51. (Zbl. **13**, p. 226.)

[SSt 38] Schouten, J.A. and Struik, D.J: *Einführung in die neuren Methoden der Differentialgeometrie*, 2. Aufl., 2. Band, Geometrie von D.J. Struik, P. Noordhoff N.V., Groningen/Batavia, 1938, xii+338 pp. (Zbl. **19**, p. 183.)

[SegB 47] Segre, B: *Sui teoremi di Bézout, Jacobi e Reiss*, Ann. Mat. Pura Appl. (4) **26** (1947), 1–26. (MR **10**, p. 397; Zbl. **30**, p. 175.)

[SegB 71] Segre, B: *Some properties of differentiable varieties and transformations: with special reference to the analytic and algebraic cases*, 2nd ed., Springer-Verlag, New York/Heidelberg/Berlin, 1971, ix+195 pp. (MR **43** #3453; Zbl. **214**, pp. 481–482); Springer-Verlag, Berlin/Göttingen/Heidelberg, 1957, viii+183 pp. (MR **16**, p. 679; Zbl. **81**, p. 374.)

[SegC 07] Segre, C: *Su una classes di superficie degli iperspazi legate colle equazioni lineari alle derivate parziali di* 2° *ordine*, Atti Accad. Sci. Torino Cl. Sci. Fis. Mat. Natur. **42** (1907), 1047–1049. (F. **38**, pp. 671–673.)

[SegC 07] Segre, C: *Su una classes di superficie degli iperspazi legate colle equazioni lineari alle derivate parziali di 2° ordine*, Atti Accad. Sci. Torino Cl. Sci. Fis. Mat. Natur. **42** (1907), 1047–1049. (F. **38**, pp. 671–673.)

[SegC 21a] Segre, C: *Le linee principali di una superficie di S_5 una proprietà caratteristica della superficie di Veronese. I, II*, Atti Accad. Naz. Lincei Rend. Cl. Sci. Fis. Mat. Natur. (5) **30** (1921), 200–203, 217–231. (F. **48**, p. 764.)

[SegC 21b] Segre, C: *Le superficie degli iperspazi con una doppia infinità di curve piane o spazial*, Atti Accad. Sci. Torino Cl. Sci. Fis. Mat. Natur. **56** (1921), 143–157. (F. **48**, pp. 764–765.)

[SegC 22] Segre, C: *Le superficie degli iperspazi con una doppia infinità di curve piane o spazial*, Atti Accad. Sci. Torino Cl. Sci. Fis. Mat. Natur. **57** (1922), 575–585. (F. **48**, pp. 764–765.)

[SR 85] Semple, J.G. and Roth, L: *Introduction to algebraic geometry*, Oxford: Clarendon Press, New York, 1985, xvii+454 pp. (MR 86m:14001; Zbl. 576.14001.)

[Sev 01] Severi, F: *Intorno ai punti doppi impropri di una superficie generale dello spazio a 4 dimensioni e a' suoi punti tripli apparenti*, Rend. Circ. Mat. Palermo **15** (1901), 33–51. (F. **32**, pp. 648–649.)

[SS 59] Shirokov, P.A. and Shirokov, A.P: *Affine differential geometry*, Gosudarstv. Izdat. Fiz.-Mat. Lit., Moscow, 1959, 319 pp. (MR **21** #7516; Zbl. **85**, p. 367.) German transl: *Affine Differentialgeometrie*, B.G. Teubner Verlagsgesellschaft, Leipzig, 1962, xi + 275 pp. (MR **27**, #660; Zbl. **106**, p. 147.)

[Shc 60] Shcherbakov, R. N: *Course of affine and projective differential geometry*, (Russian) Izdat. Tomsk. Univ, Tomsk, 1960, 194 pp. (MR **23** #A4060.)

[ShK 73] Shcherbakov, R. N. and Kruglyakov, L.Z: *On the geometry of a k-pseudofocal family of d-planes in a projective space*, (Russian) Geometry Collection, No. 13, Trudy Tomsk. Gos. Univ. **246** (1973), 43–51. (MR **50** #14254; Zbl. 481.53007.)

[Shu 64] Shulikovskii, V.I: *Projective theory of nets*, (Russian) Izdat. Kazan. Univ., Kazan', 1964, 78 pp.

[Shv 55] Shveykin, P.I: *On affinely invariant constructions on a surface*, (Russian) Uspekhi Mat. Nauk **10** (1955), no. 3 (65), 181–183. (Zbl. **64**, p. 160.)

[Shv 56] Shveykin, P.I: *On affinely invariant normalization of a surface*, (Russian) Proc. III All-Union Math. Congr. (Moskva, 1956), Vol. I, p. 175, Akad. Nauk SSSR, Moskva, 1956.

[Shv 58] Shveykin, P.I: *Invariant constructions on an m-dimensional surface in an n-dimensional affine space*, (Russian) Dokl. Akad. Nauk SSSR **121** (1958), no. 5, 811–814. (MR **20** #6712; Zbl. **85**, p. 165.)

[Shv 66] Shveykin, P.I: *Normal geometric objects of a surface in an affine space*, (Russian) Trudy Geometr. Sem **1** (1966), 331–423. (MR **35** #4836; Zbl. **185**, p. 249.)

[Si 30] Sisam, C.H: *On varieties of three dimensions with six right lines through each point*, Am. J. Math. **52** (1930), 607–610. (F. **56**, p. 574.)

[Sm 49] Smirnov, R.V: *p-conjugate systems*, Uspekhi Mat. Nauk **4** (1949), no. 4 (32), 162–163. (MR **11**, p. 397.)

[Sm 50] Smirnov, R.V: *Laplace transformations of p-conjugate systems*, Dokl. Akad. Nauk SSSR **71** (1950), no. 3, 437–439. (MR **11**, p. 616; Zbl. **40**, p. 375.)

[St 64] Sternberg, S: *Lectures on differential geometry*, Prentice-Hall, Englewood Cliffs, N.J., 1964 xii+390 pp. (MR **33** #1797); 2nd edition, Chelsea Publishing Co., New York, 1983, xi+442 pp. (MR 88f:58001; Zbl. 518.53001.)

[Sti 20] Stipa, I: *Le superficie proiettivamente applicabili*, Atti Accad. Naz. Lincei Rend. Cl. Sci. Fis. Mat. Natur. (5) **29** (1920), 127–129. (F. **47**, pp. 657–658.)

[Sto 75a] Stolyarov, A.V: *On the fundamental objects of regular hyperbands*, (Russian) Izv. Vyssh. Uchebn. Zaved. Mat. **1975**, no. 10 (161), 97–99; English transl: Soviet Math. (Iz. VUZ) **19** (1975), no. 10, 89–92. (MR **54** #8492; Zbl. 321.53016.)

[Sto 75b] Stolyarov, A.V: *Conditions for a regular hyperband to be quadratic*, (Russian) Izv. Vyssh. Uchebn. Zaved. Mat. **1975**, no. 11 (162), 106–108; English transl: Soviet Math. (Iz. VUZ) **19** (1975), no. 11, 90–92. (MR **53** #14325; Zbl. 324.53004.)

[Sto 75c] Stolyarov, A.V: *The projective differential geometry of a regular hyperband distribution of m-dimensional line elements*, (Russian) Problems in Geometry, Vol. **7**, 117–151. Akad. Nauk SSSR Vsesoyuz. Inst. Nauchn. i Tekhn. Inform., Moscow, 1975. (MR **58** #12759; Zbl. 551.530137.)

[Sto 76] Stolyarov, A.V: *Application of the theory of regular hyperbands to the study of the geometry of multidimensional surfaces of a projective space*, (Russian) Izv. Vyssh. Uchebn. Zaved. Mat. **1976**, no. 2 (165), 111–113; English transl: Soviet Math. (Iz. VUZ) **20** (1976), no. 2, 104–107. (MR **58** #18193; Zbl. 321.53008.)

[Sto 77] Stolyarov, A.V: *Dual geometry of nets on a regular hyperband*, (Russian) Izv. Vyssh. Uchebn. Zaved. Mat. **1977**, no. 8 (183), 68–78; English transl: Soviet Math. (Iz. VUZ) **21** (1977), no. 8, 51–58. (MR **57** #13733; Zbl. 383.53003.)

[Sto 78a] Stolyarov, A.V: *Differential geometry of hyperbands*, (Russian) Problems in Geometry, Vol. **10**, 25–54. Akad. Nauk SSSR Vsesoyuz. Inst. Nauchn. i Tekhn. Inform., Moscow, 1978; English transl: J. Soviet Math. **14** (1980), no. 3, 1187–1204. (MR 80e:53021; Zbl. 404.53009 & 443.53013.)

[Sto 78b] Stolyarov, A.V: *A condition for a regular hyperband to be quadratic*, (Russian) In *Differential Geometry of Manifolds of Figures* **9** (1978), 93–101. (MR 80j:53018; Zbl. 324.53004.)

[Sto 82] Stolyarov, A.V: *The dual theory of a hyperband distribution and its applications*, (Russian) In *Differential Geometry of Manifolds of Figures* **13** (1982), 95–102. (MR 85c:53023; Zbl. 528.53016.)

[Sto 92a] Stolyarov, A.V: *Dual theory of regular hyperbands and its applications*, (Russian) Chuvashsk. Gos. Univ., Cheboksary, 1992, 108 pp.

[Sto 92b] Stolyarov, A.V: *Dual theory of normalized manifolds*, (Russian) Chuvashsk. Gos. Univ., Cheboksary, 1992, 290 pp.

[Su 43] Su Buchin: *Moutard-Čech hyperquadrics associated with a point of a hypersurface*, Ann. of Math. (2) **44** (1943), 7–20. (MR **4**, p. 171; Zbl **60**, p. 731.)

[Su 73] Su Buchin: *The general projective theory of curves*, Science Press, Peking, 1973, iv+242 pp. (MR 86i:53010.)

[Su 83] Su Buchin: *Affine differential geometry*, Science Press, Bejing, Gordon & Breach Science Publishers, New York, 1983, iv+248 pp. (MR 85g:53010; Zbl. 539.53002.)

[Šv 58] Švec, A: *L'élément linéaire projectif d'une surface plongée dans l'espace à connexion projective*, Czechoslovak. Math. J. **8 (83)** (1958), no. 2, 285–291. (MR **21** #3882; Zbl. **83**, p. 377.)

[Šv 59] Švec, A: *Courbes de Darboux d'une hypersurface*, Časop. Pěstov. Mat. **9 (84)**, no. 2, 162–164. (MR **21** #4443; Zbl **92** #4443.)

[Šv 61a] Švec, A: *Les quadriques de Lie d'une surface plongée dans un espace à trois dimensions à connexion projective*, Czechoslovak. Math. J. **11** (1961), no. 3, 386–397. (MR **26** #2890; Zbl. **113**, p. 150.)

[Šv 61b] Švec, A: *Sur la géométrie différentielle d'une surface plongée dans l'espace tridimensionnel à connexion projective*, Czechoslovak. Math. J. **11 (86)**, no. 1, 134–142. (MR **25** #4441; Zbl. **101**, p. 145.)

[Šv 61c] Švec, A: *Sur l'exposition de la théorie des espaces à connexion*, Časopis Pěst. Mat. **11 (86)** (1961), no. 4, 425–432. (MR **24** #A3600; Zbl. **101**, p. 400.)

[Šv 65] Švec, A: *Projective differential geometry of line congruences*, Czechoslovak. Acad. Sci., Prague, 1965, 208 pp. (MR **33** #7949; Zbl. **187**, p. 190.)

[Svo 67] Svoboda, K: *Sur la déformation projectif des pseudocongruences complètement focales, I*, Arch. Mat. **3** (1967), no. 2, 83–98. (MR **36** #7047; Zbl. **212**, p. 540.)

[Svo 69] Svoboda, K: *Sur une classe de congruences de droites dans un espace projectif de dimension paire*, Arch. Mat. **5** (1969), no. 3, 157–165. (MR **43** #5396; Zbl. 241.53006.)

[Tei 53] Teixidor, J: *Über die Umkehrung des Theorems von Reiss*, Arch. Math. **4** (1953), no. 3, 225–229. (MR **15**, p. 152; Zbl. **50**, p. 383.)

[Te 73] Terentjeva, E.I: *Invariant framing of an $(n-2)$-dimensional regular hyperband Γ_{n-2} of the projective space P_n*, (Russian) Kaliningrad. Gos. Univ. Differentsial'naya Geom. Mnogoobraz. Figur, Vyp. 3 (1973), 133–142. (MR **58** #7432.)

[To 28] Togliati, E.G: *Sulle V_3 di S_5 coincidenze di tangenti principali*, Atti. Reale Istituto Veneto **87** (1928), 1373–1421. (F. **54**, p. 774.)

[Tr 59] Trushin, G.I: *λ-conjugacy of three directions on p-dimensional submanifolds in an n-dimensional projective space*, (Russian) Dokl. Akad. Nauk SSSR **129** (1959), no. 1, 37–39. (MR **22** #942; Zbl. **100**, p. 177.)

[Tr 60] Trushin, G.I: *Third order conjugate systems and the problem of their focal transformations*, (Russian) Dokl. Akad. Nauk SSSR **131** (1960), no. 2, 265–268; English transl: Soviet Math. Dokl. **1** (1960), no. 2, 245–248. (MR **22** #8447; Zbl. **100**, p. 177.)

[Tr 61a] Trushin, G.I: *The problem of focal transformations of conjugate systems of third order*, (Russian) Vestnik Moskov. Univ. Ser. I Mat. Mekh. **1961**, no. 2, 3–9. (MR **24** #A3581; Zbl. **135**, p. 218.)

[Tr 61b] Trushin, G.I: *Holonomic conjugate systems of third order*, (Russian) Vestnik Moskov. Univ. Ser. I Mat. Mekh. **1961**, no. 1, 9–16. (MR **24** #A3580; Zbl. **113**, p. 150.)

[Tr 61c] Trushin, G.I: *Conjugate systems of third order*, (Russian) Vestnik Moskov. Univ. Ser. I Mat. Mekh. **1961**, no. 6, 26–33. (MR **23** #A4077; Zbl. **99**, p. 170.)

[Tr 62a] Trushin, G.I: *Third order conjugate systems from semi-asymptotic lines*, (Russian) Vestnik Moskov. Univ. Ser. I Mat. Mekh. **1962**, no. 1, 16–23. (MR **25** #3444; Zbl. **134**, p. 171.)

[Tr 62b] Trushin, G.I: *A λ-conjugate system*, (Russian) Izv. Vyssh. Uchebn. Zaved. Mat. **1962**, no. 4 (29), 161–169. (MR **26** #2960; Zbl. **129**, p. 138.)

[Tz 24] Tzitzéica, G: *Géométrie différentielle projective des réseaux*, Paris, 1924, iii+291 pp. Roumanian transl: Editura Acad. R.P.R., Bucuresti, 1956, 285 pp. (MR **18**, p. 928; Zbl. **74**, p. 375.)

[Ud 63] Udalov, A.S: *On the theory of curves and surfaces in affine and projective spaces*, (Russian) Izv. Vyssh. Uchebn. Zaved. Mat. **1963**, no. 4 (35), 158–164. (MR **29** #2736; Zbl. **129**, p. 135.)

[Ud 68] Udalov, A.S: *An invariant normalization of surfaces in a multidimensional projective spaces*, (Russian) Izv. Vyssh. Uchebn. Zaved. Mat. **1968**, no. 2 (69), 102–105. (MR **37** #2109; Zbl. **155**, p. 304.)

[V 61] Vangeldère, J: *Sur les V_3 de S_5 qui possédent deux systèmes distinct d'asymptotiques doubles*, Bull. Soc. roy. sci. Liège, **30** (1961), no. 3–4, 110–126. (MR **25** #512; Zbl. **102**, p. 373.)

[V 64a] Vangeldère, J: *Quatre théorèmes généraux relatifs aux doubles systèmes conjugués de première espèce d'une V_3 de S_5*, Bull. Soc. Roy. Sci. Liège, **33** (1964), no. 5–6, 299–308. (MR **30** #523; Zbl. **124**, p. 143.)

[V 64b] Vangeldère, J: *Recherches sur la géométrie projective différentielle des V_3 de S_5*, Mém. Soc. Roy. Sci. Liège, **10** (1964), no. 1, 117 pp. (MR **29** #2730; Zbl. **125**, p. 389.)

[Va 70a] Vasilyan, M.A: *The invariant framing of a hyperband*, (Russian) Akad. Nauk Armyan. SSR Dokl. **50** (1970), 65–70. (MR **44** #4658; Zbl. **194**, p. 225.)

[Va 70b] Vasilyan, M.A: *Quadratic hyperbands of rank $n-2$*, (Russian) Akad. Nauk Armyan. SSR Dokl. **50** (1970), 193–197. (MR **44** #4659; Zbl. **197**, p. 476.)

[Va 71] Vasilyan, M.A: *Projective theory of multidimensional hyperbands*, (Russian) Izv. Akad. Nauk Armyan. SSR Ser. Mat. **6** (1971), no. 6, 477–481. (MR **47** #9445.)

[Va 73] Vasilyan, M.A: *Affine connections that can be induced by the normalization of a hyperband*, (Russian) Akad. Nauk Armyan. SSR Dokl. **57** (1973), 200– 205. (MR **49** #9765; Zbl. 311.53017.)

[Va 77] Vasilyan, M.A: *On projective differential geometry of one-parameter hyperbands*, (Russian) In *Differential Geometry*, Kalinin. Gos. Univ., Kalinin, 1977, 38–45. (MR 82e:53021.)

[Vas 87] Vasilyev, A.M: *Theory of differential-geometric structures*, (Russian) Moskov. Gos. Univ., Moscow, 1987, 192 pp. (MR 89m:58001; Zbl. 656.53001.)

[Ve 49] Verbitsky, L.L: *On the metric differential geometry of hypersurfaces of second order*, Trudy Sem. Vektor. Tenzor. Anal. **7** (1949), 319–340. (MR **12**, p. 204; Zbl. **40**, p. 86.)

[Vi 37a] Villa, M: *Le varietà a quatro dimensioni, che possegono ∞^7 quasi-asintotiche $\gamma_{1,3}$*, Atti Accad. Naz. Lincei Rend. Cl. Sci. Fis. Mat. Natur. (6) **25**, 434–439. (Zbl. **17**, p. 223.)

[Vi 37b] Villa, M: *Proprieta differenziali dei coni di Veronese*, Atti Accad. Naz. Lincei Rend. Cl. Sci. Fis. Mat. Natur. (6) **25** (1937), 691–694. (Zbl. **17**, p. 224.)

[Vi 38] Villa, M: *Sopra una classe di V_k situate sui coni di Veronese*, Atti Accad. Naz. Lincei Rend. Cl. Sci. Fis. Mat. Natur. (6) **27** (1938), 217–221. (Zbl. **19**, p. 138.)

[Vi 39a] Villa, M: *Proprieta differenziale caratteristica dei coni proiettanti le varietà che representano la totalita delle quadriche di uno spazio lineare*, Atti Accad. Naz. Lincei Rend. Cl. Sci. Fis. Mat. Natur. (6) **28** (1938), 3–12. (Zbl. **19**, p. 367.)

[Vi 39b] Villa, M: *Sulle varietà situate sui coni proiettanti la $V_r^{2^*}$ che rappresentano la totalita delle quadriche di S_r. I*, Atti Accad. Naz. Lincei Rend. Cl. Sci. Fis. Mat. Natur. (6) **28** (1939), 365–370. (Zbl. **20**, pp. 391–392.)

[Vi 39c] Villa, M: *Sulle varietà situate sui coni proiettanti la $V_r^{2^*}$ che rappresentano la totalita delle quadriche di S_r. II*, Atti Accad. Naz. Lincei Rend. Cl. Sci. Fis. Mat. Natur. (6) **28** (1939), 395–402. (Zbl. **20**, pp. 391–392.)

[Vi 39d] Villa, M: *Ricerche sulle curve quasi-asintotiche. I*, Atti Accad. Naz. Lincei Rend. Cl. Sci. Fis. Mat. Natur. (6) **28** (1939), 246–253. (Zbl. **20**, p. 258.)

[Vi 39e] Villa, M: *Ricerche sulle curve quasi-asintotiche. II*, Atti Accad. Naz. Lincei Rend. Cl. Sci. Fis. Mat. Natur. (6) **28** (1939), 302–310. (Zbl. **20**, p. 258.)

[Vi 39f] Villa, M: *Ricerche sulle varietà V_k che possegono $\infty^\delta E_2$ di $\gamma_{1,3}$ con particolare riguardo al caso $k = 4, \delta = 8$*, Mem. R. Accad. Naz. Lincei **7** (1939), 373–427. (MR **1**, p. 168; Zbl. **22**, p. 168.)

[Vi 39g] Villa, M: *Nuove ricerche nella teoria delle curve quasi-asintotiche*, Ann. Mat. Pura Appl. (4) **18** (1939), 275–308. (MR **1**, p. 268; Zbl. **22**, p. 169.)

[Vi 40] Villa, M: *Sulle superficie quasi-asintotiche della V_4^6 si S_8 che rappresenta le coppie di punti di due piani*, Atti Accad. Ital. Cl. Sci. Fis. Mat. Nat. (7) **1** (1940), 228–237. (MR **1**, p. 268; Zbl. **24**, p. 173.)

[Vi 52] Villa, M: *Varietà quasi-asintotiche e transformazioni puntuale*, Ann. Univ. Ferrara Sez. VII (N.S.) **1** (1952), 17–21. (MR **14**, p. 680; Zbl. **48**, p. 395.)

[VV 50] Villa, M. and Vaona, G: *Varietà quasi-asintotiche a più indice e curve caratteristiche di una transformazione puntuale*, Atti Accad. Naz. Lincei Rend. Cl. Sci. Fis. Mat. Natur. (8) **8** (1950), 470–476. (MR **14**, p. 205; Zbl. **39**, p. 385.)

[Vil 72] Vilms, J: *Submanifolds of Euclidean space with parallel second fundamental form*, Proc. Amer. Math. Soc. **32**, no. 1, 263–267. (MR **44** #7482; Zbl. 229.53045.)

[Vo 64] Vorontsova, N.S: *An intrinsic normalization of degenerate surfaces of rank r with an $(n-r-1)$-dimensional vertex in an $(n+1)$-dimensional projective space*, (Russian) Chelyabinsk. Gos. Ped. Inst. Trudy **2** (1964), 13–16. (MR **35** #6063.)

[Vy 49] Vygodsky, M.Ya: *Differential geometry*, (Russian) Moscow/ Leningrad, 1949, 512 pp. (MR **12**, p. 127; Zbl. **41**, p. 288.)

[Wa 50] Wagner, V.V: *The theory of a field of local hyperbands*, Trudy Sem. Vektor. Tenzor. Anal. **8** (1950), 197–272. (MR **13**, p. 778; Zbl. **41**, pp. 300–302.)

[Wak 65] Waksman, V.S: *Focal properties of ruled surfaces in N-dimensional projective space*, (Russian) Moskov. Inst. Inzh. Zheleznodor. Transporta Trudy Vyp. 222 (1965), 5–12. (MR **35** #6061; Zbl. **148**, p. 155.)

[Wal 68] Walter, R: *Zur projektiven Kinematik konjugierter Systeme des n-dimensionalen Raumes*, Arch. Math. **19** (1968), no. 3, 313–324. (MR **37** #4729; Zbl. **157**, p. 517.)

[We 38] Weise, K: *Der Berührungstensor zweier Flächen und die Affingeometrie der F_p im A_n, I, II*, Math. Z. **43** (1938), 469–480; **44** (1939), 161–184. (Zbl. **18**, p. 187 & **19**, p. 185.)

[Wi 06] Wilczynski, E.J: *Projective differential geometry of curves and surfaces*, Teubner, Leipzig, 1906, viii+298 pp. (F. **37**, p. 620.) Reprint of 1st edition, Chelsea Publ. Comp., New York, 1962. (MR **24** #A1085.)

[Wo 82] Wood, J.A: *An algebraization theorem for local hypersurfaces in projective space*, Ph. D. Dissertation, Univ. of California, Berkeley, 1982, 87 pp.

[Wo 84] Wood, J.A: *A simple criterion for local hypersurfaces to be algebraic*, Duke Math. J. **51** (1984), no. 1, 235–237. (MR 85d:14069; Zbl. 584.14021.)

[Ya 53] Yanenko, N.N: *Some questions of the theory of embeddings of Riemannian metrics into Euclidean spaces*, (Russian) Uspekhi Mat. Nauk **8** (1953), no. 1 (53), 21–100. (MR **14**, p. 1122; Zbl. **51**, p. 128.)

[Y 84] Yang, K: *Equivalence problems in projective differential geometry*, Trans. Amer. Math. Soc. **282** (1984), no. 1, 319–334. (MR 85k:53020; Zbl. 558.53018.)

[Y 85] Yang, K: *Deformation of submanifolds of real projective space*, Pacific. J. Math. **120** (1985), no. 2, 469–492. (MR 87c:53109; Zbl. 543.53008 & 573.53008.)

[Y 92] Yang, K: *Exterior differential systems and equivalence problems*, Kluwer Academic Publishers, Dordrecht-Boston-London, 1992, xi+196 pp.

SYMBOLS FREQUENTLY USED

The list below contains many of the symbols whose meaning is usually fixed throughout the book.

A_i^j: first Laplace transform, 101
A_{ik}^{jl}: second Laplace transform, 101
A^n: affine space of dimension n, 21
$b_{ij\ldots k}^{i_{q-1}}$: qth fundamental tensor, 38–39, 44–45, 48
$C_{(q)}$: asymptotic cone of order q, 48
C_x: cone with vertex at a point x, 279
$C(m,l)$: Segre cone, 28–29, 52
C^n: conformal space of dimension n, 23
$\mathbf{C}^n$: n-dimensional complex space, 5
$\Delta^p(M^n)$: p-dimensional distribution on M^n, 11
$\Delta_x^p(M^n)$: element of Δ^p at a point $x \in M^n$, 11
δ: symbol of differentiation with respect to secondary parameters, 6
δ_{ij}, δ_j^i: Kronecker symbol, 22, 176
E^l: l-dimensional generator of V_r^m, 116
E^n: Euclidean space of dimension n, 22, 134
F: focus variety, 117
Φ: focus cone, 119
$\Phi_{(q)}$: fundamental form of order q, 39, 44, 48
$\mathbf{GL}(n)$: general linear group, 1
$G(m,n)$: Grassmannian of m-dimensional subspaces in P^n, 26
$GW(d,n,r)$: Grassmann d-web of codimension r in M^{nr}, 275
γ: affine connection, 16
$\gamma(V^m)$: Gauss mapping of V^m 38
γ_n: connection in the normal bundle of a normalized V^m, 183
γ_t: affine connection in the tangent bundle of a normalized V^m, 179
H^m: m-dimensional hyperband, 254
H^n: hyperbolic space of dimension n, 134
$\mathcal{K}$: correlative transformation, 116
l_x: second normal of V^m at a point x, 174
L^n: vector space of dimension n, 1
M: differentiable manifold, 5
M^n: n-dimensional differentiable manifold, 5
m_x: first subnormal of V^m at a point x, 174
$N_x^{(q)}$: normal subspace of V^m of order q at a point x, 36, 43, 48
$\widetilde{N}_x^{(q)}$: reduced normal subspace of order q of V^m at a point x, 42, 45
n_x: first normal of V^m at a point x, 174
∇_δ: operator of covariant differentiation relative to secondary parameters, 6
$\nu(x)$: index of relative nullity at a point x, 133

$\Omega(m,n)$: image of the Grassmannian $G(m,n)$, 27
$\mathcal{P}_m(P^n)$: projectivization of P^n with the center P^m, 23
P^n: projective space of dimension n, 17
P^{n*}: dual space of P^n, 20
Q: hyperquadric, 130
$\mathcal{R}(X)$: frame bundle over X, 1, 5
$\mathcal{R}^1(X)$: bundle of first order frames over X, 15, 35
$\mathcal{R}^2(X)$: bundle of second order frames over X, 15
R^i_{jk}: torsion tensor of an affine connection, 17
R^i_{jkl}: the curvature tensor of an affine connection, 17
$\{R^\alpha_{\beta ij}, R^0_{\beta ij}\}$: curvature tensor of the connection γ_n, 184,
$\mathbf{R}^n$: n-dimensional real space, 5
Σ_k: conjugate net of order k, 73, 76
$\mathbf{SL}(n)$: special linear group, 18
$S(m,l)$: Segre variety, 29
S^n: elliptic space of dimension n, 134
$s_1, s_2, \ldots$: characters, 13
$T(M^n)$: tangent bundle of M^n, 5
$T^*(M^n)$: cotangent bundle of M^n, 6
$T_x(M^n)$: tangent space to M^n at a point x, 5
$T_x(V^m) = T_x^{(1)}$: tangent subspace to V^m at a point x, 33
$T_x^*(M^n)$: cotangent space to M^n at a point x, 6
$T_x^{(q)}(V^m)$: osculating subspace of order q of V^m at a point x, 41, 45, 46
V_d^r: r-dimensional algebraic variety of degree d, 278
$V(m)$: Veronese variety of dimension m, 30
V^m: submanifold of dimension m, 33
V_r^m: tangentially degenerate submanifold of dimension m and rank r, 96, 113
V^n: Riemannian manifold of dimension n, 132
V_c^n: Riemannian manifold of dimension n and constant curvature c, 134
$v^{(2)}$: Veronese mapping, 62
y^i_α: normalizing object of V^m the 1st kind, 193
z_i: normalizing object of V^m the 2nd kind, 193
z_α: normalizing object of V^m defining the subnormal m_x, 193
$\wedge$: symbol of exterior multiplication, 8

Index